THE FUNDAMENTALS OF ENGINEERING PHYSICS

THE FUNDAMENTALS OF ENGINEERING PHYSICS

By

DR. PURNIMA SWARUP KHARE
Head of the Physics Department,
Rajiv Gandhi Technological University,
Bhopal (M.P.)

DR. ABHAYA SWARUP
Energy Consultant
World Bank Projects
Professor Management Institute,
Bhopal (M.P.)

UNIVERSITY SCIENCE PRESS
(An Imprint of Laxmi Publications Pvt. Ltd.)

BANGALORE • CHENNAI • COCHIN • GUWAHATI • HYDERABAD
JALANDHAR • KOLKATA • LUCKNOW • MUMBAI • PATNA
RANCHI • NEW DELHI

UNIVERSITY SCIENCE PRESS
(*An Imprint of Laxmi Publications Pvt. Ltd.*)
113, Golden House, Daryaganj,
New Delhi-110002
Phone : 011-43 53 25 00
Fax : 011-43 53 25 28
www.laxmipublications.com
info@laxmipublications.com

Price : **Rs. 295.00** *Only.*

First Edition : 2006, *Second Edition* : 2008
Third Edition : 2011

OFFICES

✆ **Bangalore**	080-26 75 69 30	✆ **Chennai**	044-24 34 47 26
✆ **Cochin**	0484-237 70 04, 405 13 03	✆ **Guwahati**	0361-251 36 69, 251 38 81
✆ **Hyderabad**	040-24 65 23 33	✆ **Jalandhar**	0181-222 12 72
✆ **Kolkata**	033-22 27 43 84	✆ **Lucknow**	0522-220 99 16
✆ **Mumbai**	022-24 91 54 15, 24 92 78 69	✆ **Patna**	0612-230 00 97
✆ **Ranchi**	0651-221 47 64		

UEP-9476-295-ENGINEERING PHYSICS-KHA
C. 2213 /010/09
Typeset at : Kalyani Computer Services, New Delhi.
Printed at Mehra Offset Press, Delhi

CONTENTS

Preface to the Third Edition

We are pleased to present the third edition of this book. The warm reception which the previous editions and reprints of this book have enjoyed all over India has been a great satisfaction to us.

The entire book has been thoroughly revised as per the latest syllabus prescribed by R.G.P.V., Bhopal.

Any suggestions for the improvement of this book will be thankfully acknowledged.

—AUTHORS

Preface to the First Edition

Physics is the 'Queen' of sciences and forms the foundation of engineering and technology. Therefore, for a real understanding of new discoveries, the latest developments in the technical field and a sound knowledge of basic physics and its principles are of paramount importance.

The main intention in writing this book is to expose engineering students to the fundamentals of physics, allowing them to gain insight into the subject. During the past twenty years of teaching, we have noticed that students lack appreciation of the basic concepts. Their comprehension is confined to only the mathematics of the problem, and they often display an astonishing ignorance of the physics which is intricately involved in it. We sincerely hope that the present book will inspire students and young teachers to think about the mysteries of physics and discuss the ideas behind its basic principles. This book will surely make some contribution in developing a conceptual foundation for the upcoming generation of engineers.

Though we make no claims of originality, we have relied on our rich experience gained from teaching basic physics courses over the last two decades. We are grateful for all constructive suggestions and comments that will provide improvement of the book.

—AUTHORS

Chapter 1

Quantum Physics

1.1 OVERVIEW

The universal activities are governed with the two types of basic, identities, *i.e.*, 'Matter' and 'Energy'. All the visible things are made up of matter, and energy is the invisible identity which causes physical activities. The two fundamental identities matter and energy can be interchangeable into each other's forms and can be quantified by the Einstein mass energy relation

$$E = mc^2.$$

Light is energy. It is an invisible universal identity which causes physical activities. Light also exhibits many optical phenomenon like reflection, refraction, interference, diffraction, polarization, etc. All these phenomenon can be explained with the electromagnetic theory of light. But this wave theory of light fails to explain the photoelectric effect and Compton effect. These phenomenon could be explained by the quantum theory of light. According to Einstein, the energy of light is concentrated into small regions. This represents the smallest quantity of energy known as a photon. This photon is an energy particle. Hence light shows its wave nature on the one hand and particle nature on the other hand. This shows that light has both wave nature as well as particle nature. This nature of light is known as dual nature and the property is known as wave particle duality.

The particle nature of light is governed by the quantum theory. In the beginning of the 20th century, physicists accepted the dual nature of electromagnetic radiations with the acceptance of the quantum theory. Two decades later, in 1924, french physicist Louis de Broglie put forward the bold suggestion theoretically that matter, which is normally considered as made up of discrete particles *e.g.*, protons, electrons, atom, possesses wave properties under appropriate conditions. Thus matter also has a dual nature, particle as well as wave, like radiation. This revolutionary hypothesis of de Broglie's shook the very foundations of mechanics and opened the door for the solution of many baffling problems concerning the structure of atoms and their interaction with electromagnetic radiation. These waves associated with particles are called matter waves.

According to de Broglie, matter waves are generated by the motion of a body, be it a planet, a stone, a particle of dust, or an electron. Like electromagnetic waves these waves can also propagate in a vacuum. Hence they are not mechanical waves like sound waves. And unlike the electromagnetic waves, they are also associated with the motion of electrically neutral bodies; hence they are not electromagnetic waves.

1.2 WAVE AND PARTICLE DUALITY OF RADIATION

To understand the wave and particle duality, it is necessary to know what is a particle and what is a wave. The concept of a particle is easy to grasp. It has mass, it is located at some definite point, it can move from one place to another, it gives energy when slowed down or stopped. Thus the particle is specified by (*i*) mass m, (*ii*) velocity v, (*iii*) momentum p, and (*iv*) position (x, y).

The concept of a wave is a bit more difficult than that of a particle. A wave is spread out over a relatively large region of space, it cannot be said to be located just here and there, and it is hard to think of mass being associated with a wave. Actually, a wave is nothing but a spread out disturbance. A wave is specified by its (*i*) frequency, (*ii*) wavelength, (*iii*) phase or wave velocity, (*iv*) amplitude, and (*v*) intensity.

Considering the above facts, it appears difficult to accept the conflicting ideas that radiation has a dual nature, *i.e.*, radiation is a wave which is spread out over space and also a particle which is localized at a point in space. However, this acceptance is essential because radiation sometimes behaves as a wave and at other times as a particle as explained below:

- Radiations, including visible light, infrared, ultraviolet, X rays, etc., behave as waves in experiments based in interference, diffraction etc. This is due to the fact that these phenomena require the presence of two waves at the same position at the same time. Obviously, it is difficult for the two particles to occupy the same position at the same time. Thus we conclude that radiations behave like waves.
- Plank's quantum theory was successful in explaining black body radiation, the photoelectric effect, the Compton effect, etc. and had clearly established that the radiant energy, in its interaction with matter, behaves as though it consists of corpuscles. Here radiation interacts with matter in the form of photons or quanta. Thus we conclude that radiation behaves like a particle.

Radiation thus, sometimes behaves as waves and at other times as particles, *i.e.*, it has a wave particle dualism. Here it should be remembered that radiation cannot exhibit its particle and wave properties simultaneously.

1.3 WAVE PARTICLE DUALITY OF MATTER

With the confirmation of de Broglie's matter waves, the distinction between waves and particles broke down. Light waves behaved like particles and particles behaved like waves. Whereas in good old classical physics, waves were waves and particles were particles, we are now confronting waves possessing properties of particles and particles possessing properties of waves.

It has been observed that when an intense beam of electrons is passed through a tiny hole and allowed to impinge a photographic plate, we get a group of light and dark rings (Figure 1). The distribution curve of electrons on the photographic plate is wave-like in shape. A similar wave-form is exhibited by the intensity of the diffraction pattern due to light and to X rays, which are definitely waves. The electron wave is a bit different. It is in fact a probability wave. It determines the probability of an electron impinging upon some point on the photographic plate.

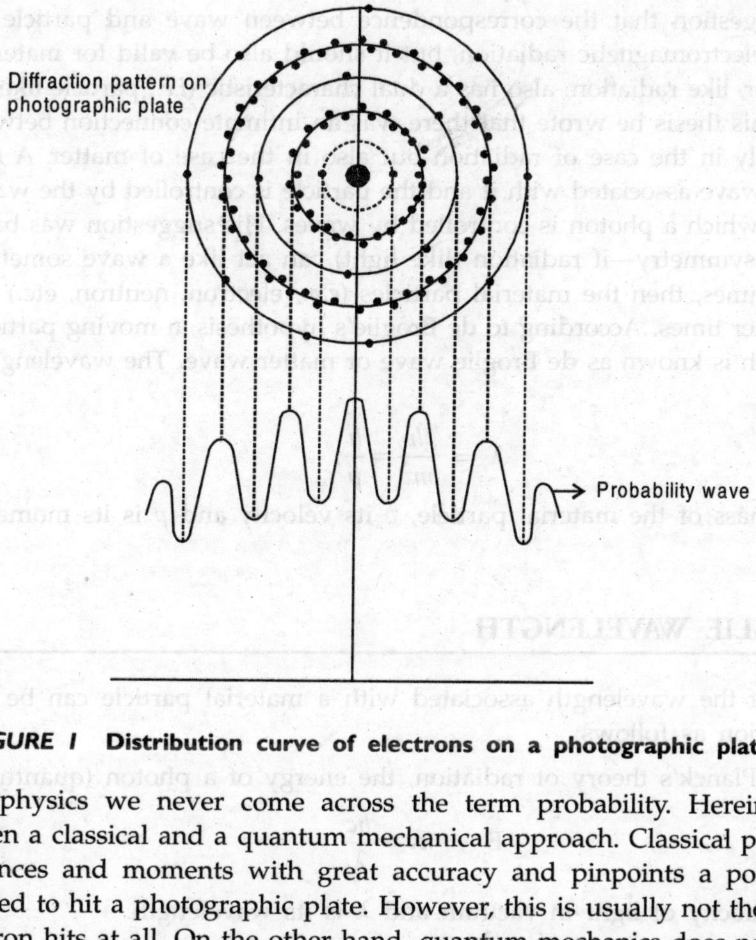

FIGURE 1 **Distribution curve of electrons on a photographic plate.**

In classical physics we never come across the term probability. Herein lies the basic difference between a classical and a quantum mechanical approach. Classical physics calculates the angles, distances and moments with great accuracy and pinpoints a position where an electron is expected to hit a photographic plate. However, this is usually not the position where a particular electron hits at all. On the other hand, quantum mechanics does not seek the exact position. It merely calculates the probability of the electron hitting the plate at a particular position.

This law of probability is then the 'wave' law which governs the de Broglie matter waves. It seems necessary, therefore, to give up the idea that waves and particles are distinctly separate entities and different laws should apply to them. We should now regard the entire universe as basically a single indivisible unit. Its two-faced character becomes distinct only in certain extreme limits like the classical limit. de Broglie's discovery has thus blended the world of physical phenomena into a coherent whole and bridged the gap between particles and waves.

At the microscopic level, evidences obtained under different experimental conditions cannot be fitted into a single picture. The wave and the particle aspects must be regarded as *complementary*.

1.4 De BROGLIE'S HYPOTHESIS

Louis de Broglie, in 1924, extended the wave particle parallelism of optics to all the fundamental entities of physics such as electrons, protons, neutrons, atoms and molecules, etc. de Broglie made a bold suggestion that the correspondence between wave and particle should not be confined only to electromagnetic radiation, but it should also be valid for material particles, as for example matter, like radiation, also has a dual characteristic (*i.e.*, particle-like and wave-like) characteristic. In his thesis he wrote that there was an intimate connection between waves and corpuscles not only in the case of radiation but also in the case of matter. A moving particle has always got a wave associated with it and the particle is controlled by the wave in a manner similar to that in which a photon is controlled by waves. His suggestion was based on the fact that nature loves symmetry—if radiation (like light) can act like a wave sometimes and like a particle at other times, then the material particles (*e.g.*, electron, neutron, etc.) should also act like a wave at other times. According to de Broglie's hypothesis, a moving particle is associated with a wave which is known as de Broglie wave or matter wave. The wavelength of the matter wave is given by

$$\lambda = \frac{h}{mv} = \frac{h}{p},$$

where m is the mass of the material particle, v its velocity and p is its momentum.

1.5 De BROGLIE WAVELENGTH

The expression of the wavelength associated with a material particle can be derived on the analogy of radiation as follows:

Considering Planck's theory of radiation, the energy of a photon (quantum) is given by

$$E = h\nu = \frac{hc}{\lambda} \tag{1}$$

where c is the velocity of light in vacuum and λ is its wavelength.

According to Einstein, the energy mass relation is given by

$$E = mc^2. \tag{2}$$

From equations (1) and (2), we get

$$mc^2 = \frac{hc}{\lambda} \quad \text{or} \quad \lambda = \frac{hc}{mc^2}$$

or

$$\lambda = \frac{h}{mc}, \tag{3}$$

where $mc = p$ (momentum associated with photon).

If we consider the case of a material particle of mass m and moving with a velocity v, (Figure 2) *i.e.*, momentum mv, then the wavelength associated with this particle is given by

$$\lambda = \frac{h}{mv} = \frac{h}{p}, \tag{4}$$

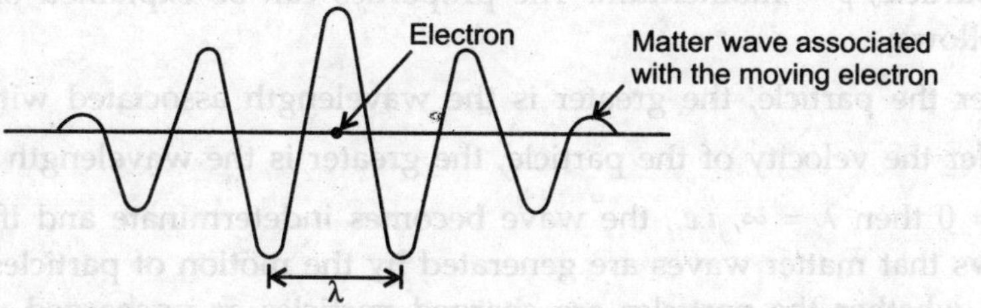

FIGURE 2 **Schematic diagram of a de Broglie wave.**

If E is the kinetic energy of the material particle, then

$$E = \frac{1}{2}mv^2 = \frac{1}{2}\frac{m^2v^2}{m} = \frac{p^2}{2m}$$

or

$$p = \sqrt{(2mE)}$$

∴ de Broglie wavelength

$$\lambda = \frac{h}{\sqrt{(2mE)}}. \tag{5}$$

1.6 De BROGLIE'S WAVELENGTH ASSOCIATED WITH ELECTRONS

Let us consider the case of an electron of rest mass m_0 and charge e which is accelerated by a potential V volts from rest to velocity v, then

$$\frac{1}{2}m_0v^2 = eV \quad \text{or} \quad v = \sqrt{\left(\frac{2eV}{m_0}\right)}.$$

Now,

$$\lambda = \frac{h}{m_0v} = \frac{h}{m_0}\sqrt{\frac{m_0}{2eV}} = \frac{h}{\sqrt{2eVm_0}}$$

or

$$\lambda = \frac{6.625 \times 10^{-34}}{\sqrt{[(2 \times 1.632 \times 10^{-19}\,V \times 9.1 \times 10^{-31})]}}$$

$$= \frac{12.26}{\sqrt{V}}\ \text{Å}$$

If

$$V = 100 \text{ volts}$$

then

$$\lambda = 1.226\ \text{Å}. \tag{6}$$

Equation (6) shows that the wavelength associated with an electron accelerated by 100 volts is 1.226 Å.

1.7 PROPERTIES OF MATTER WAVES

The waves associated with the particle are known as matter waves. The wavelength of these waves is given by the de Broglie's equation $\lambda = \frac{h}{p} = \frac{h}{mv}$ where m = mass of the particle, v = velocity of the particle, p = momentum. The properties can be explained on the basis of this expression as follows:

1. The lighter the particle, the greater is the wavelength associated with it.
2. The smaller the velocity of the particle, the greater is the wavelength associated with it.
3. When $v = 0$ then $\lambda = \infty$, *i.e.*, the wave becomes indeterminate and if $v = \infty$ then $\lambda = 0$. This shows that matter waves are generated by the motion of particles. These waves are produced whether the particles are charged particles or uncharged particles ($\lambda = h/mv$ is independent of charges). This fact reveals that these waves are not electromagnetic waves but that they are a new kind of wave (electromagnetic waves are produced only by motion of charged particles).
4. The velocity of a matter wave depends on the velocity of a matter particle *i.e.*, it is not a constant, while the velocity of a electromagnetic wave is a constant.
5. The velocity of a matter wave is greater than the velocity of light.

   The velocity of a matter wave (de Broglie wave velocity) is greater than the velocity of light. This can be proved:

   A particle in motion with an associated matter wave has two different velocities, one referring to the mechanical motion of the particle represented by v, and the second related to the propagation of the wave represented by w.

   We know that $\quad E = h\nu \quad$ and $\quad E = mc^2$

   $$h\nu = mc^2 \quad \text{or} \quad \nu = \frac{mc^2}{h}$$

   The wave velocity (w) is given by

   $$w = \nu \times \lambda = \frac{mc^2}{h} \times \frac{h}{mv} \quad \because \lambda = \frac{h}{mv}$$

   or

   $$w = \frac{c^2}{v}$$

   As particle velocity v cannot exceed c (velocity of light), therefore w is greater than the velocity of light. We can understand this unexpected result by considering the wave velocity also known as phase velocity and group velocity.
6. The wave and particle aspects of moving bodies can never appear together in the same experiment. What we can say is that waves have particle-like properties and particles have wave-like properties and the concepts are separately linked. Matter-wave representation is only a symbolic representation.
7. The wave nature of matter introduces uncertainty in the location or the position of the particle because a wave cannot be said to be exactly at this point or exactly at that point. However, where the wave is large (strong) there is a good chance of finding the particle while where the wave is small (weak), there is very small chance of finding the particle.

1.8 EXPERIMENTAL VERIFICATION OF De BROGLIE HYPOTHESIS

Davisson and Germer Experiment

This experiment was performed by Davisson and Germer in 1927. Using diffraction methods, they succeeded not only in detecting de Broglie waves but also in measuring their wavelength for slow electrons.

Experimental Arrangement

The experimental set up used is shown in Figure 3(*a*). A beam of electrons is produced from an electron gun. This contains a tungsten filament which is heated to eject electrons by thermionic emission. The emitted electrons are accelerated towards an anode and collimated by suitable slits to emerge as a fine beam. Any desired velocity can be imparted to the electrons by suitably adjusting the anode voltage.

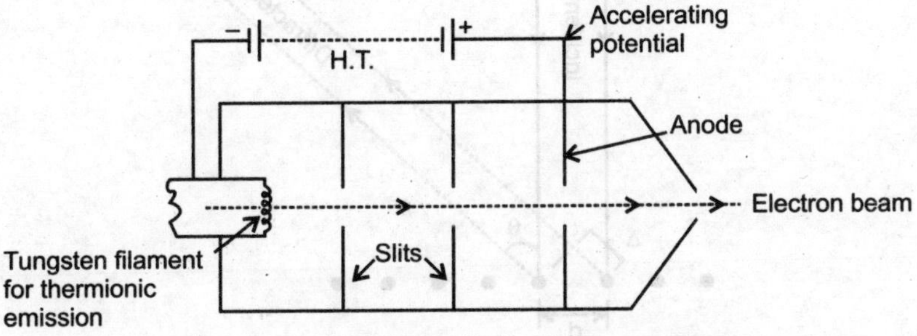

FIGURE 3(a) **Electron gun.**

On coming out of the electron gun, the beam was allowed to fall perpendicularly on the surface of the nickel crystal. The surface lattice of the crystal acts as a grating and the electron beam is diffracted off in certain preferred directions. Since the penetration of the electrons into the crystal can be neglected, the diffraction grating can be regarded as two dimensional.

Method

The crystal target was mounted at different azimuths. For each azimuth, the diffracted beam was collected by a detector scanning the beam over a circular scale (Figure 3(*b*)). The detector thus records the positions of diffraction maxima. The presence of the diffraction pattern vindicates the existence of de Broglie matter waves.

Calculation of Wavelength

The position of diffraction maxima is given by the condition

$$\text{path difference } \Delta = n\lambda$$

From Figure 3(*c*) it is clear that the path difference in question is given by

$$\Delta = d \sin \theta$$

where d is the two dimensional lattice constant and θ the angle of diffraction.

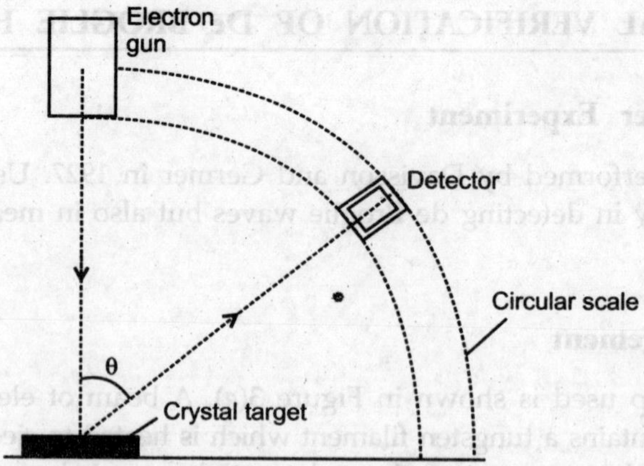

FIGURE 3(b) **Detector scanning of a diffracted electron beam.**

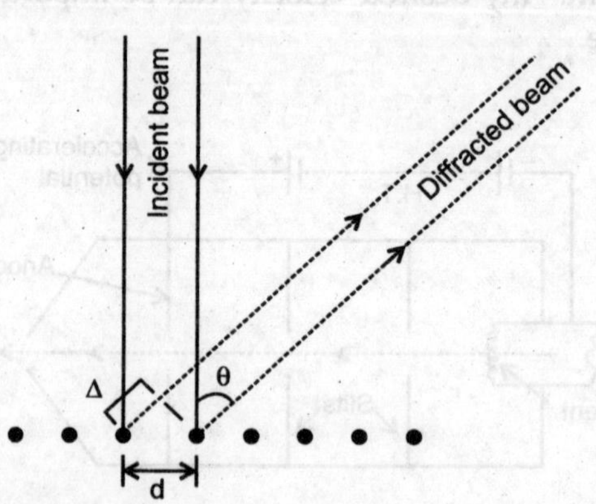

FIGURE 3(c) **Diffraction of an electron beam by a crystal lattice.**

Thus the diffraction maxima are given by

$$d \sin \theta = n\lambda$$

or

$$\lambda = \frac{d \sin \theta}{n}$$

where n is the order of the maximum.

Thus the wavelength of the de Broglie electron waves can be determined.

Verification

The wavelength of electron waves as determined above can be compared with the one obtained from de Broglie's ideas. If the electron beam in the above experiment is accelerated through a potential V volts, its kinetic energy would be V electron volts. For the electron we have

$$\text{charge } e = 1.6 \times 10^{-19} \text{ coulombs}$$

$$\text{mass } m = 9.1 \times 10^{-31} \text{ kg}$$

Now one electron volt $= 1.6 \times 10^{-19}$ joules.

∴ The energy of the electron accelerated through V volts will be

$$E = 1.6 \times 10^{-19} \text{ V joules.}$$

The momentum p of the electron is then

$$p = \sqrt{(2mE)}$$

$$= \sqrt{(2 \times 9.1 \times 10^{-31} \times 1.6 \times 10^{-19}\,V)}$$

$$= \sqrt{(29.1 \times 10^{-50}\,V)}$$

Thus the de Broglie wavelength of these electrons will be

$$\lambda = \frac{h}{p} = \frac{6.6 \times 10^{-34}}{\sqrt{29.1 \times 10^{-50}\,V}}$$

$$= \sqrt{\frac{150}{V}} \times 10^{-10} \text{ meters}$$

$$= \sqrt{\frac{150}{V}}\ \text{Å} = \frac{12.2}{\sqrt{V}}\ \text{Å}$$

In this way, knowing V, the de Broglie wavelength can be calculated and compared with the one obtained experimentally. The results of the Davisson and Germer experiment completely confirmed de Broglie's concept of matter waves.

Thomson's Experiment

This experiment demonstrates the wave nature of fast electrons.

Experimental Arrangement

In this experiment a beam of cathode rays is produced in a discharge tube. The beam is passed through a diaphragm tube A to obtain a fine pencil of electrons. It then falls on the gold foil F. The thickness of the foil is of the order of 10^{-6} cm so that the electrons do not get absorbed in it. A photographic plate P receives the beam of electrons after it has traversed the foil (Figure 4).

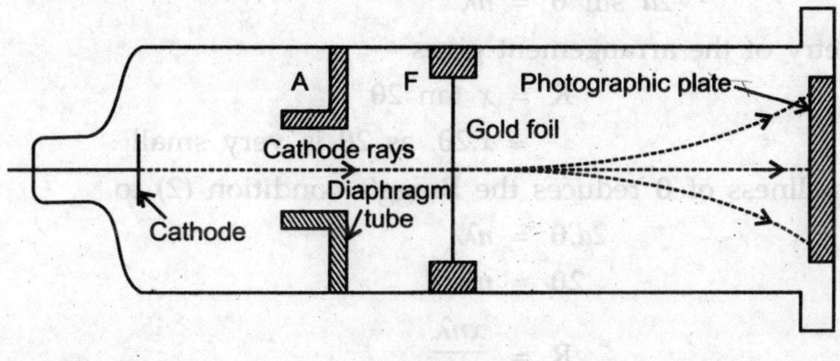

FIGURE 4(a) **Diffraction of cathode rays fast moving electron beam by thin gold foil.**

Method and Theory

The beam of electrons which has passed through the thin foil is allowed to fall on the photographic plate for some time. The plate is then developed. If the electrons here behaved like particles, the foil would merely scatter them on to the photographic plate and we would obtain a continuous patch of uniform intensity. However, in this experiment Thomson got a symmetrical pattern consisting of alternate dark and bright concentric rings. These rings can be interpreted as the diffraction pattern of the electron beam. Such a pattern is seen, for example, when one looks at a street lamp through a dusty glass. The diffraction of electrons is a definite indication of their wave character.

Quantitative Verification

In the thin film of gold there will be some crystals oriented in the right position to give a Bragg reflection. If there are enough crystals distributed at random, the result of such reflections will be a series of rings arising from the intersection of the cones of diffraction with the photographic plate [Figure 4(*b*)].

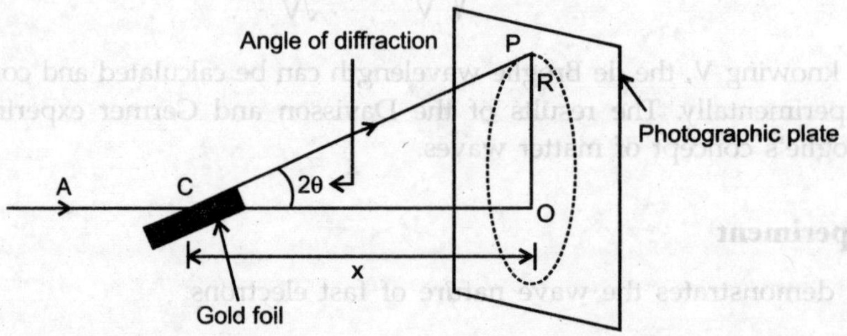

FIGURE 4(b) **Analytic treatment of the diffraction of cathode rays.**

Let AC be the incident beam passing through the foil at C and let CP be the beam which has suffered a Bragg reflection at some crystal in the foil. The diffracted beam falls at the point P of the photographic plate at a distance R from the central point O. Let the distance CO from the crystal to the photographic plate be x. The angle PCO is 2θ, where θ is given by the Bragg relation

$$2d \sin \theta = n\lambda \tag{1}$$

Now the geometry of the arrangement gives

$$R = x \tan 2\theta$$

$$\approx x.2\theta, \text{ as } 2\theta \text{ is very small} \tag{2}$$

The extreme smallness of θ reduces the Bragg's condition (2) to

$$2d.\theta = n\lambda$$

$$2\theta = n\lambda/d$$

hence we obtain $$R = \frac{xn\lambda}{d}$$

or $$\lambda = \frac{Rd}{xn} \tag{3}$$

Measuring the radius R of the nth ring and knowing x and d, the wavelength λ of the electron waves was calculated. The value thus attained was compared with the one obtained from the de Broglie relation.

$$\lambda = \frac{h}{mv} = \sqrt{\frac{150}{V}}\ \text{Å}$$

The experimental and theoretical values agreed extremely well. The de Broglie matter waves are thus confirmed.

1.9 THE DEVELOPMENT OF QUANTUM THEORY

The de Broglie concept of matter waves brought the breakthrough in the development of quantum theory in 1925. In classical mechanics, Newton's laws describe the motion of particles. The future history of the particle is completely determined by its initial position and momentum together with the forces that act upon it. But the concept of matter waves introduced the uncertainty in the position and momentum of a particle. Moreover, according to the de Broglie theory, a material particle is associated with the wave. So the mathematical reformation using a wave function associated with matter waves is needed. Such a mathematical formation known as quantum mechanics was developed in 1926 by Schrödinger. Schrödinger described the amplitude of matter waves by a complex quantity $\psi(x, y, z, t)$ known as wave function or the state of the system. It describes the particular dynamical system under observation.

1.10 WAVE FUNCTION AND IT'S PHYSICAL SIGNIFICANCE

The wave and particle properties of electrons and other particles can be explained only by considering the concept of a wave packet or a wave group as shown in Figure 5.

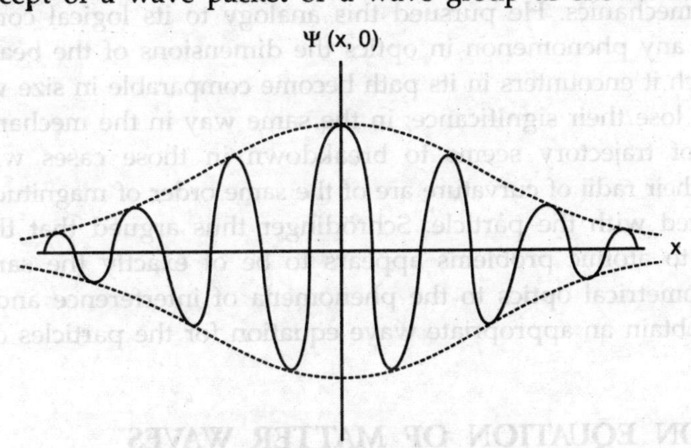

FIGURE 5 **Wave function.**

Let us try to understand what a wave function is

We know that in a light wave the electromagnetic field varies in space and time whereas in sound a wave pressure varies in space and time. Similarly, in a case of de Broglie waves the variable quantity is called the wave function ψ (Psi). Its value associated with a moving body at a given point in space and time t is related to the likelihood of finding the body there at that time. Thus, it deals with the probability P of finding the body at that time. The amplitude of any wave may be positive as well as negative. Now negative probability is meaningless. Therefore ψ cannot be an observable quantity and it has no physical significance. But $|\psi|^2$, square of the absolute value of the wavefunction which is known as a probability density, has some meaning. The probability of finding the body described by the wavefunction ψ is proportional to $|\psi|^2$ at that time. If $|\psi|^2$ is large, the chances of the body's presence might be remembered, that as long as $|\psi|^2$ has a value, however small it may be, there is a definite chance of detecting the body there. If the wavefunction ψ is complex, with both real and imaginary parts, the probability density is given by $\psi^*\psi$ when ψ^* is the complex conjugate of ψ.

$$\psi = C + iD$$

$$\psi^* = C - iD$$

or

$$\psi^*\psi = C^2 - i^2D^2$$

$$= C^2 + D^2$$

Thus $\psi^*\psi$ is always a positive real quantity.

1.11 SCHRÖDINGER WAVE MECHANICS

Schrödinger worked from a more mathematical point of view, trying to find a beautiful theory for describing atomic events. He extended de Broglie's idea of matter waves to get a very elegant equation, known as Schrödinger's wave equation, for describing atomic processes. Schrödinger got this equation by pure thought, looking for some beautiful generalization of de Broglie's idea without any regard to the experimental developments of the time.

The basic idea from which Schrödinger started was the analogy existing between geometrical optics and classical mechanics. He pursued this analogy to its logical conclusion. He argued that just as when in any phenomenon in optics the dimensions of the beam of light or of the optical obstacles which it encounters in its path become comparable in size with the wavelength, the geometrical laws lose their significance; in the same way in the mechanics of electrons and atoms the concept of trajectory seems to breakdown in those cases where dimensions of mechanical orbits or their radii of curvature are of the same order of magnitude as the wavelength of the wave associated with the particle. Schrödinger thus argued that the inapplicability of ordinary mechanics to atomic problems appears to be of exactly the same character as the inapplicability of geometrical optics to the phenomena of interference and diffraction etc. He therefore set out to obtain an appropriate wave equation for the particles of the atomic world.

1.12 THE MOTION EQUATION OF MATTER WAVES

Schrödinger Time Independent Wave Equation

According to de Broglie theory, a particle of mass is always associated with a wave whose wavelength is given by $\lambda = h/mv$. If the particle has wave properties, it is expected that there should be some sort of wave equation which describes the behavior of the particle. Let x, y, z

be the coordinates of the particle and ψ, the wave displacement for the de Broglie at any time t. The classical differential equation of a wave motion is given by

$$\frac{\partial^2\psi}{\partial t^2} = v^2\left(\frac{\partial^2\psi}{\partial x^2}+\frac{\partial^2\psi}{\partial y^2}+\frac{\partial^2\psi}{\partial z^2}\right) = v^2\nabla^2\psi \qquad (1)$$

where

$$\nabla^2 = \frac{\partial^2}{\partial x^2}+\frac{\partial^2}{\partial y^2}+\frac{\partial^2}{\partial z^2}$$

(∇^2 being Laplacian operator) and v is the wave velocity.

The solution of equation (1) gives ψ as a periodic displacement in terms of time, *i.e.*,

$$\psi(x, y, z, t) = \psi_0(x, y, z)\, e^{-i\omega t} \qquad (2)$$

where $\psi_0(x, y, z)$ is a function of (x, y, z) and gives the amplitude at the point considered. The equation (2) can also be expressed as

$$\psi(\vec{r}, t) = \psi_0(\vec{r})\, e^{-i\omega t} \qquad (3)$$

Differentiating equation (3), twice we get

$$\frac{\partial^2\psi}{\partial t^2} = -\omega^2\psi_0(\vec{r})\, e^{-i\omega t} = -\omega^2\psi$$

Substituting the values of $\frac{\partial^2\psi}{\partial t^2}$ in equation (1), we have

$$\frac{\partial^2\psi}{\partial x^2}+\frac{\partial^2\psi}{\partial y^2}+\frac{\partial^2\psi}{\partial z^2}+\frac{\omega^2}{v^2}\psi = 0 \qquad (4)$$

But

$$\omega = 2\pi\nu = 2\pi\frac{v}{\lambda}$$

or

$$\frac{\omega}{v} = \frac{2\pi}{\lambda}$$

Substituting the value of $\frac{\omega}{v}$ in equation (4), we have

$$\frac{\partial^2\psi}{\partial x^2}+\frac{\partial^2\psi}{\partial y^2}+\frac{\partial^2\psi}{\partial z^2}+\frac{4\pi^2}{\lambda^2}\psi = 0$$

or

$$\nabla^2\psi + \frac{4\pi^2}{\lambda^2}\psi = 0 \qquad (5)$$

Now from de Broglie relation

$$\lambda = \frac{h}{mv}$$

$$\nabla^2\psi + \frac{4\pi^2}{h^2}m^2v^2\psi = 0 \qquad (6)$$

If E and V are the total and potential energies of the particle respectively, then its kinetic energy $\frac{1}{2}mv^2$ is given by

$$\frac{1}{2}mv^2 = E - V$$

$$m^2v^2 = 2m\,(E - V) \tag{7}$$

From equations (6) and (7), we have

$$\nabla^2\psi + \frac{4\pi^2}{h^2} \times 2m\,(E - V)\,\psi = 0$$

or
$$\nabla^2\psi + \frac{8\pi^2 m}{h^2}(E - V)\,\psi = 0 \tag{8}$$

Equation (8) is known as Schrödinger time independent wave equation. Substituting $\hbar = \frac{h}{2\pi}$ in equation (8), the Schrödinger wave equation can be written as

$$\nabla^2\psi + \frac{2m}{\hbar^2}(E - V)\psi = 0 \tag{9}$$

For free particle V = 0, hence the Schrödinger wave equation for the free particle can be expressed as

$$\boxed{V^2\,\Psi + \frac{2mE}{\hbar^2}\Psi = 0} \tag{10}$$

The Schrödinger Time Dependent Wave Equation

The Schrödinger time dependent wave equation may be obtained from the time independent wave equation by eliminating E.

Differentiating equation (3) with respect to t, we have

$$\frac{\partial\psi}{\partial t} = -i\omega\psi_0\,(\vec{r})\,e^{-i\omega t}$$

$$= -i\,(2\pi\nu)\,\psi_0\,(\vec{r})\,e^{-i\omega t}$$

$$= -2\pi\,i\,\nu\,\psi$$

$$= -2\pi i\frac{E}{h}\psi \qquad \left(\because E = h\nu \text{ or } \nu = \frac{E}{h}\right)$$

$$= -\frac{iE}{\hbar}\Psi$$

or
$$E\psi = i\hbar\frac{\partial\psi}{\partial t} \tag{11}$$

Substituting the value of Eψ in the Schrödinger time independent wave equation, we get

$$\nabla^2 \Psi + \frac{2m}{\hbar^2}\left[i\hbar \frac{\partial \Psi}{\partial t} - V\Psi \right] = 0$$

or
$$\nabla^2 \psi = -\frac{2m}{\hbar^2}\left[i\hbar - \frac{\partial \psi}{\partial t} - V\psi \right]$$

or
$$\boxed{-\frac{\hbar^2}{2m}\nabla^2 \psi + V\psi = i\hbar \frac{\partial \psi}{\partial t}} \tag{12}$$

This equation is known as Schrödinger time dependent wave equation.

Equation (12) can be written as

$$\left(-\frac{\hbar^2}{2m}\nabla^2 + V \right)\psi = i\hbar \frac{\partial}{\partial t}\psi$$

or
$$H\psi = E\psi \tag{13}$$

where
$$H = \left(-\frac{\hbar^2}{2m}\nabla^2 + V \right)$$

$$E = i\hbar \frac{\partial}{\partial t}$$

Equation (13) describes the motion of a non-relativistic material particle.

Simple/Applications of Schrödinger wave equation

After establishing that particles sometimes behave likes waves and that they can be described by wave functions. Let us discuss automatic analysis of particles in bound stares (such as electrons in atoms), including finding their possible wave functions and energy levels.

Our analysis involves funding solutions of the Schrödinger equation. For any quantum-mechanical system, the wave functions for any state of definite energy must be solutions of the Schrödinger equation for that system. Solving the Schrödinger equation gives us the probabilities finding a particle in various regions.

Particle in a Box

Now we will learn how to use the Schrödinger equation to determine the possible energy levels and the corresponding wave functions for various systems. We refer to the states of definite energy as stationary states. For simplicity, we begin with problems in which the particle can move only along the x-axis. The Schrödinger equation for such one dimensional problems is

$$-\frac{\hbar^2}{2m}\frac{d^2\Psi(x)}{dx^2} + U(x)\Psi(x) = E\Psi(x) \tag{14}$$

This equation is really a statement that the kinetic energy [represented by the term $-(\hbar/2m)d^2\Psi(x)/dx^2$] plus the potential energy [represented by the term $U(x)\,\Psi(x)$] is equal to the total energy [represented by the term $E\,\Psi(x)$]. The fundamental problem is then the following: For a given potential energy functions $U(x)$, what are the possible stationary-state wave functions $\Psi(x)$, and what are the corresponding energies E?

We have earlier solved this problem for the case $\Psi(x) = 0$, corresponding to a free particle. The allowed wave functions and corresponding energies are

$$\Psi(x) = Ae^{ikx} \quad E = \frac{\hbar^2 k^2}{2m} \qquad \text{(free particle)} \tag{15}$$

The wave number k is equal to $2\pi/\lambda$, where λ is the wavelength. We found that k can have any real value, so the energy E of a free particle can have any value from zero to infinity. Furthermore, the particle can be found with equal probability at any value of x from $-\infty$ to the ∞.

Now let's look at a simple model in which a particle is bound so that it cannot escape to infinity, but rather is confined to a restricted region of space. Our system consists of a particle confined between two rigid walls separated by a distance L.

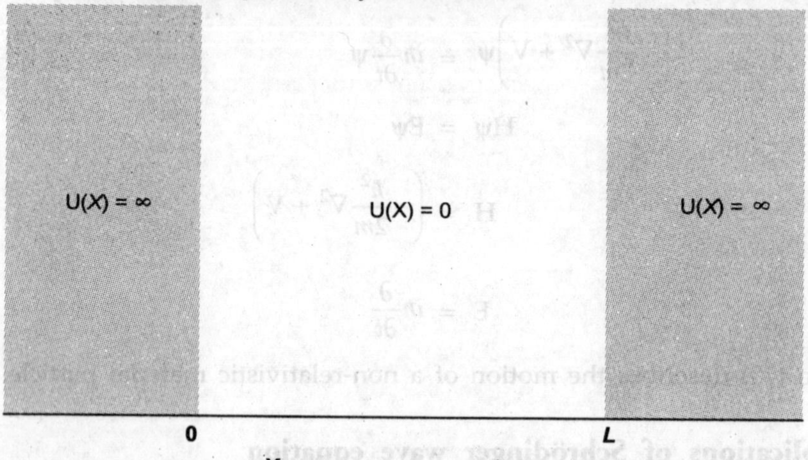

FIGURE 6 **Visualization of a potential energy function which is zero between x = 0 and x = L, and infinity elsewhere (indicated by the shaded regions). The particle cannot go into the region of infinite potential, and so is constrained to be in the region x = 0 to L. This arrangement is often called a 'particle in a box'.**

The motion is purely one-dimensional, with the particle moving along the x-axis only and the walls at $x = 0$ and $x =$ L. The potential energy corresponding to the rigid walls is infinite, so the particle cannot escape; between the walls, the potential energy is zero. This situation is often described as a "particle in a box." This model might represent an electron that is free to move within a long, straight molecule or along a very thin wire Figure.

To solve the Schrödinger equation for this system; we begin with some restrictions on the particle's wave function $\Psi(x)$. Because the particle is confined to the region $0 \geq x \leq$ L, we expect Ψ we expect $\Psi(x)$ to be zero outside that region. This agrees with the Schrödinger equation: if the term $U(x)\,\Psi(x)$ in equation (14) is to be finite, $\Psi(x)$ must be zero where $U(x)$ is infinite.

Furthermore, $\Psi(x)$ must be a continuous function to be a mathematically well behaved solution to the Schrödinger equation. If it is, then $\Psi(x)$ must be zero at the region's boundary, $x = 0$ and $x =$ L. these two conditions serve as boundary conditions for the problem. They should look familiar, because they are the same conditions that we used to find the normal modes of a vibrating string.

An additional condition is that to calculate the second derivative $d^2\Psi(x)/dx^2$ in equation (14) the first derivative $d\Psi(x)/dx$ must also be continuous except at points where the potential energy becomes infinite (as it does at the walls of the box). This is analogous to the requirement that a vibrating string cannot have a kink in it (which would correspond to a discontinuity in the first derivative of the wave function) except at the ends of the string.

We now solve for the wave functions in the region $0 \geq x \leq L$ subject to the above conditions. In this region $U(x) = 0$, so the wave function in this region must satisfy

$$-\frac{\hbar^2}{2m}\frac{d^2\Psi(x)}{dx^2} = E\Psi(x) \qquad \text{(particle in a box)} \qquad (16)$$

Equation (16) is the same Schrödinger equation as for a free particle, so it is tempting to conclude that the wave functions and energies are given by equation (15). It is true that $\Psi(x) = Ae^{ikx}$ satisfies the Schrödinger equation with $U(x)\,dx = ikAe^{ikx}$. However, this wave function does not satisfy the boundary conditions that $\Psi(x)$ must be zero at $x = 0$ and at $x = L$: at $x = 0$ the wave function in equation (15) is equal to $Ae^0 = A$, and at $x = L$ it is equal to Ae^{ikL}. (These would be equal to zero if $A = 0$, but then the wave function would be zero and there would be no particle at all!)

The way out of this dilemma is to recall Example 1, in which we found that a more general stationary-state solution to the Schrödinger equation with $U(x) = 0$ is

$$\Psi(x) = A_1e^{ikx} + A_2e^{-ikx} \qquad (17)$$

This wave function is a superposition of two waves: one traveling in the +x-direction of amplitude A_2. This is analogous to the situation for standing waves on a string which we can regard as the superposition of two sinusoidal waves propagating in opposite directions. The energy that corresponds to equation (17) is $E = \hbar^2 k^2/2m$, just as for a single wave.

To see whether the wave function given by equation (12) can satisfy the boundary conditions, let's first rewrite it in terms of sines and cosines using Euler's formula,

$$\Psi(x) = A_1(\cos kx + i\sin kx) + A_2[\cos(-kx) + i\sin(-kx)]$$
$$A_1(\cos kx + i\sin kx) + A_2[\cos kx - i\sin kx]$$
$$(A_1 + A_2)\cos kx + i(A_1 - A_2)\sin kx \qquad (18)$$

At $x = 0$ this is equal to $\Psi(0) = A_1 + A_2$, which must equal zero if we are to satisfy the boundary condition at that point. Hence $A_2 = -A_1$, and equation (18) becomes

$$\Psi(x) = 2iA_1, \sin kx = C\sin kx \qquad (19)$$

We have simplified the expression by introducing the constant $C = 2iA_1$. (We'll come back to this constant later.) We can also satisfy the second boundary condition that ? = 0 at x = L by choosing values of k such that $kL = n\pi$ (π = 1, 2, 3 ...). Hence equation (19) does give the stationary-state wave functions for a particle in a box in the region $0 \geq x \leq L$. (Outside this region, $\Psi(x) = 0$). The possible values of k and the wavelength $\lambda = 2\pi/k$ are

$$k\frac{n\pi}{L} \text{ and } \lambda = \frac{2\pi}{K} = \frac{2L}{n} \quad (n = 1, 2, 3...) \qquad (20)$$

Just as for the string in the length L of the region is an integral number of half-wavelengths.

The possible energy levels are given by $E = \hbar^2 k^2/2m = p^2/2m$, where $p = \hbar k + (h/2\pi)(2\pi/\lambda) = h/\lambda$ is the magnitude of momentum of a free particle with wave number k and wavelength λ. This makes sense, since inside the region $0 \geq x \leq L$ the potential energy is zero and the energy is all kinetic. For each value of n there are corresponding values of p, λ and E; let's call them P_m, λn, and E_n. Putting the pieces together, we get

$$P_n = \frac{h}{\lambda_n} = \frac{nh}{2L} \tag{21}$$

and

$$E_n = \frac{p^2}{2m} = \frac{n^2 h^2}{8mL^2} = \frac{n^2 \pi^2 \hbar^2}{2mL^2} \quad (n = 1, 2, 3....) \tag{22}$$

These are the possible energy levels for a particle in a box. Each energy level has its own value of the quantum number n and a corresponding wave function, which we denote by Ψ_n. When we replace k in equation (19) by $n\pi/L$ from equation (20), we find

$$\Psi_n(x) = C \sin \frac{n\pi x}{L} \quad (n = 1, 2, 3.....) \tag{23}$$

we have wave functions $\Psi_n(x)$ for n = 1,2,3,4, and 5. Note that these functions look identical to those for a standing wave on a string. The levels of successively higher energies, proportional to n^2, are spaced farther and father apart. There are an infinite number of levels because the walls are perfectly rigid; even a particle of infinitely great kinetic energy is confined within the box.

Note that the energy of a particle in a box cannot be zero. Equation (22) shows that E=0 would require n = 0, but substituting n = 0 into equation (23) gives a zero wave function. Since a particle is described by a nonzero wave function, this means that there cannot be a particle with E = 0.

Electron in an atom-size box

Example. *Find the lowest energy level for a particle in a box if the particle is in electron in a box 5.0 × 10^{-10} m across, or a little bigger than and atom.*

Solution. The lowest energy level (that is, the ground level) corresponds to n = 1 in equation (22)

From equation (22),

$$E_1 = \frac{h^2}{8mL^2} = \frac{(6.626 \times 10^{-34}\ \text{J.S})^2}{8(9.109 \times 10^{-31}\ kg)(5.0 \times 10^{-10}\ m)^2}$$

$$= 2.4 \times 10^{-10}\ \text{J} = 1.5\ eV$$

A particle trapped in a box is rather different from an electron bound in an atom, but it is reassuring that this energy is of the same order of magnitude as actual atomic energy levels.

It can easily be shown that replacing the electron with a proton or a neutron (m = 1.67 × 10^{-27} kg) in a box the width of a medium sized nucleus (L = 1.1 × 10^{-14} m) gives E_1 = 1.7 MeV. This shows us that the energies for particles in the nucleus are about a million times larger than the energies of electrons in atoms, giving us a clue as to why each nuclear fission and fusion reaction provides so much more energy than each chemical reaction.

Probability and Normalization

Several additional points need to be discussed. First is the probability interpretation of the wave function Ψ, in our one dimensional situation the quantity $|\Psi(x)|^2 dx$ is proportional to the proportional to the probability that the particle will be found within a small interval dx

about x. For a particle in a box,

$$|\Psi(x)|^2 \, dx = C^2 \sin^2 \frac{n\pi x}{L} \, dx$$

Both $\Psi(x)$ and are $|\Psi(x)|^2$ plotted in Fig. 7 for $n = 1$, 2, and 3. We note that not all positions are equally likely. This is in contrast to the situation in classical mechanics, in which all positions between $x = 0$ and $x = L$ are equally likely. We see in Fig. 7 that $|\Psi(x)|^2 = 0$ at some points, so the probability is zero of finding the particle exactly at these points. Don't let that bother you; the uncertainty principle has already shown us that we can't measure position exactly. The particle is localized only to be somewhere between $x = 0$ and $x = L$, not to be exactly at some particular value of x.

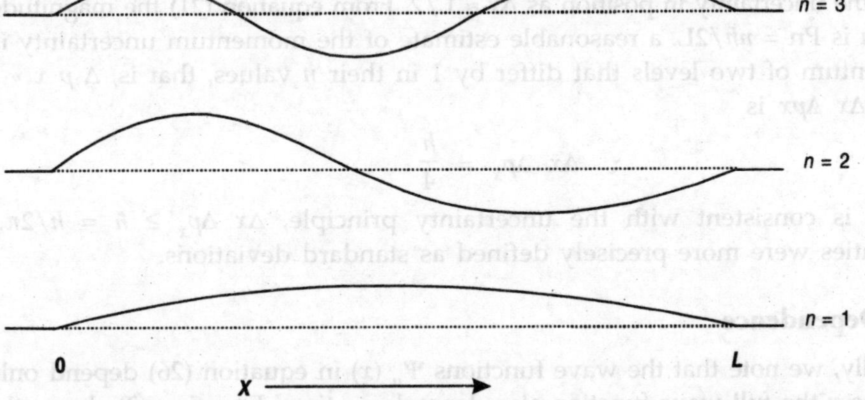

FIGURE 7 **Plots of the four lowest energy wavefunctions for the potential energy function shown in Fig. 6 (the particle in a box). The wavefunctions have been displaced vertically so that they can be seen clearly; the dotted horizontal line gives the position of $\psi = 0$ in each case. Note how the wavefunctions are zero for x less than 0 and greater than L, and how the number of nodes increases with *n*.**

We know that the particle must be somewhere in the universe, that is, between $x = -\infty$ and $x = +\infty$. So the sum of the probabilities for all the dx's every where (the total probability of finding the particle) must equal 1. That is,

$$\int_{\infty}^{\infty} |\Psi(x)|^2 \, dx = 1 \quad \text{(Normalization condition)} \tag{24}$$

A wave function is said to be normalized if it has a constant such as C in equation (23) that is calculated to make the total probability equal 1 in equation (24). The process of determining the constant is called normalization, the normalized wave function, $|\Psi(x)|^2 \, dx$ is not merely proportional to, but equals, the probability of finding the particle in dx. We call $|\Psi(x)|^2$ the probability distribution function.

Now let's normalize the wave function $\Psi(x)$ given by equation (23) for the particle in a box. Since $\Psi(x)$ is zero except between $x = 0$ and $x = L$, equation (24) becomes

$$\int_0^L C^2 \sin^2 \frac{n\pi x}{L} dx = 1 \tag{25}$$

We can evaluate this integral using the trigonometric identity $\sin^2 \theta = ½ (1 - \cos 2\theta)$; the result is $C^2 L/2$. Thus our probability interpretation of the wave function demands that $C^2 L/2 = 1$, or $C = (2/L)^{1/2}$, the constant C is not arbitrary. (This is in contrast to the classical vibrating string problem, in which C represents an amplitude that depends on initial conditions.) Thus the normalized stationary-state wave functions for a particle in a box are

$$\Psi_n(x) = \sqrt{\frac{2}{L}} \sin \frac{n\pi x}{L} \quad (n = 1, 2, 3...)\text{(particle in a box)} \tag{26}$$

Next let's check whether our results for the particle in a box are consistent with the uncertainty principal. We can write the position of the particle as $x = L/2 \pm L/2$, so we can estimate the uncertainty in position as $\Delta x \approx L/2$. From equation (21) the magnitude of momentum in state n is $Pn = nh/2L$. a reasonable estimate of the momentum uncertainty is the difference in momentum of two levels that differ by 1 in their n values, that is, $\Delta\, p\, x$ » $h/2L$. Then the product $\Delta x\ \Delta px$ is

$$\Delta x\ \Delta p_x = \frac{h}{4}$$

This is consistent with the uncertainty principle, $\Delta x\ \Delta p_x \geq \hbar = h/2\pi$, in which the uncertainties were more precisely defined as standard deviations.

Time Dependence

Finally, we note that the wave functions $\Psi_n(x)$ in equation (26) depend only on the spatial coordinate x the full wave function also depends on time. Equation (3) shows that if $\Psi(x)$ is the wave function for a state of definite energy E, the full time-dependent wave function $\Psi(x, t)$ is

$$\Psi(x, t) = \Psi(x)\, e^{-iEt/\hbar} \tag{27}$$

(time-dependent wave function for a stationary state)

From equation (26) it follows that the time-dependent wave functions for a particle in a box are

$$\Psi(x, t) = \sqrt{\frac{2}{L}} \sin\left[\frac{n\pi x}{L}\right] e^{-iE_n t/\hbar} \quad (n = 1, 2, 3....) \tag{28}$$

In this expression the energies E_n are given by equation (22).

The absolute value of $e^{-iEt/\hbar}$ is unity: $\left|e^{-iEt/\hbar}\right|^2 = \left|e^{+iEt/\hbar}\right|^2 \left|e^{-iEt/\hbar}\right|^2 = e^0 = 1$. Hence $\left|\Psi(x,t)\right|^2$ is equal to $\left|\Psi(x,t)\right|^2$, and the probability distribution function does not depend on time. We know this is the origin of the term stationary state for a state of definite energy. So in calculations with states that have definite energy, we are justified in omitting the time factor. We'll be concerned mostly with stationary states in our applications of quantum mechanics.

A potential well is a potential-energy function U(x) that has a minimum. In Newtonian mechanics a particle trapped in a potential well can vibrate back and forth with periodic motion. Our first application of the Schrödinger equation, the particle in a box, involved a rudimentary potential well with a function U(x) that is zero within a certain interval and infinite every where else. This function corresponds to a few situations found in nature, but the correspondence is only approximate.

A potential well that serves as a better approximation to several actual physical situations is a well with straight sides but finite height. Figure 6 shows a potential-energy function that is zero in the interval $0 \geq x \leq L$ and has the value U_0 outside this interval. This function is often called a square-well potential. It could serve as a simple model lf an electron within a metallic sheet with thickness L, moving perpendicular to the surfaces of the sheet. The electron can move freely inside the metal but has to climb a potential-energy barrier with height U_0 to escape from either surface of the metal. The energy U_0 is related to the work function that is similar to the photoelectric effect. In a three-dimensional application a spherical version of a potential well can be used to approximately describe the motions of protons and neutrons within a nucleus.

In Newtonian mechanics the particle is trapped (localized) in a well if the total energy E is less than U_0. In quantum mechanics, such a trapped state is often called a bound state. All states are bound for an infinitely deep well, but if E is greater than U_0 for a finite well, the particle is not bound.

For a finite square well, let's consider bound-state solutions of the Schrödinger equation, corresponding to $E < U_0$. The easiest approach is to consider separately the regions where $U = 0$ and where $U = U_0$. Where $U = 0$, the Schrödinger equation reduces to equation (16). Rearranging this equation, we find

$$\frac{d^2 \Psi(x)}{dx^2} = \frac{2mE}{\hbar^2} \Psi(x) \tag{29}$$

From equation (18) we can express the solutions of this equation as combinations of cos kx and sin kx, where $E = \hbar^2 k^2/2m$ and so $k = \sqrt{2mE/\hbar}$. Hece inside the square well ($o \geq x \leq L$) we have

$$\Psi(x) = A \cos \frac{\sqrt{2mE}}{\hbar} x + B \sin \frac{\sqrt{2mE}}{\hbar} x \tag{30}$$

Where A and B are constants. So far, this looks a lot like the particle-in-a-box analysis. In the regions $x < 0$ and $x > L$ we use equation (14) with $U = U_0$. Rearranging, we get

$$\frac{d^2 \Psi(x)}{dx^2} = \frac{2m(U_0 - E)}{h^2} \Psi(x) \tag{31}$$

The quantity $U_0 - E$ is positive, so the solutions of this equation are exponential. Using k (the Greek letter "kappa") as positive in the abbreviation $k = [2m\ (U_0 - E)]^{1/2}\ 1\hbar$, we can write the solutions as

$$\Psi(x) = Ce^{kx} + De^{-kx} \qquad \text{(outside the well)} \tag{32}$$

Where C and D are constants with different values in the $x < 0$ and $x > L$ regions.

We see that the bound-state wave functions for this system are sinusoidal inside the well and exponential outside it. We have to use the positive exponent in the $x < 0$ region and the negative exponent in the $x > L$ region. That is, $D = 0$ for $x < 0$ and $C = 0$ for $x > L$. without this choice of constants, Ψ would approach infinity as $|x|$ approached infinity, and the normalization condition, equation (24) could not be satisfied. We also have to match the wave functions so that they satisfy the boundary conditions that we mentioned $\Psi(x)$ and $d\Psi(x)/dx$ must be continuous at the boundary points ($x = 0$ and $x = L$). If the wave function Ψx or the slope $d\Psi(x)\ ldx$ were to change discontinuously at a point, the second derivative $d^2\Psi(x)/dx^2$ would be infinite at that point. But that would violate the Schrödinger equation, which says that at every point, $d^2\Psi(x)/dx^2$ is proportional to U – E. in our situation, U – E is finite everywhere, so $d^2\Psi(x)/dx^2$ must also be finite everywhere.

Comparing Finite and Infinite Square Wells

Matching the sinusoidal and exponential functions at the boundary points so that they join smoothly is possible only for certain specific values of the total energy E, so this requirement determines the possible energy levels of the finite square well. There is no simple formula for the energy levels as there was for the infinitely deep well. Finding the levels is a fairly complex mathematical problem that requires solving a transcendental equation by numerical approximation; we won't go into the details. The most striking features of the wave function are the "exponential tails" that extend outside the well into regions that are forbidden by Newtonian mechanics (because in those regions the particle would have negative kinetic energy). We see that here is some probability for finding the particle outside the potential well, despite the fact that in classical mechanics this is impossible. This penetration into classically forbidden regions is a quantum effect that has no classical analog for particles.

1.13 WAVE VELOCITY AND GROUP VELOCITY

According to the de Broglie concept, matter has a dual nature *i.e.*, sometimes it behaves like a particle and sometimes like a wave. Now the question is how these contradictory statements are explained. First of all Schrödinger explained the dual nature in terms of a wave packet. A wave packet comprises a group of waves slightly differing in velocity and wavelength, with phases and amplitude such that they interfere constructively over a small region of space where the particle can be located, and outside space they interfere destructively so that the amplitude reduces to zero. Such a wave packet (Figure 8) moves with its own velocity called the group velocity. The individual waves forming the wave packet have an average velocity which is called as *phase velocity*. It can be shown that the velocity of material particle v is the same as group velocity.

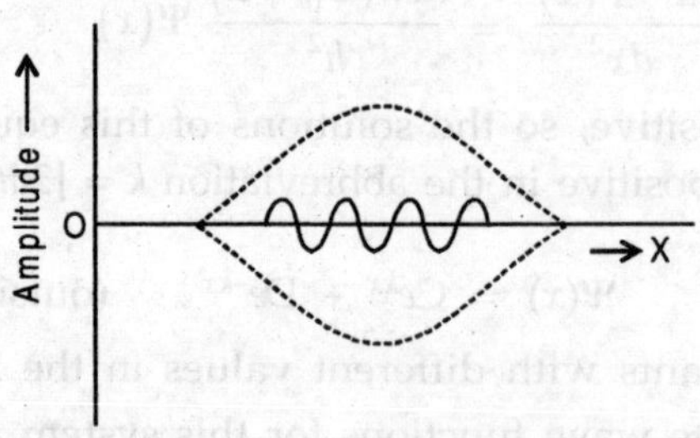

FIGURE 8 Wave packet.

Let us consider the case of two wave trains having the same amplitude but slightly different angular frequencies (ω and ω') and phase velocities (u and u'). The waves can be represented as

$$y_1 = a \sin(\omega t - kx) \quad (1)$$

and

$$y_2 = a \sin(\omega' t - k' x) \quad (2)$$

where k and k' are propagation constants defined as $(2\pi/\lambda)$. The resultant wave is given by

$$y = y_1 + y_2 = a \sin(\omega t - kx) + a \sin(\omega' t - k'x)$$

$$= 2a \cos\left[\left(\frac{\omega - \omega'}{2}\right)t - \left(\frac{k - k'}{2}\right)x\right] \times \sin\left[\left(\frac{\omega + \omega'}{2}\right)t - \left(\frac{k + k'}{2}\right)x\right]$$

$$\because \quad \sin A + \sin B = 2 \sin\left(\frac{A + B}{2}\right)\cos\left(\frac{A - B}{2}\right)$$

$$\approx 2a \cos\left[\left(\frac{d\omega}{2}\right)t - \left(\frac{dk}{2}\right)x\right] \sin(\omega t - kx) \quad (3)$$

where, $\dfrac{\omega + \omega}{2} = \omega$ and $\dfrac{k + k'}{2} = k$ (approx.)

Equation (3) represents a wave of angular frequency ω and propagation constant k. The phase velocity u of the resultant wave is given by $u = \dfrac{\omega}{k}$ (approx.) = same as that of each comprising wave.

Amplitude of the resultant wave is modified. The amplitude is given by

$$2a \cos\left[\left(\frac{d\omega}{2}\right)t - \left(\frac{dk}{2}\right)x\right]$$

or

$$2a \cos \frac{d\omega}{2}\left[t - \frac{dk}{d\omega}x\right] = 2a \cos \frac{d\omega}{2}\left[t - \frac{x}{G}\right]$$

where G is known as group velocity. This is given as

$$G = \frac{d\omega}{dk} = \frac{\omega - \omega'}{k - k'} \quad (4)$$

But $\omega = uk$ or $d\omega = du\, k + u\, dk$

$$\therefore \quad \frac{d\omega}{dk} = u + k\frac{du}{dk} \quad (5)$$

From equations (4) and (5), we get

$$G = u + k\frac{du}{dk}$$

or

$$G = u + k\frac{du}{d\lambda} \times \frac{d\lambda}{dk} \quad (6)$$

Since $k = \dfrac{2\pi}{\lambda}$; hence $\dfrac{d\lambda}{dk} = -\dfrac{2\pi}{k^2}$

Now equation (6) becomes,

$$G = u - \lambda\frac{du}{d\lambda} \quad (7)$$

Equation (7) represents a relationship between group velocity G and phase velocity *u*. Equation (7) can also be written in the following form:

$$G = \lambda^2\left[\frac{u}{\lambda^2} - \frac{1}{\lambda}\left(\frac{du}{d\lambda}\right)\right] = -\lambda^2 \frac{d}{d\lambda}\left(\frac{u}{\lambda}\right) = -\lambda^2 \frac{d\nu}{d\lambda}$$

or

$$\frac{1}{G} = -\frac{1}{\lambda^2}\frac{d\lambda}{d\nu} = \frac{d}{d\nu}\left(\frac{1}{\lambda}\right) \tag{8}$$

If E and V represent the total and potential energy of the particle respectively, then the kinetic energy of the particle is given by

$$\frac{1}{2}mv^2 = E - V \quad \text{where } v \text{ is the particle velocity}$$

or

$$v = \left[\frac{2(E-V)}{m}\right]^{1/2}$$

According to the de Broglie formula $\lambda = h/mv$

$$\therefore \quad \frac{1}{\lambda} = \frac{mv}{h} = \frac{m}{h}\left[\frac{2(E-V)}{m}\right]^{1/2} \tag{9}$$

Substituting the value of $(1/\lambda)$ from equation (9) in equation (8), we get

$$\frac{1}{G} = \frac{d}{d\nu}\left[\frac{m}{h}\left\{\frac{2(E-V)}{m}\right\}^{1/2}\right]$$

$$= \frac{d}{d\nu}\left[\frac{m}{h}\left\{\frac{2(h\nu-V)}{m}\right\}^{1/2}\right] \quad \because E = h\nu$$

$$= \frac{1}{h}\frac{d}{d\nu}\left[\{2m(h\nu - V)\}^{1/2}\right]$$

$$= \frac{1}{h}\cdot\frac{1}{2}\{2m(h\nu - V)\}^{\frac{1}{2}-1}.\,2mh$$

$$= \frac{m}{[2m(E-V)]^{1/2}} = \left[\frac{m}{2(E-V)}\right]^{1/2} = \frac{1}{v} \tag{10}$$

$\therefore$ Group velocity G = particle velocity *v*.

Hence we can say that a material particle in motion is equivalent to a group of waves or a wave packet.

When the material particle is considered to be a wave packet, it is necessary to postulate the existence of a guiding wave for it. The equation of a guiding wave was derived by Schrödinger. The physical significance of the Schrödinger wave equation is that it relates the amplitude of the guiding wave to the probability of finding the material particle at a point. If the amplitude of the guiding wave is zero at a certain point in space, then the probability of finding the particle at that point is infinitesimal. The mechanical process is accompanied by the wave process. On the other hand, if the amplitude of the guiding wave is large at any point in the space, then the probability of finding the particle at that point is maximum. This is equivalent to a particle in the packet which moves with group velocity v. Thus the wave packet plus the guiding wave passes the properties of a particle moving with velocity v. With this idea phenomena like interference, diffraction, etc., can be explained.

Example 1. *Calculate the de Broglie wavelength associated with the following:*

(i) A golf ball of 50 g moving with a velocity of 20 m/sec.

(ii) A proton moving with a velocity of 2200 m/sec.

(iii) An electron moving with a kinetic energy of 10 eV.

Solution. We know that $\lambda = h/mv$

(*i*) For a golf ball of 50 g and velocity 20 m/sec.

$$\therefore \quad \lambda = \frac{6.625 \times 10^{-34} \text{ joule-sec}}{0.05 \text{ kg} \times 20 \text{ m/sec}}$$

$$= 6.625 \times 10^{-34} \text{ m.}$$

This is too small to be detected.

(*ii*) In the case of a proton having mass 1.67×10^{-27} kg, we have

$$\lambda = \frac{6.625 \times 10^{-34} \text{ joule-sec}}{1.67 \times 10^{-27} \text{ kg} \times 2200 \text{ m/sec}}$$

$$= 1.91 \times 10^{-10} \text{ m.}$$

This wavelength is of the order of an X ray wavelength.

(*iii*) In case of an electron with kinetic energy 10 eV, we have

$$\lambda = \frac{h}{\sqrt{2mE}} \quad \text{where E = Kinetic energy}$$

$$= \frac{6.625 \times 10^{-34} \text{ joule-sec}}{\sqrt{[2\,(9.11 \times 10^{-31}\text{kg})\,(10\text{eV}) \times 1.602 \times 10^{-10}\text{joule/eV}]}}$$

$$= 3.9 \times 10^{-10} \text{m} = \mathbf{3.9\ Å.}$$

Here it should be remembered that if the kinetic energy of the electron becomes comparable to its rest mass, we must use the relativistic expression for momentum mv.

Example 2. *What is the de Broglie wavelength of an electron which has been accelerated from rest through a potential difference of 100 V.*

Solution. We know that

$$\lambda = \frac{12.25}{\sqrt{V}} \text{ Å}$$

Here $\quad V = 100$ volts

$$\therefore \quad \lambda = \frac{12.25}{\sqrt{100}}$$

$$= \mathbf{1.225\ Å.}$$

Example 3. *Calculate the de Broglie wavelength associated with a proton moving with a velocity equal to 1/20th, of the velocity of light.*

Solution. Velocity of proton

$$v = \frac{1}{20} \times \text{velocity of light}$$

$$= \frac{1}{20} \times 3 \times 10^8 = 1.5 \times 10^7 \text{ m/sec.}$$

Mass of the proton $= 1.67 \times 10^{-27}$ kg

$$\therefore \quad \lambda = \frac{h}{mv} = \frac{6.62 \times 10^{-34}}{1.67 \times 10^{-27} \times 1.5 \times 10^7} = \mathbf{2.64 \times 10^{-14}\ m.}$$

Example 4. *Find the energy of the neutron in units of electron volt whose de Broglie wavelength is 1 Å.*

Given: *mass of the neutron* $= 1.674 \times 10^{-27}$ *kg.*

Planck's constant $h = 6.60 \times 10^{-34}$ *joule-sec.*

Solution. We know that

$$\lambda = \frac{h}{mv} = \frac{h}{\sqrt{2mE}}$$

$$E = \frac{h^2}{2m\lambda^2}$$

in the given problem $m = 1.674 \times 10^{-27}$ kg

$$\lambda = 1\ Å = 10^{-10} \text{ m}$$

$$h = 6.60 \times 10^{-24} \text{ joule-sec.}$$

$$\therefore \quad E = \frac{(6.60 \times 10^{-34})^2}{2 \times 1.674 \times 10^{-27} \times (10^{-10})^2} = 13.01 \times 10^{-21} \text{ joules}$$

$$= \frac{13.01 \times 10^{-21}}{1.6 \times 10^{-10}} = \mathbf{8.13 \times 10^{-2}\ eV.}$$

Example 5. *Calculate the energy in an electron volt of an electron wave of* $\lambda = 3 \times 10^{-2}$ *m. Given:* $h = 6.62 \times 10^{-34}$ *Js.*

Solution. We know that energy is given by

$$E = \frac{h^2}{2m\lambda^2}$$

Here $h = 6.62 \times 10^{-34}$ Js, $m = 9.1 \times 10^{-31}$ kgm and $\lambda = 3 \times 10^{-2}$ m

$$\therefore \quad E = \frac{(6.62 \times 10^{-34})^2}{2 \times 9.1 \times 10^{-31} \times (3 \times 10^{-2})} \approx 0.27 \times 10^{-33} \text{ joules}$$

$$\approx \frac{0.27 \times 10^{-33}}{1.6 \times 10^{-19}} \text{ eV} \approx \mathbf{1.68 \times 10^{-15} \text{ eV.}}$$

Example 6. *What voltage must be applied to an electron microscope to produce electrons of wavelength 0.40 Å?*

Solution. The de Broglie wavelength λ is given by

$$\lambda = \frac{h}{p} = \frac{h}{\sqrt{2mE}}$$

$$E = eV \quad \text{where V is in volts}$$

Here, $h = 6.6 \times 10^{-34}$ Js, $m = 9.1 \times 10^{-31}$ kgm

and $\lambda = 0.4 \times 10^{-19}$ m

$$\therefore \quad 0.4 \times 10^{-10} = \frac{6.6 \times 10^{-34}}{\sqrt{(2 \times 9.1 \times 10^{-31} \times 1.6 \times 10^{-19} V)}}$$

Solving, we get $V = \mathbf{960 \text{ volts.}}$

Example 7. *What would be the wavelength of a quantum of radiant energy emitted, if an electron is transmitted into radiation and converted into one quantum.*

Solution. According to Planck's theory, the energy E associated with one quantum = $h\nu$, where ν is the frequency of radiation.

When the energy of an electron is transmitted into radiation, we have

$$E = mc^2 \quad \text{(mass energy relation)}$$

$$\therefore \quad h\nu = mc^2 \quad \text{or} \quad \frac{hc}{\lambda} = mc^2$$

$$\therefore \quad \lambda = \frac{h}{mc} = \frac{6.6 \times 10^{-34}}{9.1 \times 10^{-10} \times 3 \times 10^8}$$

$$= 0.0244 \times 10^{-10} \text{ m} = \mathbf{0.0244 \text{ Å.}}$$

Example 8. *Compute the de Broglie wavelength of 10^{11} keV neutron. The mass of the neutron may be taken as 1.675×10^{-27} kg.*

Solution. Here the kinetic energy of the neutron

$$= 10^{11} \text{ keV} = 10^{14} \text{ eV}$$

$$= 10^{14} \times 1.6 \times 10^{-19} \text{ joule}$$

$$= 1.6 \times 10^{-5} \text{ joule.}$$

Now $$\frac{1}{2} mv^2 = 1.6 \times 10^{-5}$$

or $$v = \left(\frac{2 \times 1.6 \times 10^{-5}}{1.675 \times 10^{-27}}\right)^{\frac{1}{2}} \text{ m/s}$$

Again $$\lambda = \frac{h}{mv} = \frac{6.625 \times 10^{-34}}{1.675 \times 10^{-27}} \times \left(\frac{1.675 \times 10^{-27}}{2 \times 1.6 \times 10^{-5}}\right)^{\frac{1}{2}}$$

$$= \mathbf{2.86 \times 10^{-18}\ m.}$$

Example 9. *Compute the de Broglie wavelength of a proton whose kinetic energy is equal to the rest energy of an electron. The mass of a proton is 1836 times that of an electron.*

Solution. According to Einstein's energy mass relation, we have

$$E = m_0c^2 = 9.1 \times 10^{-31} \times (3 \times 10^8)^2$$

$$= 81.9 \times 10^{-15} \text{ joule}$$

$$\text{Mass of the proton} = 1836 \times 9.1 \times 10^{-31}$$

Now $$\frac{1}{2}mv^2 = 81.9 \times 10^{-15}$$

or $$v = \left(\frac{2 \times 81.9 \times 10^{-15}}{1836 \times 9.1 \times 10^{-31}}\right)^{\frac{1}{2}}$$

Again $$\lambda = \frac{h}{mv}$$

$$= \frac{6.62 \times 10^{-34}}{1836 \times 9.1 \times 10^{-31}} \times \left(\frac{1836 \times 9.1 \times 10^{-31}}{2 \times 81.9 \times 10^{-15}}\right)^{\frac{1}{2}}$$

$$= 4 \times 10^{-14}\ \text{m} = 4 \times 10^{-4}\ \text{Å} = \mathbf{0.004\ Å.}$$

1.14 THE UNCERTAINTY PRINCIPLE

It may sound very strange that one of the basic pillars of modern physics is the principle of uncertainty. Heisenberg proposed it in 1927. It was arrived at from the consideration of the dualistic nature of electrons. The very definition of the corpuscle implies that it has a definite momentum and occupies a definite position in space. Unless we can determine simultaneously both the momentum and position in space we cannot say that the particle has been observed in the real sense. According to de Broglie's concept of matter waves, the electron is somewhere within the wave packet. By the wave packet is meant a wave-like disturbance whose amplitude is different from zero only in a bounded region. The statement that the position of the electron is known within an accuracy Δx at a time t can be pictured by imagining a wave packet in the proper position with an approximate extension Δx (Figure 9). This wave packet moves and changes its size and shape. The velocity of the electron is the velocity of the wave packet. But since the wave packet is not very narrow and there is some dispersion within it, there will be a certain indeterminacy in the velocity of the electron also. This is the essence of the uncertainty principle.

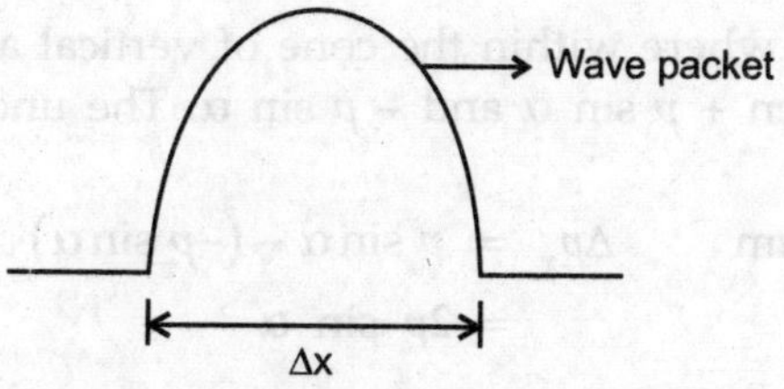

FIGURE 9 **Uncertainty in the position of the particle in a wave packet.**

Statement of the Uncertainty Principle

Heisenberg's uncertainty principle can be stated as:

It is impossible to measure precisely and simultaneously the values of both members of particular pairs of physical variables that describe the behavior of an atomic system.

These pairs of variables, called the "canonical conjugates", are:

- a rectangular coordinate x of a particle and the corresponding momentum component p_x;
- the energy E of a particle and the time t at which it is measured;
- the angular momentum L and the angular position ϕ, etc.

Quantitatively the principle of uncertainty states that the order of magnitude of the product of uncertainties in the knowledge of the two variables must be at least of the order of Planck's constant h. Thus

$$\begin{aligned} \Delta x.\Delta p_x &\geq h \\ \Delta t.\Delta E &\geq h \\ \Delta\phi.\Delta L &\geq h \end{aligned} \tag{1}$$

1.15 EXPERIMENTAL ILLUSTRATIONS OF THE UNCERTAINTY PRINCIPLE

We shall now describe some idealized experiments which illustrate the uncertainty principle.

Determination of the Position by a Microscope

Let us attempt to measure the position of a small object, say an electron in the field of view in a microscope. According to the theory of the resolving power of a microscope, the distance between two points that can be resolved by the microscope is given by

Uncertainty in position = Resolving power

$$\Delta x > \frac{\lambda}{2 \sin \alpha} \tag{1}$$

where λ is the wavelength of light used and α the semi-vertical angle of the cone of light.

Now, in order to be able to observe the electron at least one photon of light must bounce off it into the field of view in a microscope (Figure 10).

This photon will be any where within the cone of vertical angle 2α. Thus the x-component of its momentum lies between $+ p \sin\alpha$ and $- p \sin\alpha$. The uncertainty in the x-component of the momentum is therefore.

Uncertainty in Momentum
$$\Delta p_x = p\sin\alpha - (-p\sin\alpha)$$
$$= 2p\sin\alpha$$
$$= 2\frac{h}{\lambda}\sin\alpha \qquad (2)$$

In the bouncing of the electron, the photon will transfer some momentum to the electron. But due to conservation of momentum, the uncertainty in the measurement of the x-component of the momentum of the electron will be the same as for the photon. Thus for the electron also we have

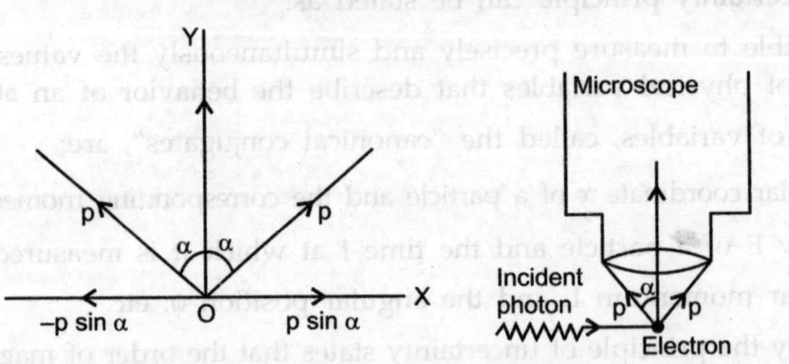

FIGURE 10 **Observation of an electron through a microscope.**

$$\Delta p_x = 2\frac{h}{\lambda}\sin\alpha \qquad (3)$$

Combining equations (1) and (3), we get

$$\Delta x\, \Delta p_x \geq \frac{\lambda}{2\sin\alpha}.2\frac{h}{\lambda}\sin\alpha$$
$$\Delta x\, \Delta p_x \geq h$$

This shows that the product of uncertainties in the determination of position and momentum of an electron by a microscope is of the order of Planck's constant.

Diffraction by a Single Slit

When a beam of electrons passes through a narrow slit and falls on a photographic plate a diffraction pattern is produced on the plate. If Δy be the width of the slit, the positions of minima of intensity are given by

$$\Delta y \sin\theta = n\lambda$$

where θ is the angle of diffraction, λ the wavelength of the electron beam and n an integer. The position of the first minimum of the pattern will be obtained by putting $n = 1$ in the preceding equation *i.e.,*

$$\Delta y \sin\theta = \lambda \qquad (4)$$

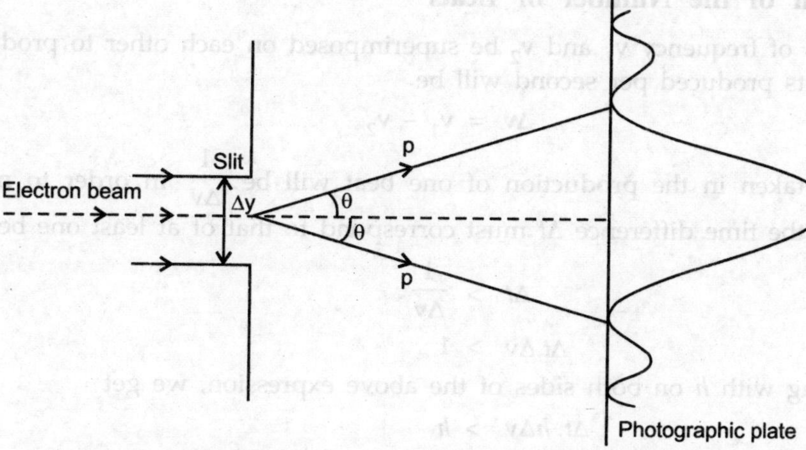

FIGURE 11 **Diffraction of an electron beam through a narrow slit.**

In producing the diffraction pattern on the screen, all the electrons have to pass through the slit, but the place from where an electron has to pass may be anywhere in the slit, so that the *y*-coordinate of the position of the electron has an uncertainty Δy equal to the width of the slit which from equation (4) is

$$d \sin \theta = n \lambda => \Delta y \sin d = \lambda$$

$$\Delta y = \frac{\lambda}{\sin \theta} \qquad (5)$$

The moving electrons in the beginning have no component of momentum along the *y*-axis, as they are all moving along the *x*-axis (Figure 11). In passing through the slit the electrons deviate from their path to form a diffraction pattern. For electrons going into the central maximum, the *y*-component of the momentum can lie anywhere between $p \sin \theta$ and $-p \sin \theta$. Thus the uncertainty in the *y*-component of the momentum is

$$\Delta p_y = p \sin \theta - (-p \sin \theta) \qquad \lambda = \frac{\lambda}{p}$$

$$= 2p \sin \theta \qquad p = \frac{h}{\lambda}$$

$$= 2\frac{h}{\lambda} \sin \theta \qquad [\because p = h/\lambda]$$

Combining this with equation (5), we get

$$\Delta y . \Delta p_y = \frac{\lambda}{\sin \theta} . \frac{2h}{\lambda} \sin \theta$$

$$= 2h$$

$$\therefore \qquad \Delta y . \Delta p_y > h$$

This experiment again illustrates that the product of uncertainties in position and momentum is of the order of Planck's constant.

Determination of the Number of Beats

Let two waves of frequency ν_1 and ν_2 be superimposed on each other to produce beats. The number of beats produced per second will be

$$\Delta\nu = \nu_1 - \nu_2$$

The time taken in the production of one beat will be $\frac{1}{\Delta\nu}$ in order to make a precise measurement; the time difference Δt must correspond to that of at least one beat. Thus

$$\Delta t > \frac{1}{\Delta\nu}$$

or

$$\Delta t.\Delta\nu > 1$$

Multiplying with h on both sides of the above expression, we get

$$\Delta t.\, h\Delta\nu > h$$

or

$$\Delta t.\Delta(h\nu) > h$$

or

$$\Delta t.\Delta E > h$$

We thus see that the product of uncertainties in the energy and time interval is of the order of Planck's constant.

1.16 INTERPRETATION OF THE UNCERTAINTY PRINCIPLE

The connection between wave intensity and probability on the one hand, combined with the connection between wavelength and momentum on the other hand, leads to the uncertainty principle. This principle is one of the most important results of the wave-particle duality and underlines the inherent limitation on the concepts of classical determinism. The term "uncertainty principle" is somewhat of a misnomer. A better term would be "the principle of limited determinism in the structure of matter". Let us examine what happens when we see a particle with an appropriate microscope.

From the relation $p = \frac{h}{\lambda}$ it is seen that as the wavelength of the photon decreases, the momentum increases rapidly. With very short waves we can define the position of a moving particle sharply, but will disturb its momentum greatly. On the other hand, with very long waves we can determine its undisturbed momentum but become rather uncertain about its positions due to pronounced diffraction effects that occur with long wavelengths.

Classical physics never had to deal with dilemmas like this. For macroscopic bodies the position and momentum could always be determined simultaneously with absolute accuracy. But in the microworld our instruments interfere directly with the phenomenon under observation and alter its natural course to an extent that we have no way of separating out the phenomenon in pure form. Thus the uncertainty principle puts an upper limit on the purity of an observation.

According to Heisenberg, we must realize that the uncertainty principle is very much a fundamental law of nature. We have been demanding that our instrument should measure the electron momentum and position simultaneously, whereas the fact is that these two quantities are not basically related in anyway. Why the uncertainty principle failed to show itself in the classical measurements is attributable to the extremely small magnitude of the Planck's constant h which determines the order of magnitude of the product of uncertainties. In the atomic world individual motions and events are not quite firmly predetermined. Photons and particles like electrons and protons move over a 'range' under the guidance of waves and this guidance is performed in a probabilitic rather than a strictly deterministic manner. We can measure only the probability that a photon will strike a given point on a screen or that a material particle will be found in a given place at a given instant.

The uncertainty principle has thus shown that the wave and particle ideas are mutually complementary ways of describing nature. It has extricated us from the wave-partical paradox.

1.17 COMPTON'S EFFECT

While discussing the scattering of X rays in the last section, we assumed that the scattered X rays had the same wavelength as that of the incident radiations. This seemed to be a reasonable supposition because the incident waves will set the electrons in the scattering material into forced vibrations of the same frequency as that of the incident wave; the oscillating electrons will then emit rays of frequency equal to their own. The scattered rays should thus be identical in wavelength with the incident waves. Scattering of this type is called Thomson or classical scattering. But in 1923 A. H. Compton discovered with the help of Bragg's crystal spectrometer that, in addition to a component of the same wavelength, the rays scattered by a light element also consisted of a component of wavelength slightly greater than that of the incident beam. This phenomenon according to which the wavelength of the radiation scattered by an element is greater than that of the incident beam is called the Compton effect.

This increase in wavelength can be accounted for by considering the scattering process as an elastic collision between a photon of incident radiation and an electron of scattering material, making use of the quantum hypothesis. In such a collision, the total energy and momentum remain conserved. As represented in Figure 12, let the X rays consist of photons or quanta of energy hf where, f is the frequency. Let a photon of energy hf strike a free electron at rest. From Einstein's mass energy relationship a photon of energy hf has a mass given by $\frac{hf}{c^2}$. Hence the momentum of the incident photon, being given by the product of mass and velocity, becomes $\frac{hf}{c^2}\,c = \frac{hf}{c}$. Let m_0 be the rest mass of the electron and thus m_0c^2 is rest mass energy.

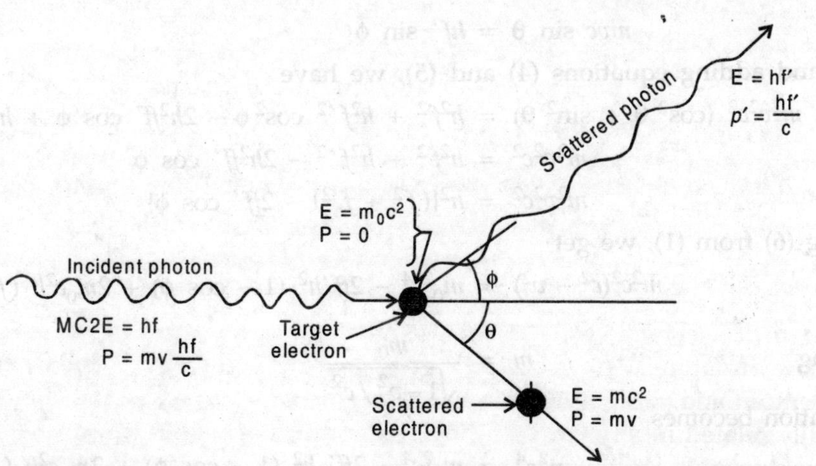

FIGURE 12 An X ray photon being scattered by an electron.

As the electron is supposed to be at rest, its initial momentum is zero. During the collision, a part of the energy of the photon is imparted to the electron which acquires a velocity v, along a direction making an angle θ, with the direction of motion of the incident photon. After having lost some energy, the photon is left with smaller energy hf' (and hence a lower frequency f') and gets scattered along a direction making an angle θ with its original direction. The electron moving with a velocity v after the impact, has an increased mass given by the relativistic equation

$$m = \frac{m_0}{\sqrt{1-\frac{v^2}{c^2}}}$$

The electron thus has energy equal to mc^2 and momentum mv.

From the principle of the conservation of energy, we have

$$hf + m_0c^2 = hf' + mc^2$$

or

$$mc^2 = m_0c^2 + h(f - f')$$

Squaring both sides,

$$m^2c^4 = m_0{}^2c^4 + h^2(f^2 + f'^2 - 2ff') + 2m_0c^2h(f - f') \quad (1)$$

Unlike energy, momentum is a vector quantity, involving both magnitude and direction. The momentum must, therefore, remain conserved both along the original direction of motion of the photon, and also the direction perpendicular to it in the plane containing the electron and the scattered photon.

Applying the principle of the conservation of momentum in the original direction of the incident photon, we have

$$\frac{hf}{c} + 0 = \frac{hf'}{c}\cos\phi + mv\cos\theta \quad (2)$$

Applying this law along the perpendicular direction, we get

$$0 = \frac{hf'}{c}\sin\phi - mv\sin\theta \quad (3)$$

From (2), $$mvc\cos\theta = hf - hf'\cos\phi \quad (4)$$

From (3), $$mvc\sin\theta = hf'\sin\phi \quad (5)$$

Squaring and adding equations (4) and (5), we have

$$m^2v^2c^2(\cos^2\theta + \sin^2\theta) = h^2f^2 + h^2f'^2\cos^2\phi - 2h^2ff'\cos\phi + h^2f'^2\sin^2\phi$$

or

$$m^2v^2c^2 = h^2f^2 + h^2f'^2 - 2h^2ff'\cos\phi$$

or

$$m^2v^2c^2 = h^2\{(f^2 + f'^2) - 2ff'\cos\phi\} \quad (6)$$

Subtracting (6) from (1), we get

$$m^2c^2(c^2 - v^2) = m_0{}^2c^4 - 2ff'h^2(1 - \cos\phi) + 2m_0c^2h(f - f')$$

Substituting $$m = \frac{m_0}{\sqrt{1 - v^2/c^2}}$$

the above equation becomes

$$m_0^2c^4 = m_0{}^2c^4 - 2ff'h^2(1 - \cos\phi) + 2m_0c^2h(f - f')$$

or $$2ff'\,h^2(1 - \cos\phi) = 2m_0c^2h\,(f - f')$$

or $$\frac{f - f'}{ff'} = \frac{h}{m_0c^2}\,(1 - \cos\phi)$$

or $$\frac{1}{f'} - \frac{1}{f} = \frac{h}{m_0c^2}\,(1 - \cos\phi)$$

or $$\frac{c}{f'} - \frac{c}{f} = \frac{h}{m_0c}\,(1 - \cos\phi) \qquad (7)$$

If λ and λ' are the wavelengths corresponding to the frequencies f and f', we have

$$\frac{c}{f} = \lambda \text{ and } \frac{c}{f'} = \lambda'$$

Thus equation (7) becomes

$$\lambda' - \lambda = \Delta\lambda = \frac{h}{m_0c}\,(1 - \cos\phi) \qquad (8)$$

This equation was first derived by Compton and it gives the increase $\Delta\lambda$, (called the Compton shift), in the wavelength of an X ray photon when it is scattered through an angle ϕ by a particle of rest mass m_0. The quantity $\frac{h}{m_0c}$ is called the Compton wavelength of the scattering material.

Inserting $h = 6.625 \times 10^{-34}$ joule-sec, $m_0 = 0.9107 \times 10^{-27}$ gm and $c = 2.998 \times 10^{-8}$ cm the value of the Compton wavelength comes out to be 0.0242 Å.

An examination of equation (8) reveals the following significant facts:

1. The wavelength λ' of the scattered photon is greater than the wavelength λ of the incident photon by an amount equal to
$$\frac{h}{m_0c}\,(1 - \cos\phi)$$
2. The increase in wavelength is independent of the wavelength of the incident photon or the nature of the scattering material but depends upon the angle of scattering.
3. When $\phi = 90°$, $D\lambda = \frac{h}{m_0c}$ which is equal to the Compton wavelength. It is constant and is equal to 0.0242 Å.
4. The maximum change in the wavelength occurs when $\phi = 180°$ and this change then is $\frac{h}{m_0c}(1+1) = \frac{2h}{m_0c} = 0.0484$ Å which is twice the Compton wavelength.

The above results agree fairly well with experimental observations and provide strong evidence in support of quantum theory of radiation.

The Relation between the Scattering Angle (ϕ) and the Angle of Recoil (θ)

Dividing equation (4) by (5), we have

$$\cot\theta = \frac{f - f'\cos\phi}{f'\sin\phi} \qquad (9)$$

But from equation (7),
$$\frac{1}{f'} = \frac{1}{f} + \frac{h}{m_0c^2}(1-\cos\phi)$$
$$= \frac{m_0c^2 + hf(1-\cos\phi)}{f.m_0c^2}$$
or
$$f' = \frac{f.m_0c^2}{m_0c^2 + hf(1-\cos\phi)}$$

Substituting f' in equation (9), we get

$$\cot\theta = \frac{f - \dfrac{f.m_0c^2\cos\phi}{m_0c^2 + f\,h(1-\cos\phi)}}{\dfrac{f.m_0c^2\sin\phi}{m_0c^2 + f\,h(1-\cos\phi)}}$$
$$= \frac{fm_0c^2(1-\cos\phi) + hf^2(1-\cos\phi)}{f.m_0c^2\sin\phi}$$
$$= \frac{(m_0c^2 + hf)\,(1-\cos\phi)}{m_0c^2\sin\phi}$$
$$= \left(\frac{m_0c^2 + hf}{m_0c^2}\right)\frac{2\sin^2\phi/2}{2\sin(\phi/2)\cos(\phi/2)}$$
$$= \left(1 + \frac{hf}{m_0c^2}\right)\tan(\phi/2)$$
or
$$\cot\theta = \left(1 + \frac{h}{\lambda m_0c}\right)\tan\phi/2\left(\because \frac{c}{f} = \lambda\right) \quad (10)$$

which gives the relationship between ϕ and θ. It follows that if $\phi = 0$, $\theta = 90°$ and if $\phi = 180°$, $\theta = 0°$. Thus as ϕ varies from 0 to 180°, θ varies from 90° to 0°. It means that where as a photon can get scattered in all directions, the electron can get recoiled only in the forward direction at angles less than 90°.

Unmodified or Classical Scattering

As said above, the scattered X rays observed at any angle also include a substantial portion having the same wavelength as the initial rays. The reason for this unmodified scattering is that sometimes a photon, instead of striking against a free electron, collides with another which is firmly bound to the nucleus. In such a case, the electron does not get detached by the impact but remains bound so that the atom as a whole recoils. The mass m_0 of the electron in equation (8) must, therefore, be replaced by the mass of the atom which is thousands of times larger than that of a single electron. The scattered photon will thus have a negligible change in wavelength which cannot be detected and this accounts for the unmodified scattering.

EXERCISES

Objective Questions

1. The scattered electron has an energy approximately
   (a) 2×10^{-13} joule (b) 2×10^{-16} joule
   (c) 2×10^{-14} joule (d) 2×10^{-12} joule
2. The change in wavelength in the Compton scattering depends upon
   (a) Angle θ of scattering
   (b) Atomic number of the scatterer
   (c) Wavelength of the incident X ray
   (d) Nature of the scatterer
3. The momentum of the electron scattered in the process as given in question No. 5 is nearly
   (a) 6×10^{-29} kg-m. sec^{-1} (b) 2×10^{-28} kg-m. sec^{-1}
   (c) 6×10^{-21} kg-m. sec^{-1} (d) 6×10^{-23} kg-m. sec^{-1}
4. Highest energy electrons will be produced by
   (a) Visible light (b) Ultraviolet light
   (c) X rays (d) Gamma rays
5. The photoelectric effect helps us to convert
   (a) Light energy into sound energy
   (b) Sound energy into electrical energy
   (c) Electrical energy into light energy
   (d) Light energy into electrical energy
6. The momentum of a photon of energy $h\nu$ is
   (a) $h\nu/c$ (b) has no momentum
   (c) $h\nu c$ (d) $h\nu$
7. Planck's constant h, connecting the energy E with frequency γ by the relation $E = h\gamma$ has the same dimensions as
   (a) Angular momentum (b) Work
   (c) Force (d) Power
8. The mass of a photon of energy $h\nu$ is
   (a) $h\nu/c^2$ (b) It changes with time
   (c) $h\nu c^2$ (d) Infinite
9. The photons in a radio wave of wavelength 3×10^4 cm have an energy
   (a) 19.86×10^{-34} joules (b) 2.2×10^{-28} joules
   (c) 6.62×10^{-40} joules (d) 6.62×10^{-28} joules
10. A hydrogen atom of mass 1.7×10^{-27} kg moving with a velocity of 1,000 m/sec has waves associated with it. Their wavelength is
    (a) 3.8 Å (b) 3.8 m
    (c) 1.7×10^{-24} m (d) 1.7×10^{-30} m

11. Momentum of a particle is given by mass multiplied by the velocity with which it moves. Using this, the photon has a mass

(a) $h\nu/c^2$ (b) $h\nu/c$

(c) $h\nu c$ (d) $h\nu$

12. A proton and an alpha particle are accelerated through the same potential difference. The ratio of the wavelengths associated with the proton to that associated with an α particle is (given: eV = 1/2 mv^2 where V is potential difference)

(a) $\frac{1}{8^{1/2}}$ (b) $8^{1/2}$

(c) 2 (d) 4

13. An electron of mass m_e and a proton of mass m_p are accelerated through the same potential difference. The ratio of the wavelengths associated with an electron to that associated with proton is

(a) $\sqrt{m_p/m_e}$ (b) m_p/m_e

(c) m_e/m_p (d) one

14. A tiny particle of mass 10^{-13} kg moving with a velocity 10 cm per second is associated with a wave whose wavelength is

(a) 6.62×10^{-34} cm (b) 6.62×10^{-20} cm

(c) 6.62×10^{-12} Å (d) 6.62×10^{-18} cm

15. In an X ray tube when the accelerating voltage V is halved, the difference between the wavelengths of the K_α line and the minimum wavelength of continuous X ray spectrum

(a) remains constant (b) becomes more than two times

(c) becomes half (d) becomes less than two times

16. Let K_1 be the maximum kinetic energy of photoelectrons emitted by the light of wavelength λ_1 and K_1 corresponding to wavelength λ_2. If $\lambda_1 = 2\lambda_2$ then

(a) $2K_1 = K_2$ (b) $K_1 = 2K_2$

(c) $K_1 < K_2/2$ (d) $K_1 > 2K_2$

17. A potential difference of 10^3 V is applied across an X ray tube. The ratio of the de Broglie wavelength of the incident electrons to the shortest wavelength of X rays produced is

(a) 1/20 (b) 1/100

(c) 1 (d) $1/10^4$

(e/m = 1.8×10^{14} C/kg for an electron)

18. The radius of the second orbit of an electron in a hydrogen atom is 2.116 Å. The de Broglie wavelength associated with this electron in this orbit would be

(a) 6.64 Å (b) 1.058 Å

(c) 2.116 Å (d) 13.28 Å

19. The de Broglie wavelength of an electron in a ground state of an hydrogen atom is

(a) 1.06 Å (b) 1.52 Å

(c) 0.53 Å (d) 3.33 Å

Short Questions for Discussion

1. What are matter waves?
2. Define group wave and group velocity.
3. State the Heisenberg's Uncertainty Principle.

4. What do you mean by the Compton shift and on what factors does it depend?
5. What is a wave packet?
6. Define group velocity and phase velocity.
7. Explain why the Compton effect is not observed with visible light?
8. Prove that the velocity of matter waves is found to be greater than the velocity of light. How is this anomaly solved?
9. Explain the dual nature of matter (define de Broglie waves).
10. Derive the expression for the de Broglie wavelength associated with an electron accelerated by the electric potential V.

Long Questions for Discussion

1. Describe the *dual nature* of matter ? Explain how de Broglie was led to this concept.
2. Derive a relation for the wavelength of de Broglie waves. How is the existence of these waves experimentally verified?
3. State and explain Heisenberg's uncertainty principle. Illustrate it with the help of experiments.
4. Write short notes on:

   (*a*) Wave-particle duality.

   (*b*) Interpretation of the uncertainty principle.
5. What are matter waves? Explain their properties.
6. Obtain the Schrödinger time-independent and time-dependent wave equations. Give the Physical interpretation of the wave function ψ.
7. What do you mean by wave-packet? What are Group velocity and Phase velocity? Establish the relation of these velocities with the particle velocity.
8. What is the Compton effect? Indicate how it strengthened the quantum concept.
9. Derive the expression for the Compton shift and the recoil angle of an electron. Discuss the various possibilities of X ray scattering.

Chapter 2

WAVE THEORY OF LIGHT

2.1 OVERVIEW

In geometrical optics we consider that light propagates accurately in straight lines, without however, concerning ourselves with the actual nature of light. The basic principles regarding the nature of light were formulated in the latter half of the seventeenth century. During that period, everyone believed in Newton's corpuscular theory. According to this theory, light consists of a stream of minute invisible particles, called corpuscles, moving at a great speed. These corpuscles are emitted in straight lines from a luminous source and their mechanical impact on the retina stimulates the sensation of vision. Different colors were ascribed to different sized corpuscles. On the basis of this theory, the phenomena of reflection, refraction, the rectilinear propagation of light, and the fact that light could propagate through a vacuum were explained satisfactorily. But the corpuscular theory could not explain the phenomenon of interference and diffraction of light. These phenomena could be explained only when the wave theory of light was considered.

The wave theory of light was first proposed by Huygens in 1678. On the basis of his wave theory, Huygens explained satisfactorily the phenomena of reflection, refraction and total internal reflection. According to his theory a luminous body is a source of disturbance in a hypothetical medium called ether. This medium pervades all space. The disturbance from the source is propagated in the form of waves through space and the energy is distributed equally in all directions. When these waves carrying energy are incident on the eye, the optic nerves are excited and the sensation of vision is produced. Huygens' theory predicted that the velocity of light in medium shall be less than the velocity of light in free space, which is just converse of the prediction made from Newton's corpuscular theory. The experimental evidence for the wave theory in Huygen's time was very small. In 1801 however, Thomas Young obtained evidence that light could produce wave effects. Very shortly, diffraction was explained by Fresnel and Fraunhoffer. While the transverse nature of light was explained by polarization is called physical optics or wave optics and should be explained by using the wave theory of light.

2.2 HUYGENS' PRINCIPLE OF WAVE PROPAGATION

Huygens' originated the concept of wavefront and secondary wavelet in order to account for the rapid speed of light. Huygens' theory is based on a geometrical construction which allows us to determine the shape of the wavefront at any time, if the shape of the wavefront at an earlier time is known.

According to Huygens' principle, each point source is a center of disturbance from which waves spread out in all directions, setting the other particles in vibration. The continuous focus of all the particles vibrating in the same phase is called the wavefront. Each point of the wavefront is a source of secondary disturbance and the wavelets (called secondary wavelets) emanating from these points spread out in all directions with the velocity of light. The envelope of these wavelets in the forward direction give the shape of the new wavefront. In Figure 1 AB represents the shape of a wavefront in a homogeneous isotropic medium at an instant $t = 0$, originating from the point source S. To find the position of the wavefront at time t using Huygens' principle, the following constructions are made:

- Consider the points *a, b, c,* on the wavefront AB. All these points are the source of secondary wavelets and send out secondary wavelets in all directions.
- Taking each point (*a, b, c,*) as center, draw circles of radii vt. These circles represent sections of secondary spherical wavelets at time t.
- Draw a common tangent A′B′ to these circles in the forward direction. The forward envelope A′B′ gives the new position of the wavefront at the required time t.

The direction of propagation of the disturbance are called the rays. In a homogeneous medium rays are normal to the wavefront as shown by the arrows in Figure 1.

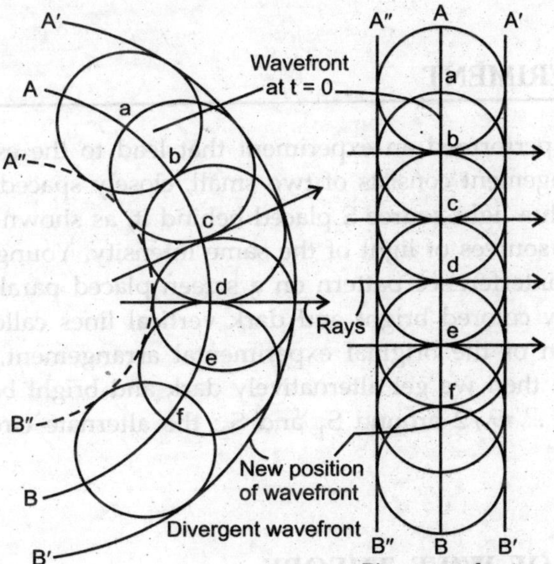

FIGURE 1

Huygens' principle does not explain why the forward wavefront is to be taken and the backward wavefront (shown by dotted lines) is to be neglected. The presence of the backward wavefront is avoided by assuming that the amplitude of the secondary wavelets is not uniform in all directions. It is maximum in the forward direction and minimum in the backward direction.

On the basis of Huygens' principle we can satisfactorily explain the phenomena of reflection and refraction of light.

2.3 INTERFERENCE OF LIGHT

The phenomenon of the interference of light has proved the validity of the wave theory of light. According to it, when two light waves of the same frequency and having a constant phase difference traverse simultaneously in the same region of a medium and cross each other, then there is a modification in the intensity of light in the region of the superposition, which is in general different from the sum of intensities due to individual waves at that point. This modification in the intensity of light resulting from the superposition of two (or more) waves of light is called interference. At certain points the waves superimpose in such a way that the resultant intensity is greater than the sum of the intensities due to individual waves. The interference produced at these points is called constructive interference or reinforcement, while at certain other points the resultant intensity is less than the sum of the intensities due to individual waves. The interference produced at these points is called destructive interference. Beyond the region of superposition the waves come through completely uninfluenced by each other.

2.4 YOUNG'S EXPERIMENT

In 1801, Thomas Young performed an experiment that lead to the establishment of the wave theory of light. His arrangement consists of two small, closely spaced holes or slits, S_1 and S_2, in an opaque surface with a light source S placed behind it, as shown in Figure 2. Here S_1 and S_2 serve as two coherent sources of light of the same intensity. Young used sunlight as a light source and observed an interference pattern on a screen placed parallel to the two sources S_1 and S_2. A series of a few colored bright and dark vertical lines called fringes appear on the screen. As a modification of the original experimental arrangement, sunlight is replaced by monochromatic light and then we get alternatively dark and bright bands. Let us draw circles of radii $\lambda/2$, $2\lambda/2$, $3\lambda/2$, ..., $n\lambda/2$ around S_1 and S_2; the alternate circles represent crests and troughs.

2.5 EXPLANATION OF WAVE THEORY

The formation of interference fringes on the screen can be explained on the basis of wave theory. When light passes through S, spherical waves spread out from S. According to Huygens' principle, each point of a wavefront becomes the source of a secondary disturbance and the secondary wavelets emanating from these points spread out in all directions. Hence spherical

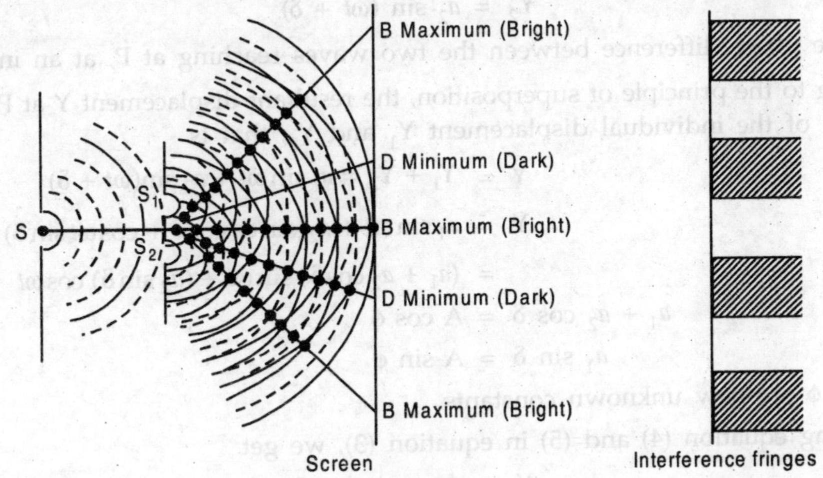

FIGURE 2

waves also spread out from the closely spaced holes S_1 and S_2. The radii of these secondary wavefronts increase as they move away from S_1 and S_2 and hence they superimpose more and more on each other. At points where a crest falls over another crest or over a trough, the vibrations are large and the resultant amplitude is greatest; hence the resultant intensity at these points is increased. These points are shown by solid dots. This is the case of constructive interference. At points where a crest of one falls over the trough of the other, the intensity is minimum. This is the case of destructive interference. The points, where the intensity is minimal, are dark and represented by hollow dots in Figure 2. Thus we get alternate bright and dark lines at the intersection of the screen with antinodal and nodal planes. These dark and bright regions of equal width are called interference fringes.

2.6 ANALYTICAL TREATMENT OF INTERFERENCE

We shall derive an expression for the resultant intensity at any point P of the screen due to the superposition of two waves of light. Let S be a narrow slit illuminated by a monochromatic light of wavelength λ (Figure 3), and S_1 and S_2 two narrow slits close together and equidistant from S. The wave of light from S arriving at S_1 and S_2 will always be in phase. Let a_1 and a_2 be the amplitudes of the two waves from S_1 and S_2 respectively. Then the displacement Y_1 due to one wave, from S_1, at any instant t is represented as

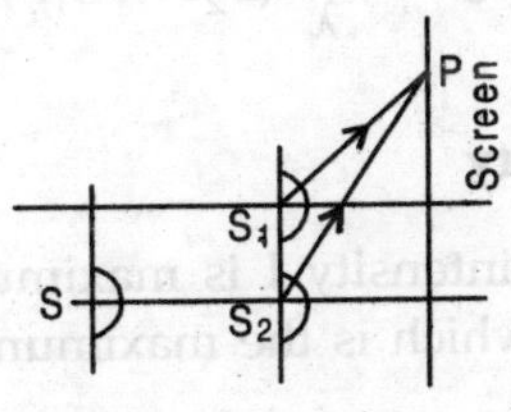

FIGURE 3

$$Y_1 = a_1 \sin \omega t \tag{1}$$

The displacement Y_2 due to other wave from S_2, at the same instant t is represented as

$$Y_2 = a_2 \sin (\omega t + \delta) \tag{2}$$

where δ is the phase difference between the two waves reaching at P, at an instant t.

According to the principle of superposition, the resultant displacement Y at P is merely the algebraic sum of the individual displacement Y_1 and Y_2, that is

$$Y = Y_1 + Y_2 = a_1 \sin \omega t + a_2 \sin(\omega t + \delta)$$

or

$$Y = a_1 \sin \omega t + a_2(\sin \omega t \cos \delta + \cos \omega t \sin \delta)$$

$$= (a_1 + a_2 \cos \delta) \sin \omega t + (a_2 \sin \delta) \cos \omega t \tag{3}$$

Suppose

$$a_1 + a_2 \cos \delta = A \cos \phi \tag{4}$$

and

$$a_1 \sin \delta = A \sin \phi \tag{5}$$

where A and ϕ are new unknown constants.

Substituting equation (4) and (5) in equation (3), we get

$$Y = A \cos \phi \sin \omega t + A \sin \phi \cos \omega t$$

or

$$Y = A \sin(\omega t + \phi) \tag{6}$$

obviously the resultant vibration at P is a simple harmonic of amplitude A and of phase ϕ.

Squaring and adding equation (4) and (5), we get

$$(a_1 + a_2 \cos \delta)^2 + (a_2 \sin \delta)^2 = A^2$$

or

$$A^2 = a_1^2 + a_2^2 \cos^2 \delta + 2a_1a_2 \cos \delta + a_2^2 \sin^2 \delta$$

or

$$A^2 = a_1^2 + a_2^2 + 2a_1a_2 \cos \delta$$

The resultant intensity at P is proportional to the square of the resultant amplitude, that is, $I = A^2$. Thus

$$I = A^2 = a_1^2 + a_2^2 + 2a_1a_2 \cos \delta \tag{7}$$

or

$$I = a_1^2 + a_2^2 + 2a_1a_2 \, (2\cos^2 \delta/2 - 1)$$

or

$$I = (a_1 - a_2)^2 + 4a_1a_2 \cos^2 \delta/2 \tag{8}$$

As a_1 and a_2 are constant, the intensity I will vary according to the variations of $\cos^2 \delta/2$. Thus we can say that the intensity variations is of the cosine square form.

The phase difference δ between the two light waves arriving at P is given by

$$\delta = \frac{2\pi}{\lambda} \text{ (path difference)}$$

or

$$\delta = \frac{2\pi}{\lambda} (S_2P - S_1P)$$

Condition for Maximum Intensity

From equation (7) it is clear that the intensity I is maximum at those points at which $\cos \delta$ is maximum, that is, where $\cos \delta = +1$, which is the maximum value of the cosine function, where phase difference

$$\delta = 2n\pi, \ n = 0, 1, 2, 3, ...$$

or $$\delta = 0, 2\pi, 4\pi, 6\pi, ..., \text{etc.}$$

or $$\text{path difference} = \frac{\lambda}{2\pi}(\delta) = \frac{\lambda}{2\pi}.2n\pi = \frac{2n\lambda}{2}$$

that is, the intensity is maximum when phase difference is an even multiple of π or the path difference is an even multiple of a half wavelength (λ/2).

The value of maximum intensity, I_{max} is given by

$$I_{max} = a_1^2 + a_2^2 + 2a_1a_2$$

[putting cos δ = +1 in equation (7)]

$$= (a_1 + a_2)^2 \quad (9)$$

Obviously $I_{max} > a_1^2 + a_2^2$. Thus, the resultant intensity is greater than the sum of the individual intensities.

Condition for Minimum Intensity

The resultant intensity I is minimum when cos δ is minimum. Minimum value of cos δ = –1, that is, where phase difference

$$\delta = (2n - 1)\pi, \ n = 1, 2, 3, ..., \text{etc.}$$

$$\text{The path difference} = \frac{\lambda}{2\pi}(2n-1)\pi = (2n-1)\lambda/2$$

That is, the intensity is minimum when the phase difference is an odd multiple of π or the path difference is an odd multiple of a half wavelength (λ/2).

The value of the minimum intensity, I_{max}, is given by

$$I_{min} = a_1^2 + a_2^2 - 2a_1a_2$$

[putting cos δ = –1 in equation (7)]

$$I_{min} = (a_1 - a_2)^2 \quad (10)$$

Obviously $I_{min} < a_1^2 + a_2^2$. Thus, the resultant intensity is less than the sum of the individual intensities.

Special case: When $a_1 = a_2$

$$I_{max} = a^2 + a^2 + 2a.a = 4a^2 \quad (11)$$

and $$I_{min} = a^2 + a^2 - 2a.a = 0 \quad (12)$$

Intensity Distribution Curve

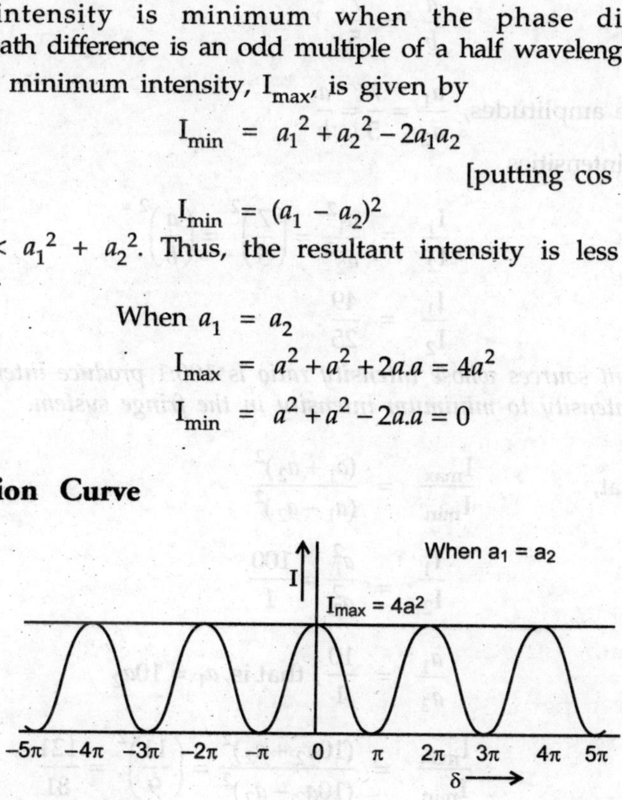

FIGURE 4

A curve showing the variation of intensity with phase is called the intensity distribution curve. From equations (11) and (12) it is clear that the intensity is maximum or bright at the point on which $\delta = 2n\pi$, where $n = 0, 1, 2, 3, \ldots$ etc. and is $4a^2$, and is minimum or dark at points on which $\delta = (2n - 1)\pi$, where $n = 1, 2, 3, \ldots$ etc. and is zero. Figure 4 shows the variation of intensity as a function of δ. The formation of fringes in the interference pattern is in accordance with the law of conservation of energy. In the interference fringes the energy is simply redistributed. It is transfered from the points of distructive interference (minima) to the points of constructive interference (maxima).

Example 1. *If in an interference pattern, the ratio between maximum and minimum intensities is 36:1, find the ratio between the amplitude and intensities of the two interfering waves.*

Solution. Let the amplitude ratio of the two interfering waves be $a : 1$. If I_{max} and I_{min} be the maximum and minimum intensities, then

$$\frac{I_{max}}{I_{min}} = \frac{(a+b)^2}{(a-b)^2}, \text{ where } a > 1$$

Here
$$\frac{I_{max}}{I_{min}} = \frac{36}{1} = \frac{(a+b)^2}{(a-b)^2}$$

or
$$\frac{(a+b)}{(a-b)} = \frac{6}{1} \quad \text{or} \quad (a + b) = 6(a - b)$$

or
$$\frac{a}{b} = \frac{7}{5}$$

$\therefore$ The ratio of the amplitudes, $\dfrac{a_1}{a_2} = \dfrac{7}{5} = \dfrac{a}{b}$

The ratio of their intensities,

$$\frac{I_1}{I_2} = \frac{a_1^2}{a_2^2} = \left(\frac{7}{5}\right)^2 = \left(\frac{a}{b}\right)^2$$

or
$$\frac{I_1}{I_2} = \frac{49}{25}.$$

Example 2. *Two coherent sources whose intensity ratio is 100:1 produce interference fringes. Deduce the ratio of maximum intensity to minimum intensity in the fringe system.*

Solution. We know that,
$$\frac{I_{max}}{I_{min}} = \frac{(a_1 + a_2)^2}{(a_1 - a_2)^2}$$

Given
$$\frac{I_1}{I_2} = \frac{a_1^2}{a_2^2} = \frac{100}{1}$$

$$\therefore \quad \frac{a_1}{a_2} = \frac{10}{1}, \text{ that is, } a_1 = 10a_2$$

$$\therefore \quad \frac{I_{max}}{I_{min}} = \frac{(10a_2 + a_2)^2}{(10a_2 - a_2)^2} = \left(\frac{11}{9}\right)^2 = \frac{121}{81}$$

$$\therefore \quad \frac{I_{max}}{I_{min}} = \frac{121}{81}.$$

Example 3. *Two sources of intensities I and 4I are used in an interference experiment. Obtain intensities at points where the waves from two sources superimpose with a phase difference of (i) 0, (ii) π/2, and (iii) π.*

Solution. The resultant intensity at the point at which the phase difference between the waves from two coherent sources δ is

$$I_\delta = a_1^2 + a_2^2 + 2a_1a_2 \cos\delta$$
$$= I_1 + I_2 + 2\sqrt{(I_1I_2)}\cos\delta$$

where $I_1 = a_1^2$ and $I_2 = a_2^2$ are the intensities due to individual waves

(i) Here, $\delta = 0$, $I_1 = I$ and $I_2 = 4I$

$$\therefore \quad I_\delta = I + 4I + 2\sqrt{(I)(4I)}\cos 0 = 9I$$

(ii) Here, $\delta = \pi/2$, $I_1 = I$ and $I_2 = 4I$

$$\therefore \quad I_\delta = I + 4I + 2\sqrt{(I)(4I)}\cos \pi/2 = 5I$$

(iii) Here, $\delta = \pi$, $I_1 = I$ and $I_2 = 4I$

$$\therefore \quad I_\delta = I + 4I + 2\sqrt{(I)(4I)}\cos \pi = 5I - 4I = I.$$

Example 4. *Two coherent beams of wavelength 5000 reaching a point would individually produce intensities 1.44 and 4.00 units. If they reach here together, the intensity is 0.90 units. Calculate the lowest phase difference with which the beams reach that point.*

Solution. The resultant intensity at a point due to interfering waves of amplitudes a_1 and a_2 with phase difference δ is given by

$$I = a_1^2 + a_2^2 + 2a_1a_2\cos\delta$$

or
$$I = I_1 + I_2 + 2\sqrt{(I_1I_2)}\cos\delta$$

where I_1 and I_2 are the intensities of coherent waves of amplitudes a_1 and a_2 respectively.

Here $I_1 = 1.44$, $I_2 = 4.00$ and $I = 0.90$

$$\therefore \quad 0.90 = 1.44 + 4.00 + 2\left(\sqrt{1.44 + 4.0}\right)\cos\delta$$

$$0.90 = 5.44 + 4.8\cos\delta$$

or
$$\cos\delta = \frac{0.90 - 5.44}{4.8} = -0.9458$$

or
$$\cos(180° - \delta) = 0.9458$$

or
$$(180° - \delta) = \cos^{-1}(0.9458) = 19°$$

$$\therefore \quad \delta = 180° - 19° = 161°.$$

Example 5. *Two coherent sources of intensity ratio α interfere. Prove that in the interference pattern*

$$\frac{I_{max} - I_{min}}{I_{max} + I_{min}} = \frac{2\sqrt{\alpha}}{1+\alpha}$$

Solution. Let I_1 and I_2 be the intensities, and a_1 and a_2 the amplitudes of the disturbances from the two coherent sources. Since the intensity of the wave is proportional to the square of the amplitude, we can write

$$\frac{a_1^2}{a_2^2} = \left(\frac{I_1}{I_2}\right) \quad \text{or} \quad \sqrt{\left(\frac{I_1}{I_2}\right)} = \frac{a_1}{a_2} = \sqrt{\alpha}$$

We know that $I_{max} = (a_1 + a_2)^2$ and $I_{min} = (a_1 - a_2)^2$

$$\frac{I_{max} - I_{min}}{I_{max} + I_{min}} = \frac{(a_1+a_2)^2 - (a_1-a_2)^2}{(a_1+a_2)^2 + (a_1-a_2)^2} = \frac{2a_1a_2}{a_1^2 + a_2^2}$$

Dividing RHS throughout by a_2^2

or
$$\frac{I_{max} - I_{min}}{I_{max} + I_{min}} = \frac{2\left(\frac{a_1}{a_2}\right)}{\left[\left(\frac{a_1}{a_2}\right)^2 + 1\right]} = \frac{2\sqrt{\alpha}}{\alpha + 1}.$$

2.7 COHERENT SOURCES

Two sources of light are said to be coherent if they emit light which always has a constant phase difference between them. It means that the two sources must emit radiations of the same wavelength. The two independent sources cannot be coherent because of the fact that independent sources cannot maintain a constant phase difference between them. For experimental purposes, two virtual sources obtained from a single parent source can act as coherent. In such case all the random phase changes occurring in the parent source are repeated in the virtual sources also, thus maintaining a constant phase difference between them. Since the wavelength of light waves is extremely small, the two sources must be narrow and must also be close to each other.

No Interference by Two Independent Light Sources

Sustained interference can never be obtained with two independent sources of light, such as two bulbs or two candles, due to the fact that any two independent beams of light are always incoherent. Actually a beam of light is built up of waves radiated from millions of excited atoms or molecules, whose vibrations are completely independent of each other, so that the initial phase of the beam is absolutely governed by chance. After a time interval of 10^{-9} sec, this phase will be randomly changed because of the fact that the excited atoms which are responsible for vibrations are replaced by other excited atoms. Thus every source undergoes haphazard changes of phase in every billionth of a second. Hence the resultant intensity on the screens vary rapidly with time and two incoherent sources cannot produce any stationary interference patterns.

2.8 CONDITION FOR SUSTAINED INTERFERENCE OF LIGHT

To obtain well defined interference patterns, the intensity at points corresponding to destructive interference must be zero, while intensity at the point corresponding to constructive interference must be maximum. To accomplish this the following conditions must be satisfied.

- The two interfering sources must be coherent, that is, they must keep a constant phase difference.
- The two interfering sources must emit the light of the same wavelength and time period. This condition can be achieved by using a monochromatic common original source, that is, the common source emits light of a single wavelength.
- The amplitudes or intensities of the interfering waves must be equal or very nearly equal so that the minimum intensity would be zero.
- The separation between the two coherent sources must be as small as possible so that the width ($D\lambda/2d$) of the fringes is large and are separately visible.
- The two sources must be narrow or they must be extremely small. A broad source is equivalent to a large number of fine sources. Each pair of fine sources will give its own pattern. The fringes of different interference patterns will overlap.
- The distance between the two coherent sources and the screen must be as large as possible so that the width of fringes ($D\lambda/2d$) is large and are separately visible.
- The two interference waves must be propagated along the same direction so that their vibrations are along a common line.
- If the interfering waves are polarized, they must be in the same state of polarization.

2.9 TYPES OF INTERFERENCE

The phenomenon of interference may be grouped into two categories:

(*i*) *Division of Wavefront*: Under this category, the coherent sources are obtained by dividing the wavefront, originating from a common source, by employing mirrors, biprisms or lenses. This class of interference requires essentially a point source or a narrow slit source. The instruments used to obtain interference by division of wavefront are the Fresnel biprism, Fresnel mirrors, Lloyd's mirror, lasers, etc.

(*ii*) *Division of Amplitude:* In this method, the amplitude of the incident beam is divided into two or more parts either by partial reflection or refraction. Thus we have coherent beams produced by division of amplitude. These beams travel different paths and are finally brought together to produce interference. The effects resulting from the superposition of two beams are referred to as two beam interference and those resulting from superposition of more than two beams are referred to as multiple beam interference. The interference in thin films, Newton's rings, and Michelson's interferometer are examples of two beam interference and Fabry-Perot's interferometer is an example of multiple beam interference.

2.10 DIVISION OF WAVEFRONT

Theory of Interference Fringes

Let S be a narrow slit illuminated by monochromatic light of wavelength λ, and S_1 and S_2 be the two narrow slits separated by a small distance $2d$ (as in Figure 5). Let a screen XY be placed at a distance D from the coherent sources S_1 and S_2. Let O be the foot of the perpendicular drawn from C, the mid point of S_1 and S_2 on the screen. Now consider a point P on the screen at a distance x from O at which the condition for bright and dark fringes are to be determined. Draw S_1M and S_2N perpendiculars from S_1 and S_2 on the screen and join S_1P and S_2P. The path difference between the waves reaching at P from S_1 and S_2 is given by

$$\Delta = S_2P - S_1P$$

In a right angled triangle S_2NP, we have

$$(S_2P)^2 = (S_2N)^2 + (NP)^2 = [D^2 + (x + d)^2]$$

or $$S_2P = [D^2 + (x+d)^2]^{1/2} = D\left[1+\left(\frac{x+d}{D}\right)^2\right]^{1/2}$$

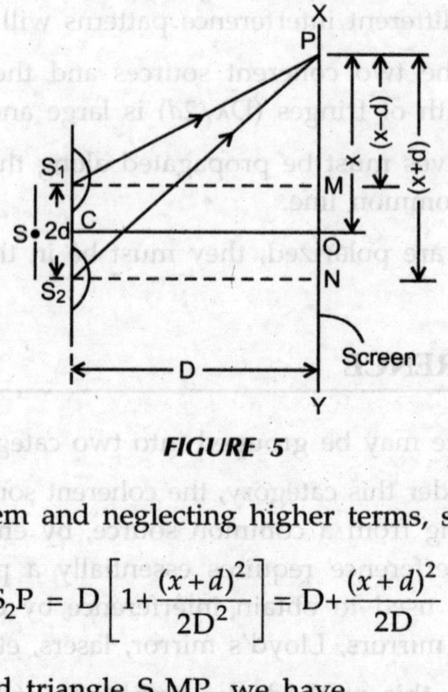

FIGURE 5

Using the binomial theorem and neglecting higher terms, since D >> x or d, we get

$$S_2P = D\left[1+\frac{(x+d)^2}{2D^2}\right] = D+\frac{(x+d)^2}{2D} \quad (1)$$

Similarly in a right angled triangle S_1MP, we have

$$S_1P = [(S_1M)^2 + (MP)^2]^{1/2} = [D^2 + (x-d)^2]^{1/2}$$

$$= D\left[1+\left(\frac{x-d}{D}\right)^2\right]^{1/2}$$

$$= D\left[1+\frac{(x-d)^2}{2D^2}\right] = D+\frac{(x-d)^2}{2D} \quad (2)$$

$\therefore$ Path difference, $$\Delta = (S_2P - S_1P) = \left(D + \frac{(x+d)^2}{2D}\right) - \left(D + \frac{(x-d)^2}{2D}\right)$$

or $$\Delta = D + \frac{(x+d)^2}{2D} - D - \frac{(x-d)^2}{2D}$$

$$= \frac{1}{2D}[(x+d)^2 - (x-d)^2]$$

$$\Delta = \frac{4xd}{2D} = \frac{2xd}{D} \qquad (3)$$

If the sources S_1 and S_2 are in the same phase as the waves reaching at P, the path differences must only be due to path difference, $\Delta = S_2P - S_1P$

$$\therefore \quad \text{Phase difference, } \delta = \frac{2\pi}{\lambda}\Delta = \frac{2\pi}{\lambda}.\left(\frac{2xd}{D}\right) \qquad (4)$$

1. *Position of Bright Fringes:* For bright fringe or maximum intensity at P, the path difference must be an even multiple of half the wavelength ($\lambda/2$) of light used, that is,

$$\Delta = 2n.(\lambda/2) = (2xd)/D, \text{ where } n = 0, 1, 2, 3, ...$$

$$\therefore \quad x = \frac{n\lambda D}{2d} \qquad (5)$$

Equation (5) gives the distance of nth bright fringe from O. Therefore writing x_n for x, we get

$$x_n = \frac{n\lambda D}{2d}, \text{ where } n = 0, 1, 2, 3, ... \qquad (6)$$

2. *Position of Dark Fringes:* For dark fringes or minimum intensity at P, the path difference must be an odd multiple of half wavelength ($\lambda/2$) of light used, that is,

$$\Delta = (2n + 1)\lambda/2, \text{ where } n = 0, 1, 2, 3, ...$$

$$\therefore \quad \Delta = \frac{2xd}{D} = (2n+1)\lambda/2$$

or $$x = \frac{(2n+1)\lambda D}{4d} \qquad (7)$$

Equation (7) gives the distance of nth dark fringe from O. Therefore writing x_n for x, we get

$$x_n = \frac{(2n+1)\lambda D}{4d}, \text{ where } n = 1, 2, 3, ... \qquad (8)$$

3. *Spacing of Fringes or Fringe Width* ω: The spacing between any two successive bright fringes or any two successive dark fringes is called the fringe width and is denoted by ω.

For bright fringes: If x_{n+1} and x_n denote the distances of $(n+1)^{th}$ and n^{th} bright fringes from O, then we have

$$x_{n+1} = (n+1)\frac{D\lambda}{2d} \quad \text{and} \quad x_n = \frac{nD\lambda}{2d}$$

∴ Fringe width,

$$\omega = x_{(n+1)} - x_n$$

$$= (n+1)\frac{D\lambda}{2d} - \frac{nD\lambda}{2d} = \frac{D\lambda}{2d}(n+1-n)$$

∴

$$\omega = \frac{D\lambda}{2d} \tag{9}$$

For dark fringes: If x_{n+1} and x_n are the distance or spacing of two consecutive dark fringes from O, then we have

$$x_{n+1} = \frac{(2n+3)\lambda D}{4d}, \; x_n = \frac{(2n+1)\lambda D}{4d}$$

∴ Fringe width,

$$\omega = x_{n+1} - x_n = \frac{(2n+3)\,\lambda D}{4d} - \frac{(2n+1)\lambda D}{4d}$$

or

$$\omega = \frac{\lambda D}{4d}[2n+3-2n-1] = \frac{\lambda D}{2d}$$

or

$$\omega = \frac{\lambda D}{2d} \tag{10}$$

Equation (9) and (10) are independent of n. Hence the separation between any two consecutive bright fringes or any two consecutive dark fringes is the same. From equation (9) or (10) it is clear that the fringe width varies directly with the slit-screen separation D, inversely with the separation of slits $2d$ and directly with the wavelength of light λ used. These conclusions are in perfect agreement with the observed interference pattern on the screen.

Shape of the Interference Fringes

The condition for maximum and minimum intensity at a point P on the screen is given by the relations

$$S_2P - S_1P = n\lambda \quad \text{and} \quad S_2P - S_1P = (2n+1)\lambda/2$$

Thus for a given value of n, the locus of points of maximum or minimum intensity is given by

$$S_2P - S_1P = \text{constant}$$

This is the equation of a hyperbola with S_1 and S_2 as foci. For a particular value of n the locus of points of maximum and minimum intensity will be a hyperboloid of revolution obtained by revolving hyperbola about the line joining S_1 and S_2 in space as shown in Figure 6.

The intersection of hyperboloids with the screen placed parallel to the line joining the sources S_1 and S_2 give the fringe pattern obtained. As λ is extremely small, the path difference ($S_2P - S_1P$) is also extremely small and hence the hyperbola appear more or less as straight. Thus if D >> $2d$, the fringe pattern will consist of nearly parallel, equidistant bright and dark lines.

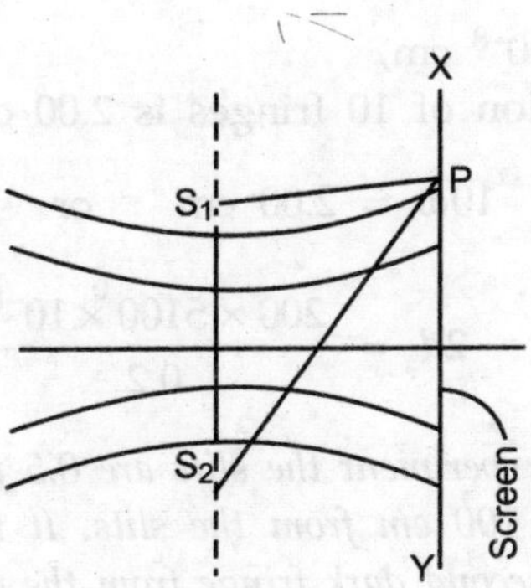

FIGURE 6

Example 6. *Two straight and narrow parallel slits 3 mm apart are illuminated by a monochromatic light of wavelength 5.9 × 10⁻⁵ cm. Fringes are obtained on a 60 cm distant screen from the slits. Find the value of the fringe width.*

Solution. In the given problem, the separation between the slits, $2d$ = 3 mm = 0.3 cm, the distance between the source and screen is D = 60 cm and $\lambda = 5.9 \times 10^{-5}$ cm.

We know that the fringe width ω is given by

$$\omega = \frac{D\lambda}{2d} = \frac{60 \times 5.9 \times 10^{-5}}{0.3}$$

$$= 11.8 \times 10^{-3} \text{ cm} = \mathbf{0.0118\ cm}.$$

Example 7. *Two coherent sources are placed 0.18 mm apart and the fringes are observed on a screen 80 cm away. It is found that with a certain monochromatic source of light the fourth bright fringe is situated at a distance of 10.8 mm from the central fringe. Find the wavelength of light.*

Solution. We know that the distance of nth bright fringe from the center of central fringe is given by

$$x_n = \frac{n\lambda D}{2d} \quad \text{or} \quad \lambda = \frac{x_n . 2d}{nD}$$

Here D = 80 cm, $2d$ = 0.18 mm = 0.018 cm, n = 4

and x_n = 10.8 mm = 1.08 cm

$$\therefore \quad \lambda = \frac{1.08 \times 0.018}{4 \times 80} \times 10^{-8} \text{ cm} = \mathbf{6075\ Å}.$$

Example 8. *Light of wavelength 5100 Å from a narrow slit is incident on a double slit. If the overall separation of 10 fringes on a screen 200 cm away is 2.0 cm, find the double slit separation.*

Solution. Let the double slit separation be $2d$, then using the fringe width relation, $\omega = \frac{D\lambda}{2d}$, we have

$$2d = \frac{D\lambda}{\omega}$$

Here λ = 5100 Å = 5100×10^{-8} cm,

D = 200 cm and the separation of 10 fringes is 2.00 cm, that is,

$$10\omega = 2.00 \text{ cm} \quad \text{or} \quad \omega = \frac{2.00}{10} = 0.2 \text{ cm}$$

$$\therefore \quad 2d = \frac{200 \times 5100 \times 10^{-8}}{0.2} = \mathbf{0.051\ cm} \text{ or } \mathbf{0.51\ mm.}$$

Example 9. *In Young's double slit experiment the slits are 0.5 mm apart and interference is observed on a screen placed at a distance of 100 cm from the slits. It is found that the 9th bright fringe is at a distance of 8.835 mm from the second dark fringe from the center pattern. Find the wavelength of light used.*

Solution. We know that the distance of nth bright fringe from the center of central fringe is given by

$$x_n = \frac{n\lambda D}{2d} = n\omega,$$

where $\omega = \left(\frac{\lambda D}{2d}\right)$ is the fringe width

For 9th bright fringe, $n = 9$, therefore, $x_9 = 9\omega$ (1)

Similarly the distance of nth dark fringe from the central fringe is given by

$$x'_n = \frac{(2n+1)\lambda D}{4d}, \qquad \text{where } n = 0, 1, 2, 3, ...$$

or

$$x'_n = \left(\frac{(2n+1)}{2}\right)\frac{\lambda D}{2d} = \frac{(2n+1)}{2}$$

For 2nd dark fringe $n = 1$, therefore, $x'_2 = \frac{3}{2}\omega$ (2)

The distance between the 9th bright fringe and 2nd dark fringe is as 8.835 mm or 0.8835 cm = $(x_9 - x'_n)$. Therefore, from equations (1) and (2) we have

$$x_9 - x'_2 = 9\omega - \frac{3}{2}\omega = \frac{15}{2}\omega = 0.8835$$

or

$$\omega = \frac{0.8835 \times 2}{15}$$

or

$$\omega = 0.1178 \text{ cm} = 1.178 \times 10^{-3} \text{ m}$$

Now the wavelength, $\lambda = \omega\frac{2d}{D}$

Here $2d$ = 0.5 mm = 0.05 cm and D = 100 cm

$$\therefore \quad \lambda = 0.1178 \times \frac{0.05}{100} = 5890 \times 10^{-8} \text{ cm} = \mathbf{5890\ A.}$$

Example 10. *In a two slit experiment with monochromatic light, fringes are obtained on a screen placed at the same distance from the slits. If the screen is moved by 5×10^{-2} m, calculate the wavelength of the light used.*

Solution. In the given problem, wavelength of light λ and the separation between the slits $2d$ is fixed. The fringe width ω changes with the distance of screen from the slit, that is, with D.

We know that the fringe width $\omega = \dfrac{\lambda D}{2d}$

If change in fringe width be Δω and change in position of the screen is ΔD, then,

$$\Delta\omega = \frac{\lambda}{2d}.\Delta D$$

Here $\Delta D = 5 \times 10^{-2}$ m, $\Delta\omega = 3 \times 10^{-5}$ m and $2d = 10^{-3}$ m

$$\therefore \quad \lambda = \frac{\Delta\omega}{\Delta D}.2d = \frac{3\times10^{-5}}{5\times10^{-2}}\times10^{-3}$$

$$= 6\times10^{-7} \text{ m} = \mathbf{6000\ Å.}$$

Example 11. *In a particular two-slit interference pattern with λ = 6000 Å, the zero order and 10th order maxima fall at micrometer readings 12.34 mm and 14.73 mm. If λ is changed to 5000 Å, deduce the positions of the zero order and 20th order fringes, the other arrangements remaining the same.*

Solution. Zero order fringe is the central fringe and independent of λ. So it falls at 12.34 mm.

We know that if the position of zero order fringe is x_0 and 10th order fringe is x_{10}, then

$$x_{10} - x_0 = 10.\frac{\lambda D}{2d} - 0.\frac{\lambda D}{2d} \qquad \left(\because x_n = \frac{n\lambda D}{2d}\right)$$

or

$$x_{10} - x_0 = \frac{10\ \lambda D}{2d}$$

or

$$\frac{\lambda D}{2d} = \omega = \frac{x_{10}-x_0}{10} = \frac{14.73-12.34}{10} = 0.239 \text{ mm}$$

Therefore for λ = 6000 Å, $\omega = 0.239 \text{ mm} = \dfrac{\lambda D}{2d}$, or $\dfrac{\omega}{\lambda} = \dfrac{D}{2d}$ (1)

We know that the fringe width changes with λ. Let the new interference pattern have ω′ fringe width, when λ′ = 5000 Å, then

$$\omega' = \frac{\lambda' D}{2d} \quad \text{or} \quad \frac{\omega'}{\lambda'} = \frac{D}{2d} \qquad (2)$$

Dividing equation (2) by equation (1), we get

$$\frac{(\omega'/\lambda')}{(\omega/\lambda)} = 1 \quad \text{or} \quad \frac{\omega'}{\lambda'} = \frac{\omega}{\lambda}$$

or

$$\omega' = \frac{\lambda'}{\lambda}\omega$$

Therefore

$$\omega' = \frac{5000}{6000}\times0.239 = 0.199 \text{ m}$$

The separation of 20 fringes = 20ω′ = 20 × 0.199 = 3.98 mm

∴ The position of 20th order fringe with λ = 5000 Å is

$x_0 \pm 3.98 = 12.34 \pm 2.98 =$ **16.32 mm, 8.36 mm.**

Example 12. *A beam of light consisting of two wavelengths 6500 Å and 5200 Å is used to obtain interference fringes in a Young's double slit experiment. The distance between the slits is 2.0 mm and the distance between the plane of slits and the screen is 120 cm. (a) Find the distance of the third bright fringe on the screen from the central maximum for the wavelength 6500 Å. (b) What is the least distance from the central maximum where the bright fringes, due to both the wavelength, coincide ?*

Solution. (*a*) The distance of nth bright fringe from the center of central fringe is given by

$$x_n = \frac{n\lambda D}{2d}.$$

Here $n = 3$, $\lambda = 6500$ Å $= 6500 \times 10^{-8}$ cm, D = 120 cm and $2d = 2$ mm $= 0.2$ cm

$$\therefore \quad x_3 = 3 \times \frac{6500 \times 10^{-8} \times 120}{0.2} = 0.117 \text{ cm} = 1.17 \text{ mm}$$

(*b*) m^{th} bright fringe of $\lambda = 6500$ Å coincides with the nth bright fringe of $\lambda' = 5200$ Å, then

$$m.\frac{\lambda D}{2d} = \frac{n\lambda' D}{2d}$$

or

$$\frac{m}{n} = \frac{\lambda'}{\lambda} = \frac{5200}{6500} = \frac{4}{5}$$

The minimum values of m and n that satisfy these equations are 4 and 5 respectively. The distance of 4th bright fringe of 5200 Å from the central fringe is

$$x_4 = \frac{4 \times 6500 \times 10^{-8} \times 120}{0.2} = \mathbf{0.156 \text{ cm.}}$$

2.11 FRESNEL'S BIPRISM

Fresnel produced the interference fringes in the laboratory by deriving two coherent sources S_1 and S_2 from a single monochromatic source S by using the deviation produced by a biprism. The experimental arrangement is shown schematically in Figure 7. The biprism consists of two acute angled prisms with their bases in contact. It is actually a single prism of an obtuse angle of about 179°. The acute angle α on both sides is about 30′ (or 1/2°). The prism is placed with its refracting edge parallel to the slits such that SE is normal to the face GF of the prism. Light from a narrow vertical slit S, illuminated with monochromatic light of wavelength λ, is allowed to fall symmetrically on the biprism. When light falls from S on the lower half of the prism, it bends upwards and appears to come from S_2. Similarly the light from S, which falls on the upper half of the prism, is bent downwards and appears to come from S_1. The virtual images S_1 and S_2, being the images of S, act as two coherent sources. The source S_1 illuminates the portion AQ of the screen while the source S_2 illuminates the portion BP. In the common shaded region the coherent beams overlap and therefore interfere. In the region AB the interference fringes have equal width, but beyond A and B the diffraction fringes of large width are obtained.

Theory

The point O on the screen is equidistant from virtual sources S_1 and S_2. Therefore at O the intensity is maximum and the central zero order fringe is bright. On both sides of O alternatively dark and bright fringes are produced. The positions of nth bright fringe and n^{th} dark fringe from the center O of the central fringe may be obtained by the relations

$$x_n = \frac{nD\lambda}{2d}, \text{ where } n = 0, 1, 2, 3, \ldots \text{ [For bright fringe]} \quad (1)$$

and
$$x_n = \frac{(2n+1)\lambda D}{4d}, \text{ where } n = 0, 1, 2, 3, \ldots \text{ [For dark fringe]} \quad (2)$$

The distance between any two successive bright fringes or any two successive dark fringes, (that is, fringe width) may be obtained by the relation

$$\omega = \frac{D\lambda}{2d} \quad (3)$$

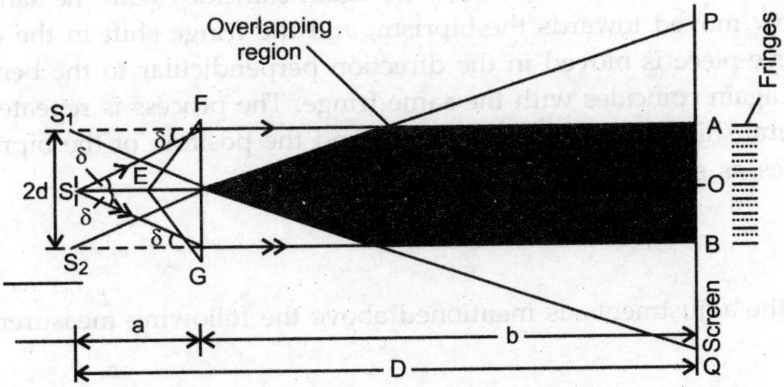

FIGURE 7

Determination of Wavelength of Monochromatic Light

Fresnel's biprism can be used to determine the wavelength of a given source of monochromatic light.

From equation (3), we have $\lambda = \omega \dfrac{2d}{D}$

Thus to find the wavelength λ, the fringes are produced and the values of ω, D, and 2*d* are measured. To find the accurate value of ω the following experimental arrangement and the adjustments are made.

Experimental Arrangement

An experimental arrangement consists of an optical bench about two meters in length. The bench carries four uprights for supporting the adjustable slit, the biprism, a high power micrometer, Ramsden's eye-piece, and a convergent lens. These uprights are capable of movement along and perpendicular to the length of bench and may be adjusted to any desired height. The slit and the biprism can also be rotated in their own planes about the axis of the upright. All the uprights are arranged at the same height above the bed of the optical bench. The slit is made narrow and is illuminated with monochromatic light whose wavelength is to be determined.

Adjustment for Getting Interference Fringes

For obtaining sharp and well defined fringes, the following necessary adjustments are made:

- The optical bench is leveled, the eye-piece is focussed on the cross-wires, and the slit, the biprism and the eye-piece are adjusted to the same height.
- The axis of the slit is made vertical.
- The line joining the obtuse edge of the biprism and the slit must be adjusted exactly parallel to the length of the bench by removing lateral shift. To remove the lateral shift the cross-wire is fixed on any one of the fringes in the center and the eye-piece is moved along the bench away from the biprism and the direction of the shift in fringes is observed. Now the biprism is moved perpendicular to the bench in a direction opposite to the shift so that the vertical cross-wire again coincides with the same fringe. The eye-piece is now moved towards the biprism, and the fringe shift in the opposite direction. Now the eye-piece is moved in the direction perpendicular to the bench till the vertical cross-wire again coincides with the same fringe. The process is repeated again and again till the lateral shift is completely removed and the position of the biprism relative to the slit becomes as shown in Figure 8.

Measurements

After making all the adjustments as mentioned above the following measurements are taken in sequence.

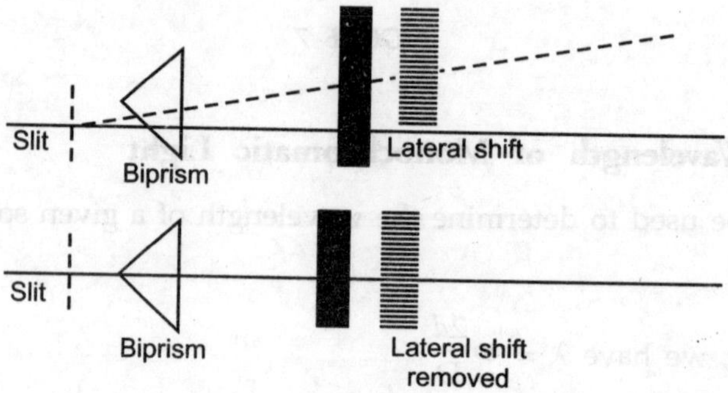

FIGURE 8

Measurement of the Fringe Width ω

The eye-piece is fixed at a suitable distance from the prism so that widely spaced and distinct fringes are visible in its principal focal plane. The cross-wire is set accurately at the center of the first bright fringe and the reading of the micrometer screw is taken. The eye-piece is then moved laterally so that the cross-wire coincides with successive bright fringes and the corresponding readings are noted down. From these readings the fringe width may be easily deduced.

Measurement of D

The distance between the eye-piece and the slit can be measured directly by reading the positions of their uprights on the optical bench and taking the difference of two readings. By applying the bench correction the accurate value of D is measured.

Measurement of (2d) the Distance between the Two Virtual Sources

The distance between two virtual sources (2*d*) is determined by the following method:

Displacement Method

In this method, the distance between the slit and the eye-piece is kept more than four times the focal length of the convex lens used so that the two positions of the lens to form the images of S_1 and S_2 are found. A convex lens of short focal length is placed between the biprism and the eye-piece (as shown in Figure 9). By moving the lens along the length of the bench, two positions L_1 and L_2 are obtained such that the real images of S_1 and S_2 are obtained in the eye-piece. Let d_1 and d_2 be the separations between real images in the positions L_1 and L_2 of the lens respectively. If u and v are the distances of the slit and the eye-piece from the lens in position L_1, then from the magnification formula we have

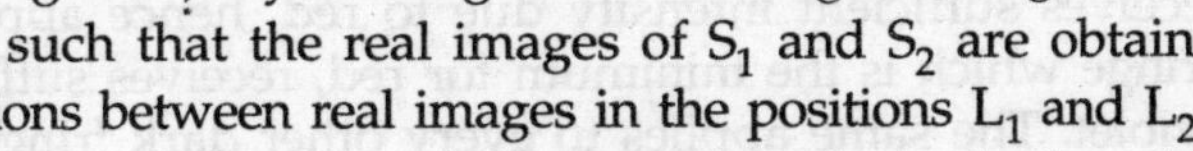

FIGURE 9

$$\frac{v}{u} = \frac{d_1}{2d} \tag{1}$$

As the two positions of the lens are conjugate, for the second position L_2 of the lens, we have

$$\frac{d_2}{2d} = \frac{u}{v} \tag{2}$$

Comparing equation (1) and (2) we get

$$\frac{d_1}{2d} = \frac{2d}{d_2} \quad \text{or} \quad 2d = \sqrt{(d_1 . d_2)}$$

Thus by measuring d_1 and d_2, 2*d* can be calculated. For an accurate measurement of 2*d*, the procedure of determining d_1 and d_2 is repeated at least three times by moving the eye-piece into different positions.

White Light Fringes and Location of Zero Order Fringe in a Biprism Experiment

When the monochromatic source of light in a biprism experiment is replaced by white light, then observed interference pattern consists of a central white fringe surrounded by few colored fringes and then a general illumination. White light consists of seven colors mixed together and the wavelengths vary from about 4000 A.U to 7800 A.U. At any point situated on the perpendicular bisector of coherent sources S_1 and S_2, the geometrical path difference is zero for the interfering light waves of all wavelengths. At this position, the condition of constructive interference is satisfied for all the wavelengths present in the white light. Thus at point O in Figure 7 we get the light of every color in exactly the same proportion as it exists in white light. Therefore, the point O will appear white; hence the central zero order fringe is white.

The separation between the successive bright or dark fringes or width of the fringe $\frac{\lambda D}{2d}$ is directly proportional to the wavelength. Therefore, the fringe width of red light is greater and that of violet light is smaller. As we move on either side of O, the path difference gradually increases from zero. At a certain point it becomes equal to half the wavelength of violet. This gives the position the first dark fringes of violet. Beyond this we obtain the first minimum of blue, green and red in the last. The inner edge of the first dark fringe, which is the first minimum for violet, receives sufficient intensity due to red, hence appears reddish. The outer edge of the first dark fringe which is the minimum for red, receives sufficient light due to violet and therefore appears violet. The same applies to every other dark fringe. Therefore we observe few colored fringes on either side of the central white fringe. The colored fringes will soon disappear because at points far away from O there will be many wavelengths in the visible region which will constructively interfere. Hence in the region of the large path difference uniform white illumination is obtained.

Location of the Zero Order Fringe

It is very difficult to locate the zero order fringe in the fringe pattern obtained by monochromatic light, as the number of fringes is very large and of the same color. Hence to locate the central fringe, we replace the monochromatic light by white light. When white light is used, in the interference pattern the zero order central fringe is white while the rest are colored. Hence we can easily adjust the cross-wires on the zero order fringe. Now white light is again replaced by monochromatic light without altering any of the arrangements. The cross-wire in this situation is still on the zero order bright fringe, which is thusly located.

Determination of the Thickness of a Thin Sheet of Transparent Material (Displacement of Fringes)

Let us now investigate the change in the interference pattern produced by introducing a thin transparent plate in the path of the two interference beams as shown in Figure 10. Let a thin transparent sheet (like mica) of thickness t and refractive index μ be introduced in the path of a wave from slit S_1. It is seen from Figure 10 that light reaching point P from source S_1 has to traverse a distance t in the sheet and a distance $(S_1P - t)$ in the air. If c and v are the velocities of light in the air and in the transparent sheet respectively, then the time taken by light to reach from S_1 to P is given by

$$= \frac{(S_1P - t)}{c} + \frac{t}{v}$$

$$= \frac{(S_1P - t)}{c} + \frac{\mu t}{c} = \frac{1}{c}[(S_1P - t) + \mu t] \qquad (\because \mu = c/v)$$

$$= \frac{1}{c}[S_1P + (\mu - 1)t] \tag{1}$$

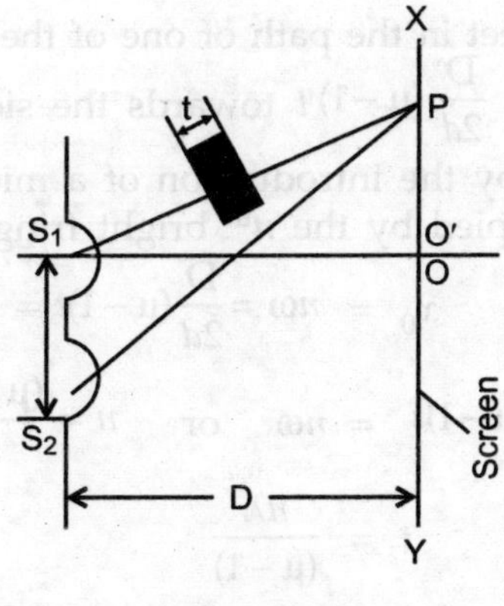

FIGURE 10

Thus by introducing a thin plate, the effective optical path in air is increased by an amount (μ – 1) *t*. In the absence of the sheet, let O be the position of the central bright fringe, which corresponds to the equal optical path from S_1 and S_2. In the presence of the sheet, the two optical paths (S_1O and S_2O) become unequal and therefore, the two waves from S_1 and S_2 do not arrive at O simultaneously. Therefore, the central fringe is shifted to a point O′ such that at O′ the two optical paths become equal. Such a shift results for all fringes.

Now the effective path difference at any point P on the screen is

$$\Delta' = S_2P - [S_1P + (\mu - 1)t]$$

$$= S_2P - S_1P - (\mu - 1)t \quad (2)$$

If the distance between the two sources S_1 and S_2 is $2d$, the distance of the screen from the sources is D and the position of the nth bright fringe is x_n, then the path difference in the absence of the sheet is

$$S_2P - S_1P = \frac{2d}{D}x_n \quad (3)$$

Thus the effective path difference

$$\Delta' = \frac{2d}{D}x_n - (\mu - 1)t$$

If the point P is now the center of the nth bright fringe, then from the condition of the bright fringe

$$\Delta' = 2n\lambda/2, \text{ when } n = 0, 1, 2, 3, \ldots \text{ we have}$$

$$\frac{2d}{D}x_n - (\mu - 1)t = n\lambda \quad \text{or} \quad x_n = \frac{D}{2d}[n\lambda + (\mu - 1)t] \quad (4)$$

In the absence of sheet $t = 0$, the distance of the nth bright fringe from O is $\frac{D}{2d}n\lambda$. Therefore, the displacement of the nth bright fringe is

$$x_n = \frac{D}{2d}[n\lambda + (\mu - 1)t] - \frac{D}{2d}n\lambda$$

$$x_0 = \frac{D}{2d}(\mu - 1)t \quad (5)$$

It is clear from equation (5) that x_0 is independent of n, that is, the displacement is the same for all bright fringes, and a similar expression can also be obtained for dark fringes. Thus the introduction of the transparent sheet in the path of one of the waves simply displaces the entire fringe system through a distance $\frac{D}{2d}.(\mu-1)\,t$ towards the side on which the sheet is placed.

Thickness of mica sheet: If by the introduction of a mica sheet, the central fringe moves through a distance formerly occupied by the nth bright fringe, then

$$x_0 = n\omega = \frac{D}{2d}(\mu-1)t = \frac{\omega}{\lambda}(\mu-1)t \qquad (\because \omega = \lambda D/2d)$$

$$\therefore \quad \frac{\omega}{\lambda}(\mu-1)t = n\omega \quad \text{or} \quad n = \frac{(\mu-1)t}{\lambda}$$

or

$$t = \frac{n\lambda}{(\mu-1)} \tag{6}$$

Thus to find the thickness of a mica sheet, the ω and hence D and $2d$, are measured by the biprism experiment and by measuring n and substituting in equation (6), the thickness of mica sheet t can be calculated.

Example 13. *A biprism is placed at a distance of 5 cm infront of a narrow slit, illuminated by sodium light ($\lambda = 5890 \times 10^{-8}$ cm) and the distance between the virtual sources is found to be 0.05 cm. Find the width of the fringes observed in an eye-piece placed at a distance of 75 cm from the biprism.*

Solution. We know that the fringe width ω is given by $\omega = \frac{\lambda D}{2d}$

In the given problem, the distance between virtual sources ($2d$) = 0.05 cm, and the distance between the slit and eye-piece, D = the distance between the biprism and narrow slit + the distance of eye-piece from the biprism *i.e.*, D = 5 cm + 75 cm = 80 cm and $\lambda = 5890 \times 10^{-8}$ cm.

$$\therefore \quad \omega = \frac{5890 \times 10^{-8} \times 80}{0.05} = \mathbf{9424 \times 10^{-5}\ cm.}$$

Example 14. *The inclined faces of a biprism of glass (μ = 1.5) make angles of 2° with the base. The slit is at 10 cm from the biprism and is illuminated by the light of wavelength 5500 Å. Find the fringe width at a distance of one meter from the biprism.*

Solution. The separation between the two virtual sources is given by

$$2d = 2a\,(\mu - 1)\alpha$$

Here

$$\alpha = \frac{2° \times \pi}{180} = \frac{\pi}{90} \text{ rad}$$

and

$$\mu = 1.5,\ a = 10 \text{ cm}$$

$$\therefore \quad 2d = 2 \times 10(1.5-1) \times \frac{3.14}{90} = 0.35 \text{ cm}$$

The fringe width, ω is given by $\omega = \frac{D\lambda}{2d}$

where D is the distance between the source and screen

i.e.,

$$D = 10 \text{ cm} + 100 \text{ cm} = 110 \text{ cm}$$

and

$$\lambda = 5500 \text{ Å} = 5500 \times 10^{-8} \text{ cm}$$

$$\therefore \quad \omega = \frac{110 \times 5500 \times 10^{-8}}{0.35} = \mathbf{0.0173\ cm.}$$

Example 15. *Interference fringes are produced by Fresnel's biprism in the focal plane of an eye-piece 200 cm away from the slit. Two images of the slit that are formed for each of the two positions of a convex lens placed between the biprism and eye-piece are found to be separated by 4.5 mm and 2.9 mm respectively. If the width of interference fringes is 0.326 mm, find the wavelength of light used.*

Solution. The wavelength of light λ is given by

$$\lambda = \frac{2d}{D}\omega$$

and the separation between the two virtual sources is $2d = \sqrt{(d_1 \times d_2)}$

where d_1 and d_2 are the image-separations for the two positions of the lens.

In the given problem,

$$d_1 = 4.5 \text{ mm} = 0.45 \text{ cm}$$

$$d_2 = 2.9 \text{ mm} = 0.29 \text{ cm}, \ D = 200 \text{ cm}$$

and $$\omega = 0.326 \text{ mm} = 0.0326 \text{ cm}$$

∴ $$2d = \sqrt{(0.45 \times 0.29)} = 0.36 \text{ cm}$$

and $$\lambda = \frac{2d}{D}\omega = \frac{0.36 \times 0.0326}{200} = \mathbf{5.87 \times 10^{-5} \ cm.}$$

Example 16. *In an experiment with Fresnel's biprism, sodium light is used and bands of 0.0196 cm in width are observed at a distance of 100 cm from the slit. A convex lens is then put between the observer and the prism so as to give an image of the source at a distance of 100 cm from the slit. The distance apart of the images is found to be 0.70 cm, the lens being 30 cm from the slit. Calculate the wavelength of sodium light.*

Solution. We know that the wavelength of light λ is given by

$$\lambda = \frac{2d}{D}\omega$$

Here $$\omega = 0.0196 \text{ cm} \quad \text{and} \quad D = 100 \text{ cm},$$

now we have to find the value of $2d$.

When the lens is at a distance of 30 cm from the slit, the separation of the two coherent images becomes 0.70 cm, that is, $d_1 = 0.70$ cm.

From the magnification formula

$$M = \frac{\text{Size of image}}{\text{Size of object}} = \frac{\text{Distance of image}}{\text{Distance of object}} \text{ or } M = \frac{I}{O} = \frac{v}{u}$$

Here $$I = 0.70 \text{ cm}, \ u = 30 \text{ cm}, \ O = 2d$$

the distance of the image from the lens, $v = 100 - 30 = 70$ cm

∴ $$\frac{0.70}{2d} = \frac{70}{30} \text{ or } 2d = \frac{0.70 \times 30}{70} = 0.30 \text{ cm}$$

Thus, $$\lambda = \frac{2d}{D}\omega = \frac{0.30}{100} \times 0.0196 = \mathbf{5880 \times 10^{-8} \ cm.}$$

Example 17. *In a biprism experiment the eye-piece was placed at a distance of 120 cm from the source. The distance between the two virtual sources was found to be equal to 0.075 cm. Find the wavelength*

of light of the source if the eye-piece has to be moved through a distance 1.888 cm for 20 fringes to cross the field of view.

Solution. The separation of 20 fringes is 1.888 cm, that is, $20\omega = 1.888$:

$$\therefore \quad \omega = \frac{1.888}{20}$$

or $$\omega = 0.944 \text{ mm} = 0.0944 \text{ cm}$$

We know that $$\lambda = \frac{2d}{D}\omega \qquad \text{Here D} = 120 \text{ cm and } 2d = 0.075 \text{ cm}$$

$$\lambda = \frac{0.075}{120} \times 0.0944 = 5900 \times 10^{-8} \text{ cm} = \mathbf{5900\ Å.}$$

Example 18. *A Fresnel biprism arrangement is set with sodium light (λ = 5893 Å) and in the field of view of the eye-piece we get 62 fringes. How many fringes shall we get if we replace the source by a mercury lamp using (a) green filter (λ = 5461 Å), (b) violet filter (λ = 4358 Å)?*

Solution. We know that the fringe width is given by

$$\omega = \frac{\lambda D}{2d} \quad \text{(for sodium light)} \tag{1}$$

(*a*) If the source is replaced by a mercury lamp (wavelength λ_1), the fringe width is given by

$$\omega_1 = \frac{D}{2d}\lambda_1 \tag{2}$$

Dividing equation (1) by equation (2), we get

$$\frac{\omega}{\omega_1} = \frac{\lambda}{\lambda_1} \tag{3}$$

The separation of 62 fringes is equal to the width of the field of view. Let the field of view contain n_1 fringes when we use a green filter.

Then the width of the field of view = $n_1\omega_1$

$$\therefore \quad n_1\omega_1 = 62\omega \quad \text{or} \quad n_1 = 62\frac{\omega}{\omega_1} \tag{4}$$

Substituting the value of $\frac{\omega}{\omega_1}$ from equation (3) in equation (4),

we get $$n_1 = 62\frac{\lambda}{\lambda_1}$$

Here $$\lambda_1 = 5461 \text{ Å} = 5461 \times 10^{-8} \text{ cm}$$

and $$\lambda = 5893 \text{ Å} = 5893 \times 10^{-8} \text{ cm}$$

$$\therefore \quad n_1 = 62 \times \frac{5893 \times 10^{-8}}{5461 \times 10^{-8}} = 67 \text{ (in whole numbers)}$$

(*b*) If the field of view contains n_2 fringes when we use a violet filter (λ_2 = 4358 Å), we have

$$n_2 = 62\frac{\lambda}{\lambda_2} = 62 \times \frac{5893 \times 10^{-8}}{4358 \times 10^{-8}} = \mathbf{83.}$$

Example 19. *In an experiment with a biprism, the reading on the optical bench of the position of the eye-piece and the two positions of the lens were respectively 100.00, 67.00 and 34.00 cm. The distances between the two images for the two positions of the lens were respectively 0.3000 and 1.2000 mm, and the width of 10 fringes was 9.720 mm. Assuming that there is no index error in any case, calculate (a) the distance between the focal plane of the eye-piece and the plane of interfering sources and (b) the wavelength of the light used.*

Solution. (*a*) The distance between the eye-piece and the second position of the lens is given by (from Figure 11)

$$v_1 = 100 - 67 = 33 \text{ cm} = u.$$

Hence, the reading of the slit on the bench is

$$= \text{Ist position of lens} - u = 34 - 33 = 1 \text{ cm}.$$

Hence, the distance between the focal plane of the eye-piece and the plane of the interfering sources is

$$D = 100 - 1 = 99 \text{ cm}$$

(*b*) The separation of the two interfering sources is $2d = \sqrt{(d_1 \times d_2)}$,

Here $d_1 = 1.2 \text{ mm} = 0.12 \text{ cm}$

and $d_2 = 0.3 \text{ mm} = 0.03\text{cm}$

$\therefore$ $2d = \sqrt{(0.12 \times 0.03)} = 0.06 \text{ cm}$

the width of 10 fringes was 9.720 mm

$\therefore$ $10\,\omega = 9.720 \text{ mm} = 0.972 \text{ cm}$

or $\omega = \dfrac{0.972}{10} = 0.0972 \text{ cm}.$

Therefore, the wavelength of light,

$$\lambda = \frac{2d}{D}\,\omega = \frac{0.06 \times 0.0972}{99} = 5891 \times 10^{-8} \text{ cm}.$$

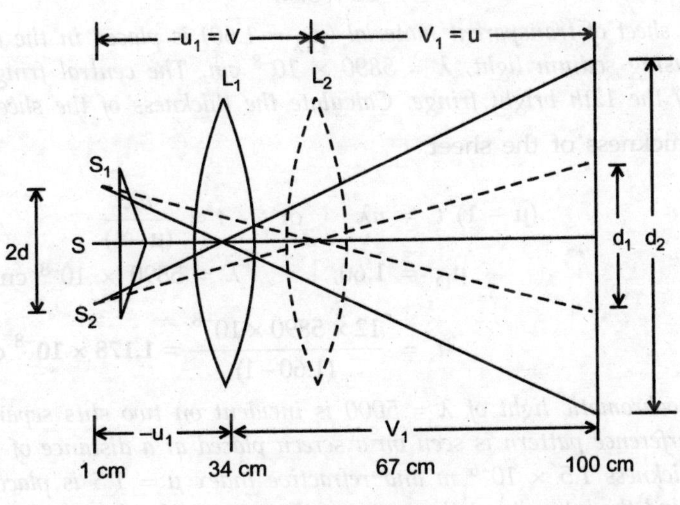

FIGURE 11

Example 20. *A two slit young's interference experiment is done with monochromatic light of wavelength 6000 Å. The slits are 2 mm apart and the fringes are observed on a screen placed 10 cm away from the slits. It is found that the interference pattern shifts by 5 mm, when a transparent plate of thickness 0.5 mm is introduced in the path of one of slits. What is the refractive index of the transparent plate?*

Solution. The shift of the interference pattern, x_0, is given by

$$x_0 = \frac{D}{2d}(\mu - 1)t \qquad (1)$$

where t is the thickness of the transparent plate.

In the given problem, $2d = 2 \text{ mm} = 0.2 \text{ cm}, D = 10 \text{ cm}$

$$\lambda = 6000 \text{ Å} = 6000 \times 10^{-8} \text{ cm}$$

$$t = 0.5 \text{ mm} = 0.05 \text{ cm and } x_0 = 5 \text{ mm} = 0.5 \text{ cm}$$

From equation (1)

$$(\mu - 1) = \frac{2d.x_0}{D.t} = \frac{0.2 \times 0.5}{10 \times 0.05} = 0.2 \quad \therefore \mu = 1 + 0.2 = \mathbf{1.2}$$

Example 21. *In a double slit interference arrangement, one of the slits is covered by a thin mica sheet whose refractive index is 1.58. The slits are 0.1 cm apart and the fringes are observed on a screen placed 50 cm away from the slit. Due to the introduction of a mica sheet the central fringes get shifted by 0.2 cm. Determine the thickness of mica sheet.*

Solution. We know that the thickness of the mica sheet t is

$$t = \frac{2d.x_0}{D(\mu - 1)}$$

Here $2d = 0.1 \text{ cm}, x_0 = 0.2 \text{ cm}$

$$D = 50 \text{ cm and } \mu = 1.58, (\mu - 1) = 0.58$$

Therefore, $$t = \frac{0.1 \times 0.2}{50 \times 0.58} = \mathbf{6.7 \times 10^{-4} \text{ cm.}}$$

Example 22. *A thin sheet of transparent material ($\mu_D = 1.60$) is placed in the interfering beams in a biprism experiment using sodium light, $\lambda = 5890 \times 10^{-8}$ cm. The central fringe shifts to a position normally occupied by the 12th bright fringe. Calculate the thickness of the sheet.*

Solution. For the thickness of the sheet, t

$$(\mu - 1)\, t = n\lambda \quad \text{or} \quad t = \frac{n\lambda}{(\mu - 1)}$$

Here $\mu_D = 1.60, \quad \lambda = 5890 \times 10^{-8} \text{ cm and } n = 12$

$$\therefore \quad t = \frac{12 \times 5890 \times 10^{-8}}{(1.60 - 1)} = \mathbf{1.178 \times 10^{-8} \text{ cm.}}$$

Example 23. *A monochromatic light of $\lambda = 5000$ is incident on two slits separated by a distance of 5×10^{-4} m. The interference pattern is seen on a screen placed at a distance of 1 m from the slits. A thin glass plate of thickness 1.5×10^{-6} m and refractive index $\mu = 1.5$ is placed between one of the slits and the screen. Find the intensity at the center of the screen. Also find the lateral shift of the central maximum.*

Solution. Shifting of the central maximum by the introduction of a thin glass plate or lateral shift of the central maximum is given by

$$x_0 = \frac{D}{2d}(\mu-1)t$$

Here D = 1 m, $2d = 5 \times 10^{-4}$ m, $\mu = 1.5$ and $t = 1.5 \times 10^{-6}$ m

$$\therefore \quad x_0 = \frac{1}{5\times 10^{-4}} \times (1.5-1)\times 1.5\times 10^{-6} = 1.5 \text{ cm}$$

The path difference created due to the introduction of a thin transparent glass plate

$$\Delta = (\mu - 1)\,t = (1.5 - 1) \times 1.5 \times 10^{-6} = 0.75 \times 10^{-6}\,\text{m}$$

$$\therefore \quad \delta = \frac{2\pi}{\lambda}\Delta = \frac{2\pi}{(5\times 10^{-7})} \times 0.75\times 10^{-6} = 3\pi$$

Intensity I, at the center of the screen $= a_1^2 + a_2^2 + 2a_1a_2\cos\delta$

Here $\quad \delta = 3\pi,\ \cos\delta = -1,\ a_1 = a_2 = a$

$$\therefore \quad I = a^2 + a^2 - 2a^2 = 0,$$

i.e., center of screen is dark.

Example 24. *In a biprism experiment the micrometer readings for a zero order fringe and a 10th order fringe are 2.37 mm and 3.55 mm respectively when light of $\lambda = 5.90 \times 10^{-5}$ cm is used. Now*

(i) What will be the position of the zero order and tenth order fringes if λ is changed to 7.5×10^{-5} cm?

(ii) If a mica sheet of thickness of 0.020 mm and $\mu_0 = 1.50$ is introduced to cover one half of the biprism, deduce the position of the zero fringe.

Solution. In the biprism experiment, the position of the zero order fringe remains unchanged as it is at equidistance from the two sources. So when the wavelength is changed from 5.90×10^{-5} cm to 7.50×10^{-5} cm, the position of the zero order fringe remains unaffected. The width of 10 fringes for $\lambda = 5.90 \times 10^{-5}$ cm is (3.55 – 2.37) mm = 0.118 cm.

Hence, $\quad 10\omega = 0.118 \quad \text{or} \quad \omega = \dfrac{0.118}{10} = 0.0118 \text{ cm}$

(i) When λ is changed to λ', ω changes to ω'

Thus $$\omega' = \frac{\lambda' D}{2d} \qquad (1)$$

For λ $$\omega = \frac{\lambda D}{2d} \qquad (2)$$

From equations (1) and (2), we have

$$\frac{\omega'}{\omega} = \frac{\lambda'}{\lambda} \quad \text{or} \quad \omega' = \frac{\lambda'}{\lambda}\omega$$

Here $\quad \lambda' = 7.50 \times 10^{-5}$ cm, $\lambda = 5.90 \times 10^{-5}$ cm,

therefore $$\omega' = \frac{7.50\times 10^{-5}}{5.90\times 10^{-5}} \times 0.0118 = \mathbf{0.015\ cm}$$

The distance between the zero order and the 10th order fringe is $10\omega' = 10 \times 0.015 = 0.15$ cm.

Thus the position of the tenth order fringe = 2.37 mm + 1.5 mm = 3.87 mm.

(*ii*) When the mica sheet is introduced to cover one half of the biprism, the shift in the fringe is given by:

$$x_0 = \frac{D}{2d}(\mu-1)t = \frac{\omega}{\lambda}(\mu-1)t$$

Here $\mu = 1.5,\ t = 0.020 \text{ mm} = 0.002 \text{ cm},\ \omega = 0.0118 \text{ cm}$

and $\lambda = 5.90 \times 10^{-5}$ cm

$$\therefore \quad x_0 = \frac{0.0118}{5.90 \times 10^{-5}} \times (1.5-1) \times 0.002$$

$$= 0.2 \text{ cm} = 2 \text{ mm}$$

Since the central fringe may shift on either side, depending upon the direction of the introduction of the sheet, the reading of the zero order fringe will be

$$= 2.37 \pm 2 = 4.37 \text{ mm}$$

or $= 0.37$ **mm.**

2.12 CHANGE OF PHASE ON REFLECTION

It is an important conclusion of Stock that when a wave of light is reflected at the surface of a denser medium, it always suffers a phase change of π or a path change of $\lambda/2$. However, no such change in phase or path is observed when the wave is reflected at the surface of a rarer medium. This phase change on reflection can be explained theoretically by using the principle of the reversibility of light.

Let a light wave AB of amplitude '*a*' incident on the interface MN, separating the two media air and glass, be partly refracted along BC and partly reflected along BD into the denser medium. Let *r* be the fraction of the amplitude reflected, and *t* the fraction transmitted, when the wave is travelling from a rarer (air) to a denser (glass) medium. Then the amplitude along BC is *ar* and that along BD is *at*, as shown in Figure 12 (*a*). If somehow or other we reverse the directions of reflected and transmitted waves, then according to the principle of reversibility we produce a wave of the original amplitude, provided there is no absorption of energy.

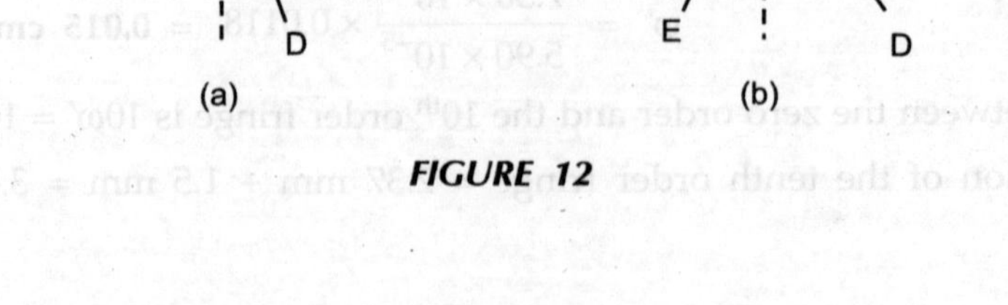

FIGURE 12

The reflected wave BC on reversal, gives a reflected wave of amplitude ar^2 along BA and a transmitted wave of amplitude art along BE [Figure 12 (*b*)]. Let r' and t' be the reflection coefficient and transmission coefficient respectively when the wave is travelling from a denser to a rarer medium. On reversing the refracted ray BD, we get a transmitted wave of amplitude att' along BA, and a reflected wave of amplitude atr' along BE. Since the reversal of ar and at must merely reproduce the amplitude a, the sum of components along BE should be zero and that along BA should be equal to a. That is,

$$art + atr' = 0 \tag{1}$$

and
$$ar^2 + att' = a \tag{2}$$

From equation (1), we get
$$r' = -r \tag{3}$$

The negative sign in equation (3) simply indicates that the displacement is in the opposite direction. This means that there is a difference of phase of π between the reflections in the rarer and denser media. In Lloyd's single mirror experiment it is observed that the central fringe is found to be dark instead of bright (as it is at equidistance from the coherent sources). This is possible only when light suffers a phase change of π when reflected at the mirror. We, therefore, conclude that a phase change of π occurs when light waves are reflected at the surface of a denser medium and no change of phase occurs when light waves are reflected at the surface of a rarer medium.

2.13 DIVISION OF AMPLITUDE—Interference in Thin Films

There are many interesting examples of interference produced in thin films. When a thin film of oil spreads on the surface of water, it is exposed to white light and beautiful colors are seen. Similar color phenomenon is observed when a soap film is illuminated with white light. These phenomena can be explained on the basis of interference between light reflected from the upper and lower surfaces of a thin film.

Interference Due to Reflected Light

To understand the phenomenon of interference in thin film due to reflected light, consider a ray SA of monochromatic light of wavelength λ incident on the upper surface of a thin transparent film of uniform thickness t and index of refraction μ, at an angle i as shown in Figure 13. The ray SA is partly reflected along AB and partly refracted along AC at an angle r. The refracted

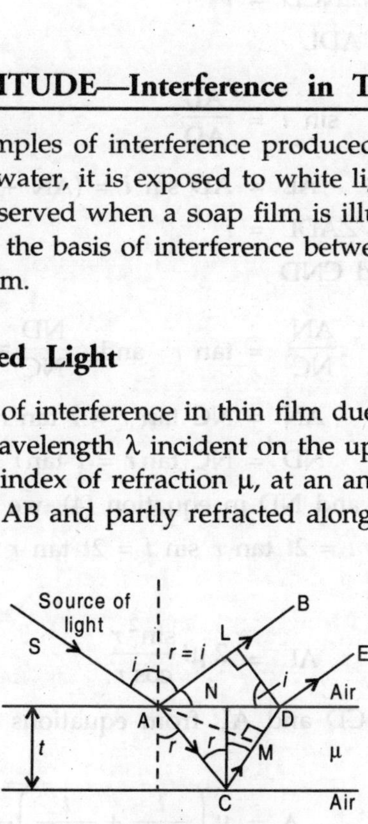

FIGURE 13

part AC is reflected from the point C on the lower surface of the film along CD and finally emerges out along DE. As the rays AB and DE are derived from the same incident ray, they are coherent. If the film is thin, the rays AB and DE will be sufficiently close to each other and in a position to interfere. To evaluate the path difference between AB and DE, the perpendiculars DL and CN are drawn on AB and AD respectively. As the paths of the rays AB and DE beyond DL are equal, the optical path difference between these two rays is given by

$$\Delta = \text{path (AC + CD) in film} - \text{path AL in air}$$

$$\Delta = \mu(\text{AC} + \text{CD}) - \text{AL}$$

$$(\because \text{ path in air} = \mu \text{ times path in the film}) \quad (1)$$

In the right angled triangle ACN, we have

$$\frac{\text{CN}}{\text{AC}} = \cos r \quad \text{or} \quad \text{AC} = \frac{\text{CN}}{\cos r} = \frac{t}{\cos r} \quad (2)$$

where $\angle\text{ACN} = r$ and CN = t, the thickness of the film.

Similarly, in the other right angled triangle CND, we have

$$\frac{\text{CN}}{\text{CD}} = \cos r \quad \text{or} \quad \text{CD} = \frac{\text{CN}}{\cos r} = \frac{t}{\cos r} \quad (3)$$

where $\angle\text{NCD} = r$

In the right angled triangle ADL

$$\sin i = \frac{\text{AL}}{\text{AD}}$$

or $\text{AL} = \text{AD} \sin i = (\text{AN} + \text{ND}) \sin i \quad (4)$

where $\angle\text{ADL} = i$

Again in triangles ACN and CND

$$\frac{\text{AN}}{\text{NC}} = \tan r \quad \text{and} \quad \frac{\text{ND}}{\text{NC}} = \tan r$$

or $\text{AN} = \text{NC} \tan r = t \tan r$

and $\text{ND} = \text{NC} \tan r = t \tan r$

Putting these values of AN and ND in equation (4), we get

$$\text{AL} = (t \tan r + t \tan r) \sin i = 2t \tan r \sin i = 2t \tan r\, (\mu \sin r)$$

(from Snell's law)

$$\text{AL} = 2\mu t \frac{\sin^2 r}{\cos r} \quad (5)$$

Substituting values of AC, CD and AL from equations (2), (3) and (5) in equation (1), we get

$$\Delta = \mu\left(\frac{t}{\cos r} + \frac{t}{\cos r}\right) - 2\mu t \frac{\sin^2 r}{\cos r}$$

or $$\Delta = \frac{2\mu t}{\cos r}[1-\sin^2 r] = \frac{2\mu t}{\cos r}.\cos^2 r = 2\mu t \cos r$$

$$\therefore \quad \Delta = 2\mu t \cos r \tag{6}$$

As the ray AB is the reflected ray from a denser medium, there occurs an additional path difference of $\lambda/2$ or phase change of π.

Thus, the total path difference $= \Delta \pm \lambda/2 = 2\mu t \cos r \pm \lambda/2$ (7)

Condition of Maximum and Minimum Intensities

(i) For constructive interference, path difference should be an even multiple of $\lambda/2$, that is, for maxima

$$2\mu t \cos r \pm \lambda/2 = 2n\lambda/2, \text{ where, } n = 0, 1, 2, 3,...$$

or $$2\mu t \cos r = (2n \pm 1)\lambda/2, \text{ where, } n = 0, 1, 2, 3,... \tag{8}$$

When condition (8) is satisfied, the film appears bright.

(ii) For destructive interference, the path difference should be an odd multiple of $\lambda/2$, that is, for minima

$$2\mu t \cos r \pm \lambda/2 = (2n \pm 1)\lambda/2, \text{ where, } n = 0, 1, 2, 3, ...,$$

or $$2\mu t \cos r = 2n\lambda/2 = n\lambda$$

Therefore, the film will appear dark if

$$2\mu t \cos r = n\lambda, \text{ where, } n = 0, 1, 2, 3, ..., \tag{9}$$

From equations (8) and (9) it is clear that the maxima and minima in the interference pattern depend upon two factors: (1) the thickness of the film t and, (2) the cosine of the angle r.

When $t = 0$, the path difference is $\lambda/2$ and the condition of minimum intensity is satisfied. Hence when $t = 0$, the film will appear dark and as the thickness is increased gradually, maxima and minima occur alternatively.

When t is kept constant and r is gradually varied, we again get an interference pattern with bright and dark bands.

Interference Due to Transmitted Light

Let a ray of light of wavelength λ be incident on the upper surface of a thin transparent film of uniform thickness t and index of refraction μ at an angle i as shown in Figure 14. The ray SA is refracted along AB at an angle r. The refracted part AB is partly reflected along BC and partly refracted along BP at an angle i. The reflected part BC is again reflected from the point C on the upper surface of the film along CD and finally emerges out along DQ. If the film is thin then the rays BP and DQ, which are coherent, will be very close to each other and in a position to interfere. To evaluate the path difference between BP and DQ, the perpendiculars

DN and CM are drawn on BP and BD respectively. As the paths of the rays BP and DQ beyond DN are equal, the optical path difference between these two rays is given by

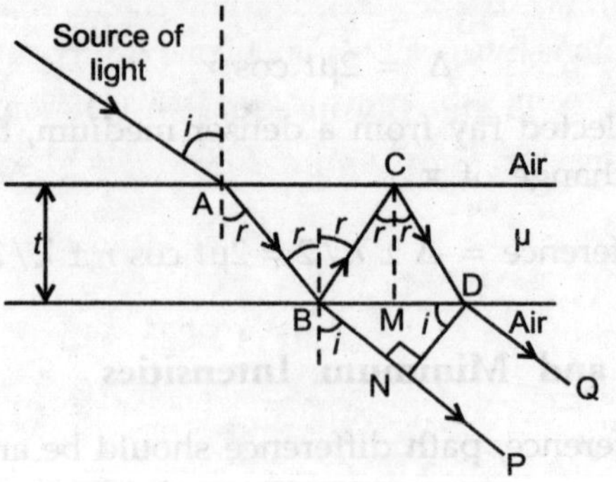

FIGURE 14

$$\Delta = \text{path (BC + CD) in film} - \text{path BN in air}$$
$$= \mu\,(BC + CD) - BN$$
$$(\because \text{ path in air} = \mu \text{ times path in the film}). \quad (1)$$

In the right angled triangle BCM, we have

$$\frac{MC}{BC} = \cos r \quad \text{or} \quad BC = \frac{MC}{\cos r} = \frac{t}{\cos r} \quad (2)$$

where $\angle BCM = r$, $MC = t$, the thickness of the film.

In other right angled triangle MDC, we have

$$\frac{MC}{CD} = \cos r \quad \text{or} \quad CD = \frac{MC}{\cos r} = \frac{t}{\cos r} \quad (3)$$

where $\angle MCD = r$

In right angled triangle BND

$$\sin i = \frac{BN}{BD}$$

or $$BN = BD \sin i = (BM + MD) \sin i \quad (4)$$

where $\angle BDN = i$

Again in right angled triangles BMC and CDM

$$\frac{BM}{CM} = \tan r \text{ and } \frac{MD}{CM} = \tan r$$

or $$BM = CM \tan r = t \tan r$$

and $$MD = CM \tan r = t \tan r$$

Putting these values of BM and MD in equation (4), we have

$$BN = (t \tan r + t \tan r) \sin i = 2t \tan r \sin i$$

But from Snell's law (sin i)/(sin r) = μ

$$\therefore \quad BN = 2t \tan r \, \mu \sin r = 2\mu t \frac{\sin^2 r}{\cos r} \qquad (5)$$

Substituting values of BC, CD and BN in equation (1) from equations (2), (3) and (4), we get

$$\Delta = \mu\left(\frac{t}{\cos r} + \frac{t}{\cos r}\right) - 2\mu t \frac{\sin^2 r}{\cos r}$$

$$= \frac{2\,\mu t}{\cos r}(1-\sin^2 r) = \frac{2\,\mu t}{\cos r}\cos^2 r$$

or
$$\Delta = 2\mu t \cos r \qquad (6)$$

Condition for Maximum and Minimum Intensities

(*i*) For constructive interference of maxima

$$\Delta = \pm 2n\lambda/2$$

or
$$2\mu t \cos r = \pm n\lambda, \text{ where } n = 0, 1, 2, 3, \ldots \qquad (7)$$

(*ii*) For destructive interference or minima

$$\Delta = (2n \pm 1)\,\lambda/2$$

or
$$2\mu t \cos r = (2n \pm 1)\lambda/2, \text{ where } n = 0, 1, 2, 3, \ldots \qquad (8)$$

Thus the condition of maxima and minima in the transmitted light are just the reverse of that in reflected light.

Colors of Thin Films

When a beam of white light from an extended source incident normally on a thin film and seen in reflected light then colored fringes will be observed. These colors arise due to the interference of the light waves reflected from the upper and lower surfaces of the film. The path difference between these interfering rays depends upon the thickness of the film t and upon the inclination of the incident rays r. At a particular point of the film (t) and for a particular position of the eye (r) the waves of only certain wavelengths satisfy the condition of maxima (*i.e.*, $2\mu t \cos r = (2n \pm 1)\,\lambda/2$). Hence only those colors (wavelengths) will present with maximum intensity. Other neighboring colors (wavelengths) will present with diminished intensity, while other wavelengths, which satisfy the condition of minima ($2\mu t \cos r = n\lambda$), will be absent from the reflected system. Hence the point of the film will appear colored. Similarly, if the same point of the film is observed with the eye, different positions or different points are observed by keeping the eye at a fixed position, a different set of colors is observed each time.

As the condition of maxima and minima for transmitted light is opposite to those for reflected light, hence the colors visible in the reflected system are absent in the transmitted system or the colors visible in reflected system will be complementary to the colors visible in the transmitted system.

Example 25. *A parallel beam of sodium light with a wavelength of 5880 Å is incident on a thin glass plate of refractive index 1.5 such that the angle of refraction in the plate is 60°. Calculate the smallest thickness of the plate which will make it appear dark by reflection.*

Solution. We know that the condition for destructive interference of light in the reflected system is given by

$$2\mu t \cos r = 2n\lambda/2$$

Here, $\mu = 1.5$, $r = 60°$, $\lambda = 5880$ Å $= 5880 \times 10^{-8}$ cm, $\cos 60° = 1/2$, for the smallest thickness $n = 1$

then
$$t = \frac{n\lambda}{2\mu \cos r} = \frac{1 \times 5880 \times 10^{-8}}{2 \times 1.5 \times 0.5}$$

$$= 3920 \times 10^{-8} \text{ cm} = \mathbf{3920 \text{ Å}}.$$

Example 26. *Calculate the thickness of the thinnest film ($\mu = 1.4$) in which interference of a violet component ($\lambda = 4000$ Å) of incident light can take place by reflection.*

Solution. We know that the condition of constructive interference in the reflected system is

$$2\mu t \cos r = (2n + 1)\lambda/2, \qquad n = 0, 1, 2, 3,...$$

For the thinnest film $n = 0$ and for normal incidence $r = 0$ or $\cos r = 1$,

Here $\lambda = 4000$ Å $= 4000 \times 10^{-8}$ cm and $\mu = 1.4$

Thus,
$$t = \frac{\lambda}{4\mu \cos r} = \frac{4000 \times 10^{-8}}{4 \times 1.4 \times 1} = \mathbf{714.3 \text{ Å}}.$$

Example 27. *A parallel beam of sodium light ($\lambda = 5890$ Å) strikes a film of oil floating on water. When viewed at an angle of 30° from the normal, an 8th dark band is seen. Determine the thickness of the film (refractive index of oil = 1.5).*

Solution. We know that the condition of destructive interference or dark band in reflected light is

$$2\mu t \cos r = 2n\lambda/2 \text{ or } \quad t = \frac{n\lambda}{2\mu \cos r} \qquad (1)$$

In this problem $\quad i = 30°, \mu = 1.5, n = 8$

From Snell's law
$$\mu = \frac{\sin i}{\sin r} \qquad \therefore \sin r = \frac{\sin i}{\mu}$$

so
$$\cos r = \sqrt{1 - \sin^2 r} = \sqrt{1 - \left(\frac{\sin i}{\mu}\right)^2}$$

$$\therefore \quad \cos r = \sqrt{\left(1 - \frac{\sin^2 30°}{(1.5)^2}\right)} = \sqrt{\left(1 - \frac{(0.5)^2}{(1.5)^2}\right)} = 0.943$$

$$\lambda = 5890 \text{ Å} = 5890 \times 10^{-8} \text{ cm}$$

Therefore, from equation (1)

$$t = \frac{8 \times 5890 \times 10^{-8}}{2 \times 1.5 \times 0.943} = \mathbf{1.665 \times 10^{-4} \text{ cm.}}$$

Example 28. *White light is incident on a soap film at an angle $\sin^{-1}\frac{4}{5}$ and the reflected light is observed with a spectroscope. It is found that two consecutive dark bands correspond to wavelengths 6.1×10^{-5} and 6.0×10^{-5} cm. If the refractive index of the film is 4/3, calculate the thickness.*

Solution. We know that the conditions for dark bands or fringe in the reflected light is

$$2\mu t \cos r = n\lambda$$

If n and $(n + 1)$ are the orders of consecutive dark bands for wavelengths λ_1 and λ_2 respectively, then

$$2\mu t \cos r = n\lambda_1 \text{ and } 2\mu t \cos r = (n+1)\lambda_2$$

or
$$2\mu t \cos r = n\lambda_1 = (n + 1)\lambda_2 \qquad (1)$$

$\therefore$
$$n\lambda_1 = (n + 1)\lambda_2 \quad \text{or} \quad n\lambda_1 = n\lambda_2 + \lambda_2$$

or
$$n(\lambda_1 - \lambda_2) = \lambda_2 \quad \text{or} \quad n = \frac{\lambda_2}{(\lambda_1 - \lambda_2)}$$

Substituting this value of n in equation (1), we get

$$2\mu t \cos r = \frac{\lambda_1\lambda_2}{(\lambda_1 - \lambda_2)} \quad \text{or} \quad t = \frac{\lambda_1\lambda_2}{(\lambda_1 - \lambda_2)} \cdot \frac{1}{2\mu \cos r} \qquad (2)$$

But
$$\cos r = \sqrt{(1 - \sin^2 r)} = \sqrt{\left[1 - \left(\frac{\sin i}{\mu}\right)^2\right]}$$

$$\left(\because \text{ From Snell's law } \mu = \frac{\sin i}{\sin r}\right)$$

Given $\mu = 4/3$ and $\sin i = 4/5$

$\therefore$
$$\cos r = \sqrt{1 - \frac{(4/5)^2}{(4/3)^2}} = \sqrt{1 - \left(\frac{3}{5}\right)^2} = \sqrt{\frac{16}{25}} = \frac{4}{5}$$

Substituting this value of $\cos r$ in equation (2) and also $\lambda_1 = 6.1 \times 10^{-5}$ cm, $\lambda_2 = 6.0 \times 10^{-5}$ cm, $\mu = 4/3$, we have

$$t = \frac{6.1 \times 10^{-5} \times 6.0 \times 10^{-5}}{(6.1 \times 10^{-5} - 6.0 \times 10^{-5}) \times 2 \times (4/3) \times (4/5)}$$

$\therefore$
$$t = \frac{6.1 \times 10^{-5} \times 6.0 \times 10^{-5} \times 3 \times 5}{0.1 \times 10^{-5} \times 2 \times 4 \times 4} = \mathbf{0.0017\ cm}.$$

Example 29. *White light is incident on two parallel glass plates separated by an air film of 0.001 cm thickness and the reflected light is examined by a spectroscope. Find the number of dark bands seen in the spectrum between the wavelength 4×10^{-5} cm and 7×10^{-5} cm when light is incident at an angle of 30° to the normal surfaces.*

Solution. We know that the condition of dark bands in the reflected light is

$$2\mu t \cos r = n\lambda \qquad (1)$$

In the given problem, $\mu = 1$, $t = 0.001$ cm, $i = 30°$ and the range of the visible spectrum is from 4×10^{-5} to 7×10^{-5} cm. Let the number of dark bands seen at $\lambda = 4 \times 10^{-5}$ cm be n_1 and at $\lambda = 7 \times 10^{-5}$ cm be n_2.

From Snell's law, $$\mu = \frac{\sin i}{\sin r} = 1, \quad \therefore \sin i = \sin r$$

or $$\cos r = \cos 30° = \frac{\sqrt{3}}{2}$$

For wavelength $\lambda = 40 \times 10^{-5}$ cm, from equation (1), we have

$$2 \times 1 \times 0.001 \times \left(\frac{\sqrt{3}}{2}\right) = 4 \times 10^{-5}\, n_1 \quad \therefore \quad n_1 = 43$$

Similarly for wavelength $\lambda = 7 \times 10^{-5}$ cm,

$$2 \times 1 \times 0.001 \times \left(\frac{\sqrt{3}}{2}\right) = 7 \times 10^{-5}\, n_2 \quad \therefore \quad n_2 = 24$$

Hence the number of dark bands observed within the given spectral range = $(n_1 - n_2)$ = 43 − 24 = **19.**

Wedge-Shaped Film

Consider a film in the shape of a thin wedge whose sides are slightly inclined to each other at an angle θ and enclosing a liquid of refractive index μ. The thickness of the film increases gradually from O to A as shown in Figure 15. When the film is illuminated by a parallel beam of monochromatic light, a system of equidistant alternate bright and dark fringes are observed due to the interference between the rays reflected at the upper and lower surfaces of the film. As the interfering rays PQ and EH are derived from the same incident ray SP, they are coherent. To evaluate the path difference between PQ and EH, the perpendiculars EK and EN are drawn on PQ and PF respectively. The optical path difference between these two rays is given by

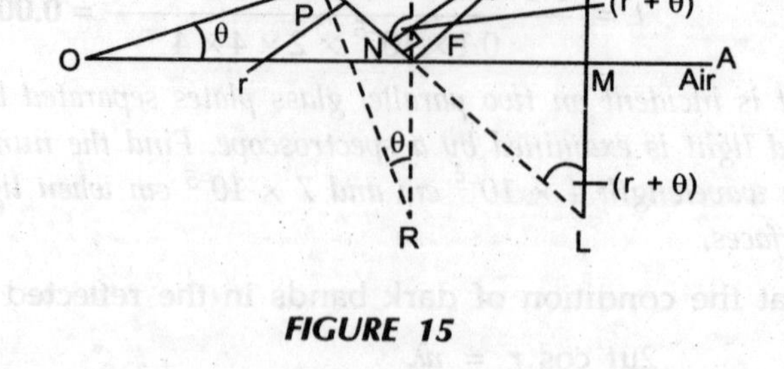

FIGURE 15

$$\Delta = (PF + FE) \text{ in film} - PK \text{ in air}$$

$$= \mu\,(PF + FE) - PK$$

($\because$ Path in air = μ times path in film)

$$= \mu(PN + NF + FE) - PK \tag{1}$$

From Figure 14 it is clear that the $\angle PEK = \angle i$ and $\angle NEP = \angle r$

$$\therefore \quad \mu \frac{\sin i}{\sin r} = \frac{(PK/PE)}{(PN/PE)} = \frac{PK}{PN} \quad \therefore\ PK = \mu PN \tag{2}$$

Substituting the value of PK from equation (2) in equation (1), we get

$$\Delta = \mu(PN + NF + FE) - \mu PN$$

or

$$\Delta = \mu\,(NF + FE) \tag{3}$$

Now draw normal EM on OA from E. When PF and EM produce, they meet at L. Triangles EFM and LFM are congruent.

$$\therefore \quad EM = ML = t \text{ and } FE = FL \tag{4}$$

Substituting the value of FE from equation (4) in equation (3), we have

$$\Delta = \mu\,(NF + FL) = \mu NL \tag{5}$$

The angle between the inclined surfaces is the same as the angle between the normals at point P and F, that is, $\angle BOA = \angle PRF = \theta$,

But $\angle FPR + \angle PRF = \text{ext. } \angle PFT$

$\therefore$ *i.e.*, $(r + \theta) = \angle PFT$

TR and EL are normals to the surface OA and PFL cuts the parallel lines TR and EL, Therefore

$$\angle PFT = \angle ELF = (r + \theta)$$

In triangle NEL, $\frac{NL}{EL} = \cos(r + \theta)$ or $NL = EL\cos(r + \theta)$

But $EL = 2EM = 2t$

$$\therefore \quad NL = 2t\cos(r + \theta) \tag{6}$$

Substituting value of NL in equation (5) from equation (6), we have the path difference,

$$\Delta = 2\mu t\cos(r + \theta)$$

As the ray PQ is the reflected ray from a denser medium, therefore there occurs an additional path difference of $\lambda/2$ or phase change of π.

Thus the total path difference = $\Delta \pm \lambda/2 = 2\mu t\cos(r + \theta) \pm \lambda/2$.

Condition for Maximum and Minimum Intensities

(*i*) For constructive interference or maxima,

$$2\mu t\cos(r + \theta) \pm \lambda/2 = 2n\lambda/2$$

or

$$2\mu t\cos(r+\theta) = (2n \pm 1)\lambda/2, \text{ where } n = 0, 1, 2, 3, \ldots \tag{7}$$

(*ii*) For destructive interference or minima,

$$2\mu t \cos(r+\theta) \pm \lambda/2 = (2n \pm 1)\lambda/2$$

or $$2\mu t \cos(r+\theta) = n\lambda, \quad \text{where } n = 0, 1, 2, 3, \ldots \tag{8}$$

The Nature of the Interference Pattern

When wedge-shaped film is exposed to a beam of monochromatic light of wavelength λ, the angle of incidence i, hence the angle of refraction r, is constant everywhere. In this case the path difference will only change due to the varying thickness of the film (as μ, r and θ are constant) from one point to the other. At a particular point, the thickness of the film is constant, thus we get a bright or dark fringe at that point due to the constant path difference. Since the locus of all points of constant thickness is a straight line, then the straight fringes parallel to the edge can be obtained.

Fringe Width

For n^{th} bright fringe, we have

$$2\mu t \cos(r+\theta) = (2n+1)\lambda/2 \tag{9}$$

Let x_n be the distance of the nth bright fringe from the edge of the film, then (from Figure 16)

$$\tan\theta = \frac{t}{x_n} \quad \text{or} \quad t = x_n \tan\theta$$

Putting this value of t in equation (9), we get

$$2\mu x_n \tan\theta \cos(r+\theta) = (2n+1)\lambda/2 \tag{10}$$

Similarly if x_{n+1} is the distance of $(n+1)^{th}$ bright fringe, then

$$2\mu x_{n+1} \tan\theta \cos(r+\theta) = [2(n+1)+1]\lambda/2 = (2n+3)\lambda/2 \tag{11}$$

Subtracting equation (10) from equation (11), we get

$$2\mu(x_{n+1} - x_n)\tan\theta \cos(r+\theta) = \lambda$$

or $$x_{n+1} - x_n = \frac{\lambda}{2\mu \tan\theta \cos(r+\theta)} = \omega$$

From normal incidence $i = r = 0$ and $\cos(r+\theta) = \cos\theta$

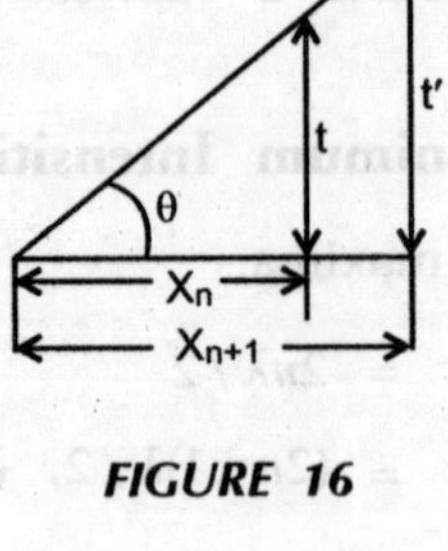

FIGURE 16

$$\omega = \frac{\lambda}{2\mu \tan\theta \cos\theta} = \frac{\lambda}{2\mu \sin\theta}$$

For the very small value of θ, sin θ = θ

$$\therefore \qquad \omega = \frac{\lambda}{2\mu\theta} \quad \text{For air film, } \mu = 1 \qquad \therefore\ \omega = \frac{\lambda}{2\theta}.$$

2.14 NECESSITY OF AN EXTENDED SOURCE

To see the interference effects over the entire film simultaneously, a broad source of light is necessary. This fact may be understood as follows:

When a thin film is illuminated with light from a point source and is viewed with a lens of small aperture, the light reflected from all corresponding points on the film does not reach the eye as shown in Figure 17(*a*). Thus only a small portion of the film will be visible. To see the whole film, the eye will have to be moved from one position to the other. Hence with a point source the entire film cannot be viewed at a glance. If we employ an extended source, the light reflected by every point reaches the eye (as shown in Figure 17(*b*)). Hence the entire film can be viewed simultaneously by keeping the eye at one place only. Hence an extended source of light is used to view a film simultaneously.

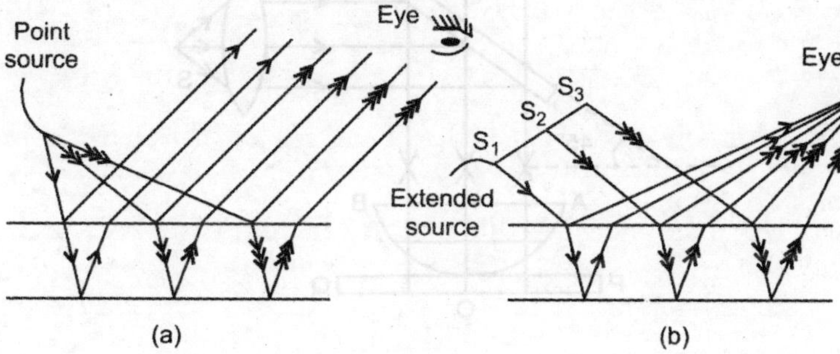

FIGURE 17

2.15 NEWTON'S RINGS

If we place a plano-convex lens of a large focal length on a glass plate, a thin film of air is formed between the curved surface of the lens and the glass plate. The thickness of the air film is zero at the point of contant O and gradually increases as we move away from the point of contact. When the monochromatic light falls normally on the surface of the lens, then the light reflected from the upper surface of the film AOB interferes with the light reflected from the lower surface of the film POQ (Figure 18). As a result, alternate dark and bright circular rings concentric around the point of contact are seen. The interference rings so formed were first investigated by Newton and hence called Newton's rings. If the incident light is white, a series of concentric colored rings is seen round about the point of contact in both the reflected and transmitted light.

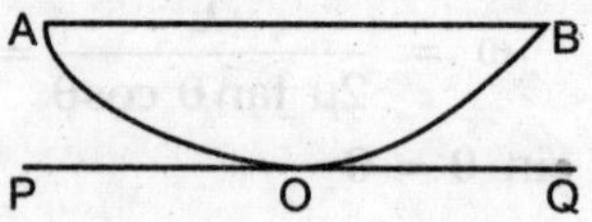

FIGURE 18

Experimental Arrangement

Newton's rings can easily be observed in the laboratory by using an apparatus as shown in Figure 19. Light from an extended monochromatic source S is rendered parallel by a lens L_1 and then falls on a 45° inclined thin transparent glass plate G, which partially reflects the light in the downward direction. These reflected beams fall normally on an air film formed between the convex surface of the plano-convex lens and glass plate. The light transmitted through the plano-convex lens on reflection from the surface of the glass plate PQ, interferes with the light reflected from the lower surface of the plano-convex lens. These reflected beams proceed upwards and enter the observer's eye through a low power travelling microscope M. On focussing a large number of Newton's rings, alternately bright and dark can be seen.

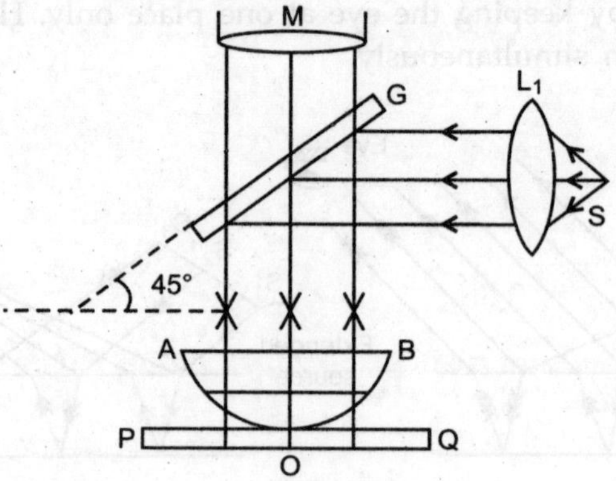

FIGURE 19

Theory (or Explanation of the Formation of Newton's Rings)

The formation of Newton's rings was satisfactorily explained by Young. According to him, these circular rings are formed due to the interference of light rays reflected from the upper and lower surfaces of the air film formed between the convex surface of a plano-convex lens and the glass plate. In Figure 20, the incident ray SF is divided into two coherent rays 1 and 2 by reflection from the upper and lower surfaces of the wedge shaped air film. The reflected rays 1 and 2 interfere and produce bright and dark circular rings around the point of contact. The effective path difference between the interfering rays in reflected light is $2\mu t \cos(r + \theta) + \lambda/2$, where μ is the refractive index of the film, t is the thickness of the film at the point of incidence, r is the angle of refraction, and θ is the angle of wedge.

For normal incidence $r = 0$ and for a very small angle of wedge

$$\cos(r + \theta) = \cos\theta = 1$$

therefore, the effective path difference = $2\mu t + \lambda/2$

At the point of contact, $t = 0$.

Therefore, the effective path difference at the point of contact = $\lambda/2$.

This is the condition of minimum intensity. Hence the central spot of the rings system is dark.

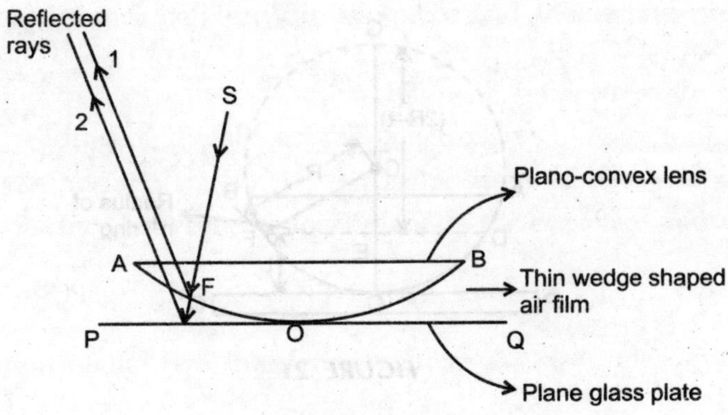

FIGURE 20

For constructive interference or for a bright ring

$$\text{The effective path difference} = 2n\lambda/2$$

$$\therefore \quad 2\mu t + \lambda/2 = n\lambda \text{ or } 2\mu t = (n - 1/2)\lambda$$

or

$$2\mu t = (2n - 1)\lambda/2,$$

$$\text{where } n = 1, 2, 3, 4,... \quad (1)$$

For destructive interference or for a dark ring

$$2\mu t + \lambda/2 = (2n + 1)\lambda/2$$

or

$$2\mu t = n\lambda, \quad \text{where } n = 0, 1, 2, 3, 4,... \quad (2)$$

Thus, equation (1) and (2) indicate that for a particular bright or dark ring, t should be constant. In the air film the locus of all the points of the same thickness is a circle with its center at the point of contact. Hence the fringes are due to equal thickness in the shape of concentric rings, with the point of contact of the film at the center of the ring.

Diameters of Bright and Dark Rings

To evaluate the diameters of bright and dark rings, consider a plano-convex lens AOB placed on a glass plate POQ. Let R be the radius of curved surface AOB of the lens and t the thickness of the film at any point F (Figure 21).

From the property of a circle,

$$EF \times DE = OE \times EG \quad (3)$$

But

$$EF = ED = r$$

(r is radius of the ring passing through the point F)

Therefore from equation (3), we have

$$r.r = t.(2R - t) = 2Rt - t^2$$

$$t << R, \text{ hence } 2Rt - t^2 = 2Rt$$

Therefore, $$r^2 = 2Rt \quad \text{or} \quad t = \frac{r^2}{2R} \qquad (4)$$

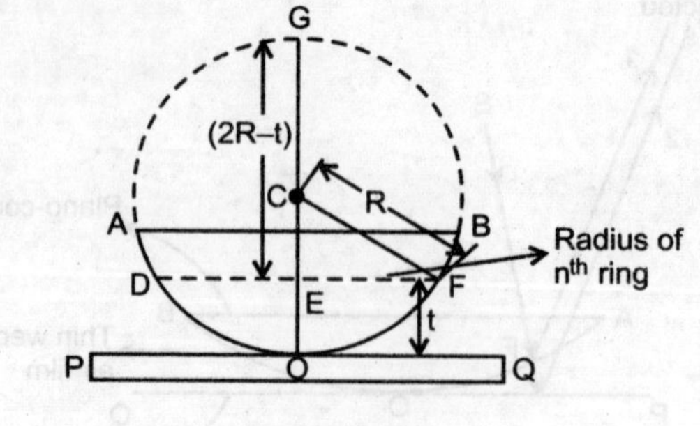

FIGURE 21

For Bright Rings

Substituting the value of t in equation (1), we have

$$2\mu\frac{r^2}{2R} = (2n - 1)\lambda/2 \quad \text{or} \quad r^2 = \frac{(2n-1)\lambda R}{2\mu}$$

If the nth ring passes through point F of the film, then r should be the radius of the nth bright ring, therefore

$$r_n^2 = \frac{(2n-1)\lambda R}{2\mu} \qquad (5)$$

If D_n is the diameter of nth bright ring, then equation (5) becomes

$$\left(\frac{D_n}{2}\right)^2 = \frac{(2n-1)\lambda R}{2\mu}$$

or $$D_n^2 = \frac{2(2n-1)\lambda R}{\mu} \qquad (6)$$

For air film $\mu = 1$, $\therefore\ D_n^2 = 2(2n - 1)\lambda R$

or $$D_n = \sqrt{[2\,(2n-1)\,\lambda R]} \quad \text{or} \quad D_n \propto \sqrt{(2n-1)}$$

or $$D_n = k\sqrt{(2n-1)}, \text{ where } k = \sqrt{(2\lambda R)} \text{ and } n = 1, 2, 3, ... \qquad (7)$$

Thus the diameters of the bright rings are proportional to the square roots of the odd natural numbers.

For Dark Rings

For the diameter of dark rings substituting the value of t from equation (4) in equation (2), we get

$$2\mu.\frac{r^2}{2R} = n\lambda \quad \text{or} \quad r^2 = \frac{n\lambda R}{\mu}$$

If r_n is the radius of the nth dark ring which passes through the point F of the film, then

$$r_n^2 = \frac{n\lambda R}{\mu}$$

or $$\left(\frac{D_n}{2}\right)^2 = \frac{n\lambda R}{\mu}, \text{ where } D_n \text{ is the diameter of } n^{th} \text{ dark ring}$$

or $$D_n^2 = \frac{4n\lambda R}{\mu} \quad (8)$$

For air film $\mu = 1$, $\therefore D_n^2 = 4n\lambda R$ (9)

or $$D_n = \sqrt{(4n\lambda R)} = k'\sqrt{n} \quad \text{or} \quad D_n \propto \sqrt{n} \quad (10)$$

where $k' = \sqrt{(4\lambda R)}$ and $n = 0, 1, 2, 3,...$

Thus the diameters of dark rings are proportional to the square roots of the natural numbers.

From equation (7) and (10), the diameters of bright and dark rings can be calculated. To find the spacing between the successive rings we have to subtract the diameters of consecutive rings. If D_{n+1} and D_n are the diameters of nth and $(n + 1)^{th}$ dark rings, then the spacing is

$$D_{n+1}^2 - D_n^2 = k\lambda R\,[(n+1)-n]$$

The spacing between consecutive dark rings 1, 2, 3, 4, and 5 is obtained by putting n = 1, 2, 3, 4,...

$$D_2^2 - D_1^2 = 4\lambda R\,(2-1) = 4\lambda R$$

$$D_3^2 - D_2^2 = 4\lambda R\,(3-2) = 4\lambda R$$

$$D_4^2 - D_3^2 = 4\lambda R\,(4-3) = 4\lambda R$$

– – – – –   – – – – – – – – –

– – – – –   – – – – – – – – –

$$D_{10}^2 - D_9^2 = 4\lambda R\,(10-9) = 4\lambda R$$

Hence, it is concluded that the spacing decreases with the order of the ring and fringes get closer and closer as their order increases. The conclusion is shown in Figure 22.

FIGURE 22

2.16 NEWTON'S RINGS BY TRANSMITTED LIGHT

In case of transmitted light, the effective path difference between the two interfering rays is $\Delta = 2\mu t$, where μ is the refractive index of the film and t is the thickness of the film at the point of incidence.

For bright rings, $2\mu t = n\lambda$, where $n = 1, 2, 3,...$ (1)

For dark rings, $2\mu t = (2n - 1)\lambda/2$, where $n = 1, 2, 3,...$ (2)

By the property of a circle, we have

$$t = \frac{r^2}{2R}, \text{ where } r \text{ is the radius of the ring or } t = \frac{D^2}{8R}$$

For the diameter of bright rings, putting the value of t in equation (1), we have

$$2\mu \frac{r^2}{2R} = n\lambda \quad \text{or} \quad r^2 = \frac{2n\lambda R}{2\mu}, \text{ for air film } \mu = 1$$

$$\therefore \quad r^2 = n\lambda R \quad \text{or} \quad D_n^2 = 4n\lambda R \quad \text{or} \quad D_n = \sqrt{(4\lambda Rn)}$$

$$\text{or} \quad D_n = k'\sqrt{n}, \qquad (3)$$

where $k' = \sqrt{(4\lambda R)}$ and $n = 1, 2, 3,...$

For the diameter of dark rings, putting the value of t in equation (2), we have

$$2\mu \frac{D^2}{8R} = (2n - 1)\lambda/2 \quad \text{or} \quad D_n^2 = \frac{2\lambda R(2n-1)}{\mu}, \text{ For air film } \mu = 1$$

$$\therefore \quad D_n^2 = 2\lambda R\,(2n-1) \quad \text{or} \quad D_n = \sqrt{[2\lambda R(2n-1)]}$$

$$\text{or} \quad D_n = k\sqrt{(2n-1)}, \qquad (4)$$

where $k = \sqrt{(2\lambda R)}$ and $n = 1, 2, 3,...$

From equations (3) and (4) it is clear that, with the transmitted light the diameters of bright rings are proportional to the square roots of natural numbers; while the diameters of dark rings are proportional to the square roots of the odd natural numbers, and the central ring is bright. The Newton's rings formed by transmitted light are shown in Figure 23.

FIGURE 23

2.17 DETERMINATION OF THE WAVELENGTH OF SODIUM LIGHT USING NEWTON'S RINGS

The apparatus for the determination of the wavelengths of monochromatic light is depicted in Figure 19. The necessary adjustments are made to observe Newton's rings. By measuring the diameters of Newton's rings with a travelling microscope, the wavelength of monochromatic light used to illuminate the wedge shaped film can be determined. Let D_n and D_{n+p} be the diameters of the nth and $(n+p)^{th}$ dark rings respectively, then

$$D_n^2 = 4n\lambda R \quad \text{and} \quad D_{n+p}^2 = 4(n+p)\lambda R$$

where p is any number, R is the radius curvature of the plano-convex curved surface.

Thus, $$D_{n+p}^2 - D_n^2 = 4(n+p)\lambda R - 4n\lambda R = 4p\lambda R$$

or $$\lambda = \frac{D_{n+p}^2 - D_n^2}{4pR}.$$

Procedure for Determination of Diameters

After focussing the eye-piece on cross-wires the microscope is moved to the extreme left of the ring system and the cross-wire is adjusted tangentially on any bright ring (say 18th) at its outer edge and the reading of the microscope is noted. The microscope is now moved to the right slowly and the reading of the micrometer screw is recorded everytime the cross-wire becomes tangential to the outer edge of every alternate (say 16th, 14th, 12th etc.) bright ring. On reaching the central spot, the cross-wire is further moved to the other side and the reading of the screw is recorded at an alternate bright ring. The cross-wire is made tangential to the inner side of the rings until we reach again the 18th bright ring on this side. By measuring the diameters of successive dark rings, a graph is plotted between D^2 and number of ring, n. This graph is a straight line as shown in Figure 24. The slope of the line gives $\frac{D_{n+p}^2 - D_n^2}{P}$. The radius of curvature of the curved surface of a plano-convex lens is measured either by a spherometer or by the Boy's optical method. Hence λ can be calculated.

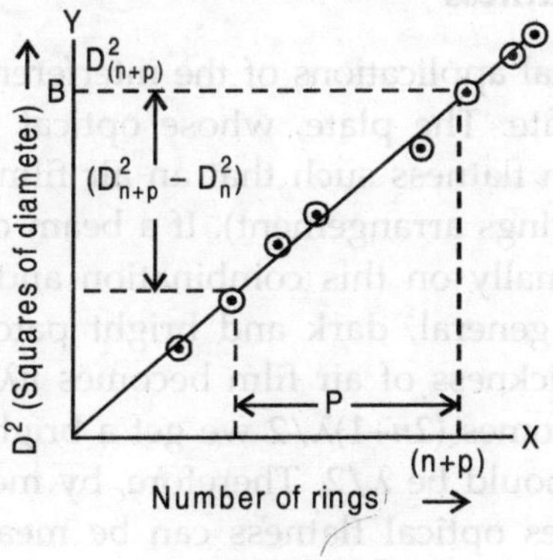

FIGURE 24

2.18 DETERMINATION OF REFRACTIVE INDEX OF A LIQUID

The necessary adjustments are made in the apparatus depicted in Figure 8 to obtain Newton's rings with air film in between the glass plate and a plano-convex lens. The diameters of nth and $(n+p)^{th}$ dark rings are measured with a low power travelling microscope in the same way as described in the method of the determination of wavelength. Let D_n and D_{n+p} be the corresponding diameters of nth and $(n+p)^{th}$ dark rings, then

$$D_n^2 = 4n\lambda R \quad \text{and} \quad D_{n+p}^2 = 4(n+p)\lambda R$$

where p is any number, λ is the wavelength of light used and R is the radius of curvature of a curved surface of a plano-convex lens.

Thus, $$D_{n+p}^2 - D_n^2 = 4\lambda pR \quad \text{or} \quad (D_{n+p}^2 - D_n^2)_{air} = 4\lambda pR \tag{1}$$

Now, a few drops of the transparent liquid whose refractive index is to be determined, is carefully introduced between the lens and the glass plate of the Newton's rings arrangement, and rings are obtained in the same manner as described above. Again the diameter of nth and $(n+p)^{th}$ dark rings are measured with a low power microscope. Let D'_n and D'^2_{n+p} be the diameters of the nth and the $(n+p)^{th}$ dark rings with a liquid medium;

then $$D'^2_n = \frac{4n\lambda R}{\mu} \quad \text{and} \quad D'^2_{n+p} = \frac{4(n+p)\lambda R}{\mu}$$

Therefore, $$D'^2_{n+p} - D'^2_n = \frac{4p\lambda R}{\mu} \tag{2}$$

From equations (1) and (2), we have

$$\mu = \frac{\left(D_{n+p}^2 - D_n^2\right)_{air}}{\left(D'^2_{n+p} - D'^2_n\right)_{liquid}}$$

Hence μ can be determined by measuring the diameters of Newton's rings in liquid and in air film.

Determination of Optical Flatness

One of the most important practical applications of the interference phenomenon is to determine the optical flatness of a glass plate. The plate, whose optical flatness is to be determined, is placed on another plate of known flatness such that an air film of varying thickness is formed between the plates (in Newton's rings arrangement). If a beam of monochromatic light, from an extended source, is incident normally on this combination and reflected light is examined by a travelling microscope, then in general, dark and bright patches will appear in the field of view. The points at which the thickness of air film becomes $n\lambda/2$ we get a dark spot, and the points on which the thickness becomes $(2n+1)\lambda/2$ we get a bright spot. The thickness of air film for two successive dark fringes should be $\lambda/2$. Therefore, by measuring the separation between successive bright and dark fringes optical flatness can be measured.

Newton's Rings with Bright Center Due to Reflected Light

Ordinarily the center of a Newton's ring is dark. At the center, the geometrical path difference is zero but the effective path difference between the two interfering rays at the point of contact is $\lambda/2$, which is the condition of minimum intensity. This path difference ($\lambda/2$) is due to a phase change π which occurs in the interfering ray reflected from the lower denser surface of the air film (that is, from glass plate) developed between the plano-convex lens and glass plate.

Hence the center of a Newton's ring is dark. To obtain a bright center, a liquid of refractive index μ is introduced between a lens of refractive index μ_1 and a plate of refractive index μ_2 such that $\mu_1 > \mu > \mu_2$. Under this condition the reflection of both the interfering rays at the point of contact becomes zero which is the condition of maximum intensity. Hence the center of Newton's rings appears bright. This situation was achieved by Young. In his experiment, he poured a few drops of sassafras oil (μ = 1.57) between a crown glass (μ =1.50) lens and a flint glass (μ = 1.65) plate and obtained a bright center of the rings system.

Newton's Rings with White Light

When an air film between the plano-convex lens and glass plate is illuminated with an extended source of white light, a few mixed colored rings around a black center are observed and beyond it a uniform illumination is obtained. This is because the diameters of the rings are a function of wavelength and white light is composed of a number of colors (wavelengths). Thus the diameters of rings of different colors will be different. As we know that $\lambda_r > \lambda_v$, therefore the diameter of violet rings of the same order will be smallest and those for red rings will be the largest, and the diameters of other colored rings shall occupy the intermediate positions. Due to the overlapping of the rings of different colors, only the first few rings will be clearly seen while the other rings cannot be observed. Hence the ring system with reflected white light will have concentric mixed colored rings from violet to red around a black center and beyond it a uniform illumination of colors is observed due to overlapping.

Example 30. *Two glass plates enclose a wedge-shaped air film, touching at one edge, and separated by a wire of 0.05 mm diameter at a distance of 15 cm from the edge. Calculate the fringe width. Monochromatic light of λ = 6,000 Å from a broad source falls normally on the film.*

Solution. In Figure 25, let θ be the angle of the wedge, then the fringe width ω in air-wedge for normal incidence is given by

$$\omega = \frac{\lambda}{2\theta}, \qquad \theta = \frac{AB}{OA} = \frac{\text{Arc}}{\text{Radius}}$$

$$\therefore \qquad \theta = \frac{0.005}{15}$$

Here $\qquad \lambda = 6000 \text{ Å} = 6000 \times 10^{-8} \text{ cm}$

$$\therefore \qquad \omega = \frac{6000 \times 10^{-8} \times 15}{2 \times 0.005} = \mathbf{0.09\ cm.}$$

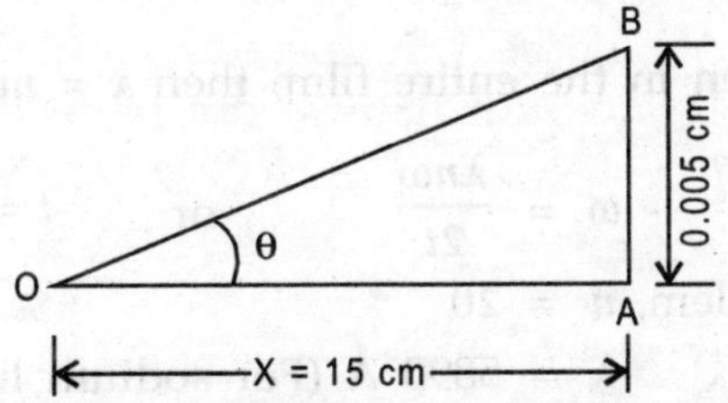

FIGURE 25

Example 31. *Light of wavelength 6000 Å falls normally on a thin wedge-shaped film of refractive index 1.35 forming fringes that are 2.0 mm apart. Find the angle of the wedge.*

Solution. If θ is the angle of the wedge formed by a medium of refractive index μ, then for normal incidence the fringe width for wavelength λ is given by

$$\omega = \frac{\lambda}{2\mu\theta} \quad \therefore \quad \theta = \frac{\lambda}{2\mu\omega}$$

In the given problem,

$$\lambda = 6000 \text{ Å} = 6000 \times 10^{-8} \text{ cm}$$

$$\mu = 1.35 \text{ and } \omega = 2.0 \text{ mm} = 0.20 \text{ cm}$$

$$\therefore \quad \theta = \frac{6000 \times 10^{-8}}{2 \times 1.35 \times 0.20} = 1.1 \times 10^{-4} \text{ rad}$$

$$= 1.1 \times 10^{-4} \times \frac{180°}{\pi}$$

or $$\theta = \mathbf{0.0063°}. \qquad \left[\because \text{ rad} = \frac{180°}{\pi}\right]$$

Example 32. *Two plane glass surfaces in contact along one edge are separated at the opposite edge by a thin wire. If 20 fringes are observed between these edges in sodium light for normal incidence, what is the thickness of the wire?*

Solution. The fringe width in air wedge for normal incidence is given by

$$\omega = \frac{\lambda}{2\theta} \qquad (1)$$

Let t be the thickness of the wire and x be the length of the glass surface from the point of contact as shown in Figure 26, then the wedge angle $\theta = t/x$, therefore equation (1)

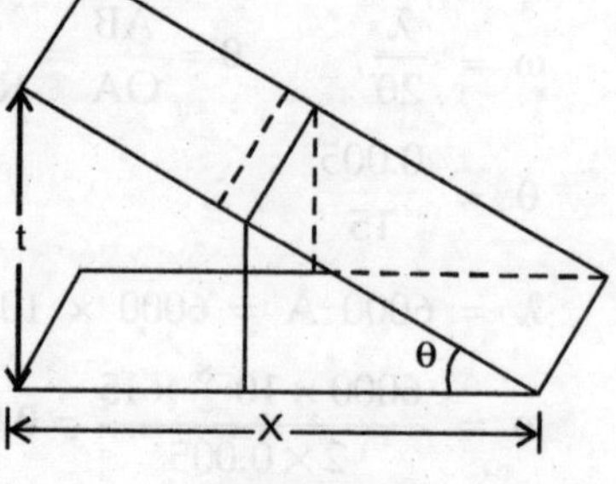

FIGURE 26

$\omega = \frac{\lambda x}{2t}$. If n fringes are seen in the entire film, then $x = n\omega$

Therefore $$\omega = \frac{\lambda n \omega}{2t} \quad \text{or} \quad t = \frac{n\lambda}{2}$$

In the given problem, $n = 20$

and $$\lambda = 5890 \text{ Å (For sodium light)}$$

$$= 5890 \times 10^{-8} \text{ cm}$$

$$\therefore \quad t = \frac{20 \times 5890 \times 10^{-8}}{2} = \mathbf{5.89 \times 10^{-4} \text{ cm.}}$$

Example 33. *Between two optically plane glass plates, at one edge a foil is introduced to get a wedge-shaped air film. Viewed in mercury green light ($\lambda = 5.46 \times 10^{-5}$) along the normal, we see 12 fringes in 0.40 cm width. Deduce (i) the angle of the wedge; (ii) the thickness of the foil if plate length is 3.0 cm in all; (iii) fringe width if water is introduced in the wedge space.*

Solution. In the given problem, the spacing between 12 fringes is 0.40 cm, therefore $12\omega = 0.40$ cm

or $$\omega = \frac{0.40}{12} = 0.33 \times 10^{-2} \text{ cm}$$

(*i*) The fringe width in air wedge for normal incidence is given by

$$\omega = \lambda/2\theta \quad \text{or} \quad \theta = \lambda/2\omega$$

Here, $$\lambda = 5.46 \times 10^{-5} \text{ cm}$$

and $$\omega = 0.33 \times 10^{-2} \text{ cm}$$

$$\therefore \quad \theta = \frac{5.46 \times 10^{-5}}{2 \times 0.33 \times 10^{-2}} = 8.2 \times 10^{-4} \text{ rad}$$

(*ii*) Let t be the thickness of the foil and l be the length of plate,

therefore $$\theta = \frac{t}{l} \quad \text{or} \quad t = \theta l$$

Given: $l = 3.0$ cm

$$\therefore \quad t = 8.2 \times 10^{-4} \times 3.0 = 2.46 \times 10^{-3} \text{ cm}$$

(*iii*) We know that if θ is the wedge formed by a medium of refractive index μ, then for normal incidence

$$\omega' = \frac{\lambda}{2\mu\theta},$$

Here $\mu = 1.33$ for water

$$\therefore \quad \omega' = \frac{5.46 \times 10^{-5}}{2 \times 1.33 \times 8.2 \times 10^{-4}} = \mathbf{0.025 \text{ cm.}}$$

2.19 HAIDINGER'S FRINGES: FRINGES OF EQUAL INCLINATION

The interference fringes can also be obtained from the films of uniform thickness. For a plane parallel sided film, the path difference $2\mu t \cos r$ between the coherent beams is due to the variation in the angle of refraction r only. A very small change in r produces a large change in path difference provided that t is large. All directions equally inclined to the normal to the film will correspond to the same path difference, so that the circular fringe concentric with the point at infinity in the direction of normal will be seen. Such like fringes are called the fringes of equal inclination or *Haidinger's fringes*. In this case all the pairs of interfering coherent rays of equal inclination pass through the plate as a parallel beam and hence meet at a point at infinity. The other pairs of different inclination meet at different points at infinity. Figure 27 shows the formation of Haidinger's fringes in a parallel sided air film. Let C be the point on the film which satisfies the condition of minimum intensity, $2t \cos \theta = nl$, where θ is the angle between normal to the film EO and EC, then the point C appears dark. Thus all points on the film having equal inclination θ lie in a circle of radius OC around O and satisfy the condition of minimum intensity. As a result, a dark ring of angular radius θ appears. As θ increases $2t \cos \theta$ decreases. Let D be another point on the film at which the condition of maximum intensity is satisfied. Thus D appears bright and all points on the film having equal inclination θ' lie in a circle of radius OD with O as center and satisfy the condition of maximum intensity, therefore a bright ring of angular radius θ' appears. In this way, as θ increases, we get alternate bright and dark rings of equal inclination at infinity and can be viewed with a telescope focussed at infinity.

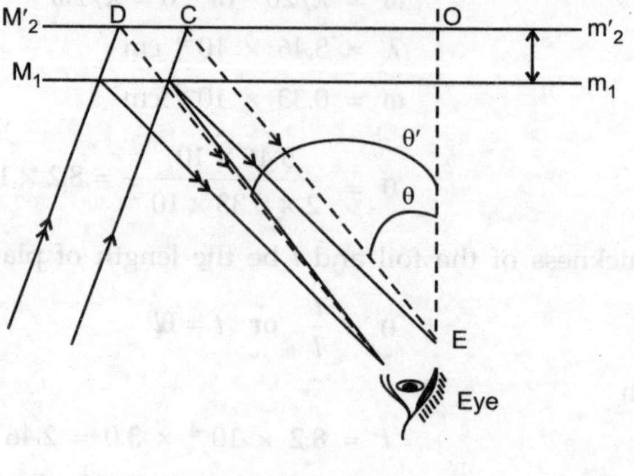

FIGURE 27

2.20 THE MICHELSON INTERFEROMETER

The Michelson interferometer, which utilizes long path lengths to produce interference fringe and based on the principle of division of amplitude, is one of the most ingenious instruments for the accurate measurement of wavelength difference between a close doublet, refractive indices of the transparent thin sheets, wavelength of monochromatic light, and to test the planeness of surface, and also to reduce the reflecting power of the lens and prism surfaces. The ingenuity of the instrument lies in the sense that both equal thickness and equal inclination interference fringes can be produced by it.

Construction

The essential parts of the instrument are schematically shown in Figure 28. It consists of two highly polished good quality plane mirrors M_1 and M_2 mutually perpendicular to each other. The planes of the mirrors can be slightly tilted with the five screws at their back. One of the mirrors M_2 is fixed and the other M_1 is capable of moving away or towards the beam splitter G_1, by means of a very fine micrometer screw capable of reading accurately a displacement of the order of 10^{-5} cm. There are two parallel glass plates G_1 and G_2 of equal thickness and of the same material. The face of beam splitter G_1 towards the compensating plate G_2 is semi-silvered so that the reflected rays and the transmitted rays are of about the same intensity. Both the plates are mounted vertically, exactly parallel to each other and inclined at 45° to the axis of the incident ray. The interference pattern is observed by a Telescope T.

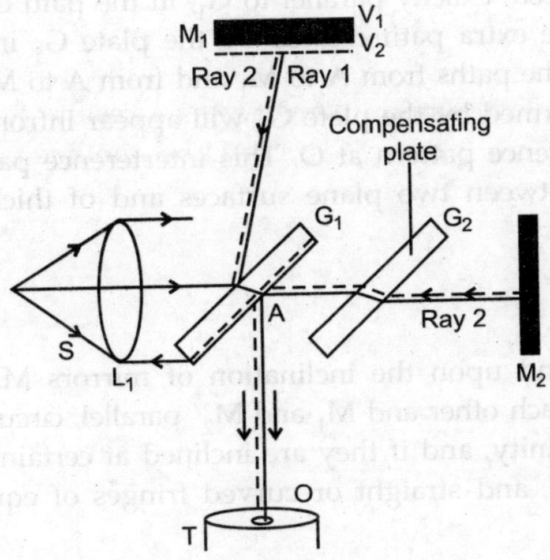

FIGURE 28

Working

Light from an extended monochromatic source S, rendered nearly parallel by the lens L_1, is incident on a semi-silvered glass plate G_1. The incident ray is equally split by the beam splitter G_1 into a reflected ray 1 and a transmitted ray 2, where rays 1 and 2 follow the paths perpendicular to each other. The ray 1 undergoes further reflection at M_1 and reflected back along its path of incidence and after being partially transmitted through G_1 enters the telescope T. The ray 2 after passing through compensating plate G_2 reflected normally from the mirror M_2 and retraces its original path. At G_1 it is reflected through 90° to enter the telescope. The two rays entering the telescope are derived from the same incident beam, hence they are coherent and very close to each other and capable of interference.

Function of the Compensating Plate G_2

The ray 1 reflected at G_1 crosses the plate G_1 twice, while the ray 2 in the absence of G_2 lies wholly in air. Thus two paths AM_1 and AM_2 are unequal and the ray 1 has to travel an extra optical path $2(\mu - 1)t$, where t is the thickness of the plate. To compensate for this additional

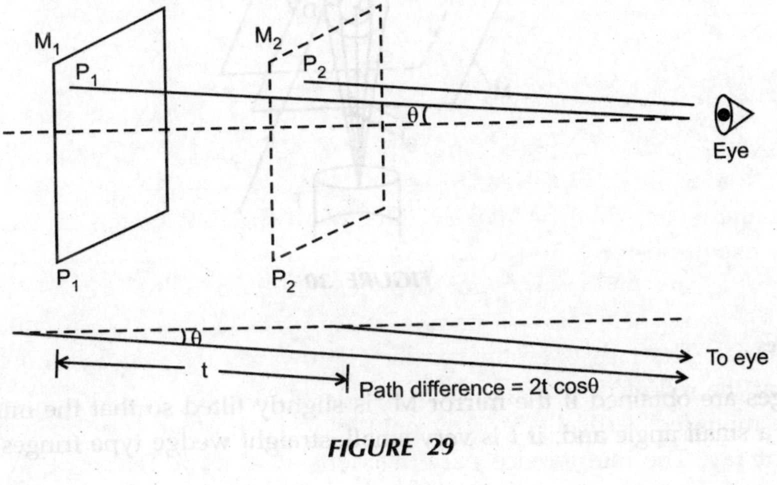

FIGURE 29

path, a plate G_2 is introduced, exactly parallel to G_1, in the path of the ray 2. Thus the function of G_2 is to compensate the extra path created by the plate G_1 in the ray 1, hence it is called the compensating plate. If the paths from A to M_1 and from A to M_2 are unequal then the virtual image M'_2 of mirror M_2 formed by the plate G_1 will appear infront of or behind the mirror M_1, which provides an interference pattern at O. This interference pattern is the same as that due to an air film enclosed between two plane surfaces and of thickness t, where $t = M_1M'_2$ as shown in Figure 29.

Types of Fringes

The types of fringes depend upon the inclination of mirrors M_1 and M_2. If the mirrors are perfectly perpendicular to each other and M_1 and M_2' parallel, circular fringes of equal inclination are obtained, located at infinity, and if they are inclined at certain angles, then the enclosed air film will be wedge-shaped, and straight or curved fringes of equal thickness will be seen.

Circular Fringes

To obtain the concentric circular fringes of equal inclination with a monochromatic extended source of light, it is essential that the mirrors M_1 and M_2 are exactly perpendicular to one another and each at 45° to the beam splitter plate G_1. Under this situation an air film of constant thickness t is enclosed between two plane surfaces M_1 and M_2', where $M_1 M_2' = t$ and a set of concentric circular interference fringes of equal inclination are formed in accordance with the equation, $2\mu t \cos\theta$ or $2t \cos\theta$ as $\mu = 1$ for air film.

At the center of the interference pattern, $\theta = 0$ so $\cos\theta = 1$. Away from the center θ increases and $\cos\theta$ decreases, at a certain value of θ all points on the film having inclination θ lie in a circle of radius OC around O and satisfy the condition of minimum intensity (Figure 30). As a result, a dark ring of angular radius θ appears. Similarly at a certain other value of $\theta = \theta'$ all the points on the film having inclination θ' lie in a circle of radius OD and satisfy the condition of maximum intensity. Therefore a bright ring of angular radius θ' appears. In this way the circular fringes of equal inclination are formed as shown in Figure 30.

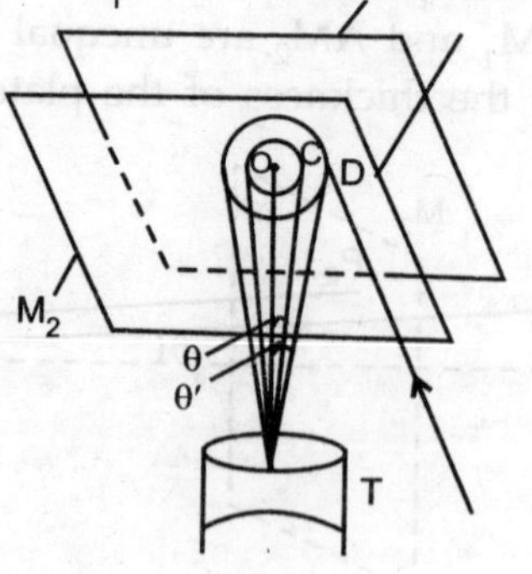

FIGURE 30

Localized Fringes

Localized fringes are obtained if the mirror M_2 is slightly tilted so that the mirrors M_1 and M_2' are inclined at a small angle and, if t is very small, straight wedge type fringes will be observed.

With a larger value of t, these fringes will become curved. The shape of the fringes depends on the thickness of the film and the angle of incidence. When the mirror M_1 is in a position as shown in Figure 31(*a*) the observed fringes are convex towards the left. As the distance between M_1 and M_2' is gradually decreased, fringes move away from the thin edge of the wedge and gradually become straighter. When M_1 intersects M'_2, the path difference becomes zero and fringes so obtained are perfectly straight as in Figure 31(*b*). When M_1 is moved further in the same direction fringes are again curved but with their convex side towards the right as in Figure 31(*c*). These fringes become invisible for large path difference.

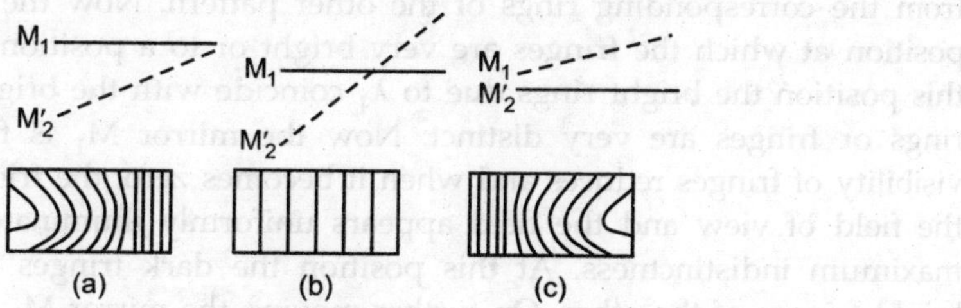

FIGURE 31

White Light Fringes

When the planes of the mirror M_1 and virtual mirror M_2' intersect in the middle, the path difference along the line of their intersection is zero for all wavelengths. If we use an extended source of white light for the film of small thickness, there will be a central achromatic bright fringe along the line of intersection, and on either side of it there will be a few colored straight fringes in the symmetrical order, the violet being nearest with the red being farthest. For the large thickness of the film uniform illumination is observed. White light fringes are useful for the determination of zero path difference in many applications.

Uses of the Michelson Interferometer

Following are the uses of the Michelson Interferometer:

(*i*) *Determination of the Wavelength of Monochromatic Light:* The monochromatic light whose wavelength is to be determined is incident on a semi-silvered glass plate G_1 of the interferometer and the mirrors M_1 and M_2 are adjusted exactly at right angles to each other so that the circular fringes are visible in the field of view. If M_1 and M_2 are equidistant from glass plate G_1, the field of view is perfectly dark. The mirror M_1 is kept fixed and the position of mirror M_1 is adjusted to obtain a bright spot at the center of the field of view. If the distance between mirror M_1 and the image M_2' of M_2 in G_1 is t and n is the order of the spot obtained then at the center, $2t = n\lambda$. Now the mirror M_1 is moved slowly and the disappearance of fringes at the center of the field of view is counted. Suppose for the monochromatic light of wavelength λ, the distance through which mirror M_1 is moved, as measured by the micrometer screw, is d cm and the number of disappearances counted is N, then

$$d = N\lambda/2$$

because a disappearance occurs each time the path difference is decreased by $\lambda/2$. Hence λ is obtained from $2d/N$.

(ii) *Determination of the Difference in Wavelength Between Two Neighboring Spectral Lines (Resolution of Spectral Lines):* The Michelson interferometer can appreciably be used in the measurement of two closely spaced wavelengths, such as sodium D lines (λ_1 = 5896 Å and λ_2 = 5890 Å). Let λ_1 and λ_2 be the wavelengths of two close doublets such as $\lambda_1 > \lambda_2$. When the interferometer is adjusted for obtaining circular fringes, each spectral line produces its own pattern, and the rings of one pattern are closely separated from the corresponding rings of the other pattern. Now the mirror M_1 is moved to a position at which the fringes are very bright or to a position of maximum visibility. In this position the bright rings due to λ_1 coincide with the bright rings due to λ_2 and the rings or fringes are very distinct. Now the mirror M_1 is further moved so that the visibility of fringes reduces and when it becomes zero, the fringes totally disappear from the field of view and the field appears uniformly illuminated. This is the position of maximum indistinctness. At this position the dark fringes of one system fall on the bright fringes of the other. On further moving the mirror M_1, the positions of maximum distinctness and maximum indistinctness are alternatively repeated.

Let x be the distance moved by the movable mirror M_1 between two consecutive positions of maximum indistinctness (or distinctness), the path difference is $2x$. During this movement of M_1, n fringes of λ_1 and $(n+1)$ fringes of λ_2 have appeared at center. Thus we have

$$2x = n\lambda_1 = (n+1)\lambda_2 \quad (1)$$

or

$$n\lambda_1 = n\lambda_2 + \lambda_2 \quad \text{or} \quad n = \frac{\lambda_2}{(\lambda_1 - \lambda_2)}$$

Substituting this value of n in equation (1), we get

$$2x = \frac{\lambda_2}{(\lambda_1 - \lambda_2)}.\lambda_1 \quad \text{or} \quad (\lambda_1 - \lambda_2) = \frac{\lambda_2\lambda_1}{2x}$$

As λ_1 and λ_2 are very nearly equal, we have

$$\lambda_1.\lambda_2 = \lambda^2 \text{ where } \lambda = \frac{\lambda_1 + \lambda_2}{2}$$

Therefore

$$\lambda_1 - \lambda_2 = \frac{\lambda^2}{2x} \quad (2)$$

Thus, the difference in wavelengths can be determined.

(iii) *Determination of Refractive Index of Thin Films:* To determine the refractive index of thin films or sheets of transparent material, the interferometer is adjusted for straight line fringes by moving mirror M_1 in such a way that M_1 intersects the image M'_2 of M_2 as depicted in Figure 30(*b*) and film is illuminated by an extended source of white light. The cross-wires of the telescope are adjusted on the central bright fringe. The thin film of transparent material, whose refractive index is to be determined, is introduced in the path of one of the interfering beams, and the light in that beam transverse the plate twice. The increase in optical path as a result of the plate over the optical path in air is $2(\mu-1)t$, where t is the thickness and μ is the refractive index of the film. The fringes are therefore displaced. The movable mirror M_1 is now moved until the fringes are brought back to their initial position so that the central bright fringe again coincides with the cross-wires. If the displacement of mirror M_1 from initial position is x, then

$$2x = 2(\mu-1)t \quad \text{or} \quad x = (\mu-1)t$$

If x and thickness t are known, the refractive index of the material μ is determined. It is possible to determine the refractive index of a gas with a Michelson interferometer by admitting the gas slowly to a hollow parallel-sided glass vessel in one of the beams.

Example 34. *Circular fringes are observed in the Michelson interferometer illuminated with light of wavelength 5896 Å. When the path difference between the mirrors M_1 and M_2 is 0.3 cm, the central fringe is bright. Calculate the angular diameter of the 7th bright fringe.*

Solution. The condition for bright circular fringes is

$$2d \cos r = n\lambda \tag{1}$$

At the center $\quad r = 0$

$$\therefore \quad 2d = n\lambda,$$

where 'n' stands for the order of the central bright fringe. The order of fringes decreases as we move away from the center. Therefore, the second bright fringe is of $(n-1)^{th}$ order, the seventh bright fringe is of $(n-6)^{th}$ order. If θ is the angular radius of the seventh bright fringe, we have

$$2d \cos \theta = (n-6)\lambda = n\lambda - 6\lambda$$

From equation (1) $\quad n\lambda = 2d$, therefore $2d(1 - \cos\theta) = 6\lambda$

or $$\cos\theta = 1 - \frac{6\lambda}{2d}$$

Here $\quad \lambda = 5896\text{Å} = 5896 \times 10^{-8}$ cm and $d = 0.3$ cm

$$\therefore \quad \cos\theta = 1 - \frac{6 \times 5896 \times 10^{-8}}{2 \times 0.3}$$

$$= 1 - 0.00059 = 0.9991$$

$$\therefore \quad \theta = \cos^{-1}(0.9991) = 2°$$

Thus the angular diameter $= 2° \times 2 =$ **4°.**

Example 35. *In moving one mirror in a Michelson interferometer through a distance of 0.1474 mm, 500 fringes cross the center of the field of view. What is the wavelength of light?*

Solution. Let x be the distance moved by the mirror when N fringes cross the field of view, then

$$x = N\frac{\lambda}{2} \quad \text{or} \quad \lambda = \frac{2x}{N}$$

In the given problem, $x = 0.1474$ mm $= 0.01474$ cm and N = 500

Therefore $$\lambda = \frac{2 \times 0.01474}{500} = 5896 \times 10^{-8} \text{ cm} = \mathbf{5896\ Å.}$$

Example 36. *For a sodium lamp, the distance traversed by the mirror between two successive disappearances is 0.289 mm. Calculate the difference in the wavelengths of the D_1 and D_2 lines of the sodium lamp. Assuming $\lambda = 5890$ Å.*

Solution. If x be the distance moved by the movable mirror between two successive disappearances, then the difference in wavelengths $(\lambda_1 - \lambda_2)$ is

$$\lambda_1 - \lambda_2 = \frac{\lambda^2}{2x}$$

In the given problem x = 0.289 mm = 0.0289 cm and λ = 5890 Å = 5890 × 10^{-8} cm

Therefore
$$\lambda_1 - \lambda_2 = \frac{5896 \times 10^{-8} \times 5896 \times 10^{-8}}{2 \times 0.0289}$$
$$= \mathbf{6 \times 10^{-8}\ cm.}$$

EXERCISES

Objective Questions

1. In the Michelson interferometer, two waves interfere after travelling a distance of
   (*a*) centimeter order (*b*) wavelength order
   (*c*) kilometer order (*d*) none of these
2. The instrument which is capable of forming the fringes of equal thickness and of equal inclination is
   (*a*) Newton's rings (*b*) Fabry-Perot interferometer
   (*c*) Michelson interferometer (*d*) Jamin refractometer
3. In the Michelson interferometer the shape of fringes depends upon
   (*a*) distance of source from compensating plate
   (*b*) distance of semi-silvered plate from compensating plate
   (*c*) the inclination between two mirrors
   (*d*) distance of semi-silvered plate from the telescope
4. In the Michelson interferometer, if a uniform sheet of thickness t and refractive index μ is introduced in one of the paths, then the additional one-way path introduced in that beam is
   (*a*) $2(\mu - 1)t$ (*b*) $(\mu - 1)t$
   (*c*) $\frac{D}{2d}(\mu - 1)t$ (*d*) μt
5. In the Fabry-Perot interferometer, the sharpness of maxima depends upon
   (*a*) the distance between the glass plates
   (*b*) the inclination between the two plates
   (*c*) the reflectivity of the glass plates
   (*d*) the resolving power of a telescope
6. In the Fabry-Perot interferometer, the width of maxima for equal separation of the plates
   (*a*) increases with the coefficient of reflection
   (*b*) decreases with the coefficient of reflection

(*c*) remains constant with the coefficient of reflection

(*d*) none of these

7. The most suitable method for the determination of the refractive index of gases and liquids is

(*a*) the Michelson interferometer

(*b*) Newton's rings

(*c*) the Fabry-Perot interferometer

(*d*) the Jamin refractometer

8. A wedge-shaped film is observed in reflected sunlight first through a red glass and then through a blue glass. The number of fringes in the later case is

(*a*) less (*b*) more

(*c*) equal in both cases (*d*) none of these

9. The loss of intensity due to reflection can be reduced substantially by coating the glass surface with a uniform film of optical thickness

(*a*) $\lambda/2$ and μ less than that of glass

(*b*) $\lambda/2$ and μ greater than that of glass

(*c*) $\lambda/4$ and μ less than that of glass

(*d*) $\lambda/4$ and μ greater than that of glass

10. Which of the following is not suitable for a non-reflection coating

(*a*) magnesium fluoride (*b*) sodium aluminum fluoride

(*c*) zinc sulphide (*d*) none of these

11. The reflectivity of a surface can be enhanced by coating the glass surface with a uniform film of optical thickness

(*a*) $\lambda/2$ and refractive index less than that of glass

(*b*) $\lambda/2$ and refractive index greater than that of glass

(*c*) $\lambda/4$ and refractive index less than that of glass

(*d*) $\lambda/4$ and refractive index greater than that of glass

12. In a Fresnel's biprism experiment bright and dark fringes are obtained using sodium yellow light. If the angle of the biprism is increased, then

(*a*) the fringe width increases

(*b*) the fringe width decreases

(*c*) the fringe width remains unchanged

(*d*) the contrast of bright and dark fringes becomes poorer and poorer

13. In a Fresnel's biprism experiment, if the width of the slits is increased, then

(*a*) the contrast between the bright and dark fringes becomes poorer and poorer

(*b*) the contrast between the bright and dark fringes becomes richer and richer

(*c*) the contrast between the bright and dark fringes remains unchanged

(*d*) the screen appears completely dark

14. In a biprism experiment interference fringes are obtained using sodium yellow light. What happens if the yellow light is replaced by blue light?

(*a*) fringes disappear (*b*) fringe width decreases

(*c*) fringe width increases (*d*) no change in fringe width

15. When monochromatic light is replaced by white light in Fresnel's biprism arrangement, the central fringe is

(*a*) dark (*b*) white

(*c*) colored (*d*) it changes color from violet to red

16. When a thin transparent sheet of thickness t and refractive index μ is placed in the path of one of the interfering waves, then the entire fringe pattern is displaced through a distance

(*a*) $(\mu - 1)t$ (*b*) $\frac{D}{2d}(\mu - 1)t$

(*c*) $\frac{2d}{D}(\mu - 1)t$ (*d*) $\frac{2d}{(\mu - 1)t}$

where $2d$ is the separation between the sources.

17. When a thin transparent sheet is placed in the path of one of the interfering waves then

(*a*) the entire fringe system shifts towards the side on which the sheet is placed

(*b*) the entire fringe system shifts toward the side on which no sheet is placed

(*c*) there is no shift in the fringe system

(*d*) the fringe system becomes narrower

18. When a thin sheet of mica is introduced in the path of one of the interfering beams, then the fringe width

(*a*) increases (*b*) decreases

(*c*) remains unchanged (*d*) none of these

19. If one of the two slits of a standard Young's double slit interference experiment is pointed over so that it transmits half the light intensity of the other

(*a*) the dark fringes will be more dark and the bright fringes will be more bright

(*b*) the dark fringes will be brighter and the bright fringes darker

(*c*) the fringe pattern will disappear

(*d*) there will be no effect

20. When light of wavelength λ falls on a thin parallel film of constant thickness t and refractive index μ, the essential condition for the production of constructive interference fringes by the rays reflected from the upper and lower edge of the film is

(*a*) $2\ \mu t\ \cos r = (2n - 1).\lambda/2$ (*b*) $2\mu t\ \cos r = 2n.\lambda/2$

(*c*) $2\mu t\ \cos\ (r + \theta) = (2n - 1).\lambda/2$ (*d*) $2\mu t\ \cos\ (r + \theta) = 2n.\lambda/2$

21. When light of wavelength λ falls on a thin film of air of varying thickness, the essential condition for the production of constructive interference by the two rays in the transmitted system is

(*a*) $2\ \mu t\ \cos r = (2n - 1).\lambda/2$ (*b*) $2\mu t\ \cos r = 2n.\lambda/2$

(*c*) $2\mu t\ \cos\ (r + \theta) = (2n - 1).\lambda/2$ (*d*) $2\mu t\ \cos\ (r + \theta) = 2n.\lambda/2$

22. In the Newton's rings arrangement with air film in reflected light the diameter of nth dark ring is D_n. If the air film is replaced by a liquid film of refractive index μ, the diameter of nth dark ring will become (*a*) $\sqrt{\mu}$ times (*b*) $1/\sqrt{\mu}$ times

(*c*) $1/\mu$ times (*d*) μ times

23. In the Newton's rings arrangement, the diameters of bright rings are

(*a*) directly proportional to the square root of natural numbers

(*b*) inversely proportional to the square root of odd natural numbers

(*c*) directly proportional to the square root of odd natural numbers

(*d*) directly proportional to the square root of even natural numbers

24. In the Newton's rings arrangement, bright and dark rings are obtained using sodium yellow light. What happens if the top surface of the glass plate on which the lens is kept is highly silvered?

(*a*) fringe width decreases (*b*) fringe width increases

(*c*) fringe width remains unchanged (*d*) fringes disappear

25. In the Newton's rings arrangement, if the distance between the lens and the plate is increased, the order of the ring at a given point

(*a*) decreases (*b*) increases

(*c*) remains unchanged (*d*) none of these

26. In the Newton's rings arrangement, bright and dark rings are obtained using sodium yellow light. If the entire arrangement is dipped into water then the diameters of rings

(*a*) increases (*b*) decreases

(*c*) fringe pattern disappears (*d*) remains unchanged

Short Questions for Discussion

1. Explain the Huygens' wave theory of light. What is a wave front?
2. What is interference of light?
3. Discuss the basic conditions of obtaining sustained interference patterns
4. Define coherent sources
5. Is it possible to achieve coherent sources from two independent sources? Explain.
6. Name the methods of obtaining coherent sources and give examples.
7. Give the expression for the fringe width of the interference pattern obtained by Fresnel's biprism.
8. Discuss the characteristics of Fresnel's biprism interference pattern.
9. Calculate the displacement of the fringes when a thin transparent sheet is introduced in the path of one of the interfering beams of a biprism.
10. Discuss the appearance of colors in thin films illuminated by an extended source of white light.
11. Discuss the phase change due to reflection of light from the surface of a denser medium.
12. Why are Newton's rings circular?
13. Explain why the fringe pattern of Newton's experiment is complimentary to each other in reflected and transmitted light.
14. What type of Newton's rings would be obtained if the monochromatic sources of light are replaced by a chromatic source.
15. Discuss the role of the compensating plate in the Michelson interferometer.

Long Questions for Discussion

1. Explain Huygens' wave theory of light. What is a wavefront?
2. What is interference of light? Describe and explain Young's experiment demonstrating interference of light.
3. What do you mean by coherent sources? Explain why two independent sources of the same wavelength cannot act as coherent sources? State the methods of realizing coherent sources from a single source.
4. Give the theory of interference for two coherent sources and deduce the expression for fringe width.
5. State the conditions of obtaining sustained interference. How can the coherent sources be achieved by a single source?
6. Explain the formation of interference fringes by means of a Fresnels biprism when a monochromatic source of light is used and derive an expression for the fringe width.
7. Describe the method for determination of the wavelength of monochromatic light by Fresnels biprism.
8. Calculate the displacement of the fringes when a thin transparent plate is introduced in the path of one of the interfering beams of a biprism. Show how this method is used to find the thickness of a thin mica sheet.
9. Discuss the phase change due to the reflection of light from the surface of a denser medium. How can it be demonstrated experimentally?
10. Explain the phenomenon of the interference of light due to thin films and find the condition of maxima and minima. Show that the interference patterns of reflected and transmitted monochromatic sources of light are complementary.
11. Why are Newton's rings formed? Obtain the expression of the diameter of nth dark fringe in reflected light.
12. What are Newton's rings? How can Newton's rings be used to determine the wavelength of light and the refractive index of a liquid?
13. Why is the center of Newton's rings dark? How can be get bright center?
14. Explain the construction and working of the Michelson interferometer with the help of a diagram. Explain the use of a compensating glass plate. Explain how the Michelson interferometer is used to determine

    (*i*) the wavelength of monochromatic light

    (*ii*) the difference in wavelengths very close the each other

Chapter 3

DIFFRACTION OF LIGHT

3.1 OVERVIEW

We have seen that the interference phenomenon arises due to two or more coherent beams of light obtained either by wavefront division or by amplitude division or both. In this chapter, we shall deal with the interference phenomenon occurring between light beams from various parts of one aperture. Ordinarily, when a plane wave of light is incident on a long narrow slit placed in front of a screen, a sharp patch of light and a shadow is obtained on the screen (as shown in Figure 1). This shows that the light travels approximately in straight lines. But a very close examination of the light distribution on the screen reveals dark and bright fringes near the edges. It is observed that, if the aperture reduces in width, the patch of light on the screen becomes distinctly wider than given by laws of geometrical optics, and the fringes become more and more distinct until they become broad and practically cover the entire patch. Similarly, when an opaque obstacle is placed in the path of light, instead of a sharp shadow, bright and dark fringes are obtained on the screen. This observation can be explained by assuming that light waves bend around an obstacle. The amount of bending, however, depends upon the size of the obstacle and the wavelength of the wave.

Thus we see when light passes through a small opening (or aperture) or by the side of a small obstacle, it bends, to some extent, into the region of geometrical shadow and its intensity falls off rapidly. This deviation is extremely small when the wavelength is small in comparison to the dimension of the obstacle. But the deviation becomes much more pronounced when the dimensions of the aperture (or opaque disc) are comparable with the wavelength of light. Thus the phenomenon of bending of light around corners of an obstacle and their spreading into the geometrical shadow of an object is called diffraction, and the distribution of light intensity resulting in dark and bright fringes (that is, with alternate maxima and minima) is called a diffraction pattern. This phenomenon was first discovered in 1665 by an Italian scientist named Grimaldi and was studied by Newton. In terms of the wave theory, the first attempt to explain the diffraction phenomenon was made by Young. He considered the diffraction as interference between the direct light waves which pass through the edge of the obstacle and the wave of light reflected from the edge of the obstacle. This idea was not accepted on the ground that the diffraction pattern does not depend upon the sharpness of the edge, the material of obstacle, and its degree of polish.

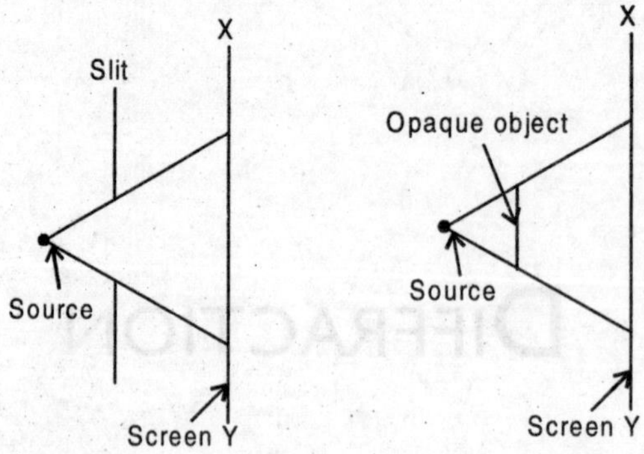

FIGURE 1 **Bending of light at edges.**

A. J. Fresnel, in 1815, gave a correct interpretation of the diffraction phenomenon on the basis of wave theory by combining Huygens' principle of secondary wavelets with the principle of interference.

3.2 FRESNEL'S DIFFRACTION

According to Fresnel, the diffraction phenomenon is caused by the interference of the innumerable secondary wavelets originating from various points of an exposed portion of the same wavefront. Fresnel not only explained diffraction but also showed that the rectilinear propagation of light is approximate.

3.3 DIFFRACTION AT A STRAIGHT EDGE

In Figure 2, let O be the sharp straight edge of an obstacle OA, placed perpendicular to the plane of the paper and parallel to a long narrow slit S, which is illuminated with a monochromatic source of light of wavelength λ. Let XY be a screen parallel to OA. WOW' is the incident cylindrical wavefront. Join SO and produce it to meet the screen at C. According to the laws of geometrical optics, we should get uniform illumination in the region CX and complete darkness in the region CY. A careful examination of the screen reveals that immediately above C, there are a series of unequal maxima and minima regions, parallel to the straight edge, called diffraction fringes. As we move above the point C, the intensity of maxima gradually decreases while that of minima gradually increases until they finally merge into a uniform illumination.

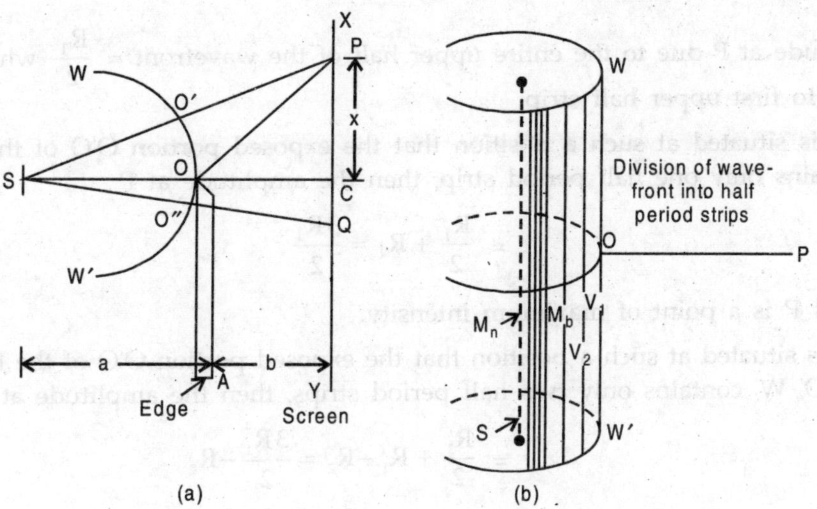

FIGURE 2

3.4 EXPLANATION OF DIFFRACTION FRINGES IN THE ILLUMINATED REGION

The wavefront emanating from the slit is cylindrical and in order to find the amplitude at any point P on the screen, the wavefront is divided into half period strips with respect to P as shown in Figures 2 and 3. With P as the center and with the radii equal to PO + $\lambda/2$, PO + $2\lambda/2$, PO + $3\lambda/2$, ..., PO + $(n-1)\lambda/2$, PO + $n\lambda/2$ etc. draw arcs. These arcs cut the wavefront at points (M_1, M_1'), (M_2, M_2'), (M_3, M_3'), ..., (M_n, M_n').

Now through the points O, M_1, M_1', M_2, M_2', ..., M_n, M_n' draw lines on the surface of the wavefront parallel to the slit. These lines divide the surface into a number of strips called half period strips. The effect of the entire wavefront WW' on point P is equal to the effect of the entire upper half of the wavefront plus the effect of the exposed portion of the lower half OW'.

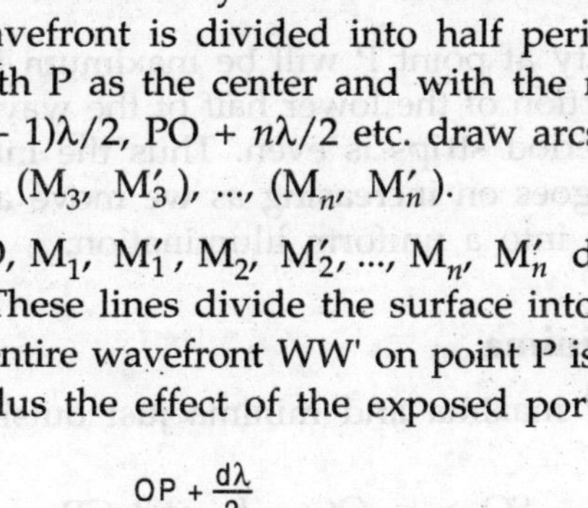

FIGURE 3

The amplitude at P due to the entire upper half of the wavefront = $\frac{R_1}{2}$ where R_1 is the amplitude due to first upper half strip.

If point P is situated at such a position that the exposed portion O'O of the lower half wavefront contains only one half period strip, then the amplitude at P

$$= \frac{R_1}{2} + R_1 = \frac{3R_1}{2}$$

Thus, point P is a point of maximum intensity.

If point P is situated at such a position that the exposed portion O'O of the lower half of the wavefront O' W' contains only two half period strips, then the amplitude at P will be

$$= \frac{R_1}{2} + R_1 - R_2 = \frac{3R_1}{2} - R_2$$

Hence in comparison to the previous position point P is a point of minimum intensity.

If the portion OO' allows three half period strips, then the amplitude at P will be

$$= \frac{R_1}{2} + R_1 - R_2 + R_3 = \left(\frac{3R_1}{2} - R_2\right) + R_3$$

Thus point P is sufficiently bright but the intensity is slightly less than the first maximum. Similarly, we get maxima at P if the lower half wavefront exposes three, five, or seven....half period strips and minima at P if the lower half wavefront exposes two, four, six, or eight......half period strips.

We conclude that the intensity at point P will be maximum if the number of half period strips allowed by the exposed portion of the lower half of the wavefront is odd, and minimum if the number of allowed half period strips is even. Thus the intensity of maxima decreases gradually, while that of minima goes on increasing as we move along the screen above point C, until finally the fringes merge into a uniform illumination.

Positions of Maxima and Minima

In order to find the positions of maxima and minima just outside the geometrical shadow, consider Figure 2. Let

$$SO = a,\ OC = b \text{ and } CP = x$$

The path difference between the waves reaching at P from O and O' is PO – PO'.

In a right angled triangle POC, we have

$$PO = \left[(OC)^2 + (CP)^2\right]^{1/2} = (b^2 + x^2)^{1/2} = b\left(1 + \frac{x^2}{b^2}\right)^{1/2}$$

$$= b\left(1 + \frac{x^2}{2b^2}\right) = b + \frac{x^2}{2b} \qquad \text{(By Binomial theorem)} \quad (1)$$

and $$PO' = SP - SO' = SP - a.$$

In a right angled triangle PSC,

we have $$SP = \left[(SC)^2 + (CP)^2\right]^{1/2}$$

$$= \left[(a+b)^2 + x^2\right]^{1/2} = (a+b) + \frac{x^2}{2(a+b)}$$

Thus $$PO' = a+b+\frac{x^2}{2(a+b)} - a = b + \frac{x^2}{2(a+b)} \qquad (2)$$

Thus the path difference

$$PO - PO' = b + \frac{x^2}{2b} - b - \frac{x^2}{2(a+b)} = \frac{x^2}{2} \cdot \frac{a}{b(a+b)}. \qquad (3)$$

For maxima

$$PO - PO' = \frac{x^2 a}{2b(a+b)} = (2n+1)\frac{\lambda}{2}$$

where $n = 0, 1, 2, 3, \ldots$

or $$x_n = \left(\frac{b(a+b)}{a}(2n+1)\lambda\right)^{1/2} \qquad (4)$$

Equation (4) represents the distance of nth maximum from C towards X.

For minima

$$PO - PO' = \frac{x^2 a}{2b(a+b)} = 2n\frac{\lambda}{2}$$

where $n = 0, 1, 2, 3, \ldots$

or $$x_n = \sqrt{\frac{b(a+b)}{a} 2n\lambda} \qquad (5)$$

Equation (5) gives the distance of nth minimum from C towards X.

If $$\sqrt{\frac{b(a+b)\lambda}{a}} = k$$

where k is constant, then equations (4) and (5) take the form

$$x_n = k.\sqrt{(2n+1)} \qquad (6)$$

and $$x_n = k.\sqrt{2n}\,.$$

Thus the separations between the successive maxima are

$$x_2 - x_1 = \left(\sqrt{3}-1\right)k = 0.73\,k$$

$$x_3 - x_2 = \left(\sqrt{5}-\sqrt{3}\right)k = 0.50\,k$$

$$x_4 - x_3 = \left(\sqrt{7}-\sqrt{5}\right)k = 0.43\,k$$

Obviously the separations go on decreasing. Similarly the separations between successive minima are

$k(\sqrt{4}-\sqrt{2}), k\left(\sqrt{6}-\sqrt{4}\right), k\left(\sqrt{8}-\sqrt{6}\right), \ldots$ which is also in decreasing order.

Thus we conclude that the diffraction fringes get closer and closer as we move above the edge of the geometrical shadow.

3.5 INTENSITY AT THE EDGE OF THE GEOMETRICAL SHADOW

If point P is at the edge of the geometrical shadow C, the pole of the wavefront O is at the edge of the obstacle. In this position the upper half wavefront OW is fully exposed and the lower half OW' is completely blocked by the obstacle. Thus the resultant amplitude at C is $R_1/2$ (due to the upper half wavefront) and the intensity at C is $R_1^2/4$. If the entire wavefront were exposed, the resultant amplitude at C would be $R_1/2 + R_1/2 = R_1$ and hence the intensity would be R_1^2. Thus the intensity at the edge of the geometrical shadow is one-fourth of the intensity of that produced in the absence of an obstacle.

Intensity within the Geometrical Shadow

If Q is a point inside the geometrical shadow and O" is its corresponding pole, then the entire lower half wavefront O"W' and the portion O"O of the upper half wavefront O"W is blocked by the straight edge and only the part OW of the upper half is exposed. If point Q is such that the unexposed portion of the upper half wavefront O"O cuts off only the first upper half period strip, then the amplitude at Q will be

$$R_2 - R_3 + R_4 + ... = R_2/2.$$

If point Q is such that the unexposed portion of the upper half wavefront O"O obstructs the first two strips of the upper half wavefront, in addition to the lower half, then the resultant amplitude at Q will be

$$R_3 - R_4 + R_5 - R_6 + ... = R_3/2.$$

Thus as the point Q moves further into the geometrical shadow, the number of strips contributing to the resultant amplitude become smaller and smaller. Therefore, the intensity falls off continuously until all the effective strips are cut off and hence the intensity becomes zero. Thus we conclude that the intensity within the geometrical shadow decreases first rapidly than slowly to a minimum zero as we move away from the edge.

The Intensity Distribution Curve

The intensity distribution curve is depicted in Figure 4. It is clear from this curve that the intensity inside the geometrical shadow decreases rapidly to minimum zero. While in the illuminated region, we get alternate maxima of decreasing intensity and minima of increasing intensity, until at a certain distance the intensity reaches a steady value resulting in a uniform illumination.

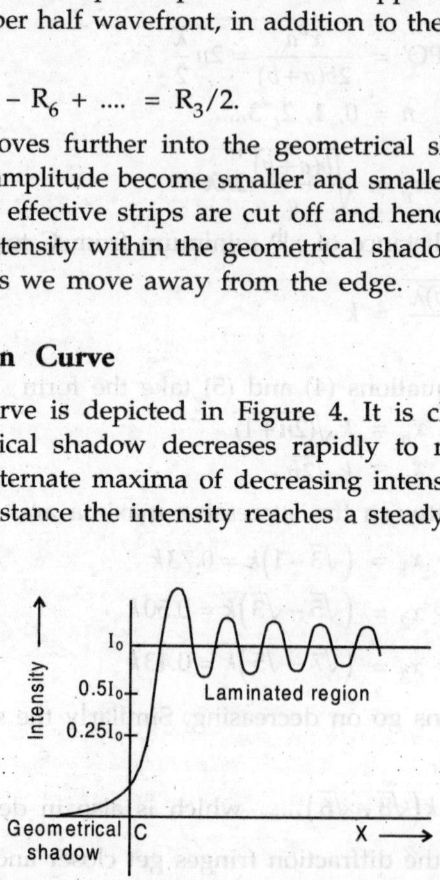

FIGURE 4 **Intensity distribution in the geometrical shadow line.**

The Determination of a Wavelength of Light by a Straight Edge

To determine the wavelength λ of monochromatic light, a thin razor blade is used as a straight edge and placed parallel to the slit in one of the uprights of the optical bench. The upright containing razor the blade is kept in between the slit upright and microscope upright. The slit, thin razor blade, and microscope are adjusted to the same height. The slit is illuminated by monochromatic light source whose wavelength is to be determined and the apparatus is adjusted to obtain the diffraction pattern. The diffraction fringe, so obtained, is observed by a travelling microscope. Now the distance between the first or second bright band and one of the most distant that can be seen distinctly is measured with the help of a micrometer screw. If the former is m^{th} and latter is n^{th} band then we have

$$x_n = \sqrt{\frac{(2n+1)\,(a+b)\,b\lambda}{a}}$$

and

$$x_m = \sqrt{\frac{(2m+1)\,(a+b)\,b\lambda}{a}}$$

$$x_n - x_m = \sqrt{\frac{b\lambda\,(a+b)}{a}}\left(\sqrt{(2n-1)} - \sqrt{(2m-1)}\right).$$

Thus by measuring a and b, λ can be calculated.

Example 1. *A narrow slit, illuminated by light of wavelength 4900 Å, is placed at a distance of 3 m from a straight edge. If the distance between the straight edge and screen is 6 m, calculate the distance between the first and the fourth band.*

Solution. The distance of nth minimum outside the geometrical shadow is given by

$$x_n = \sqrt{\frac{b(a+b)\lambda}{a}}.\sqrt{2n} \text{ or } k\sqrt{2n}$$

where

$$k = \sqrt{\frac{b(a+b)\lambda}{a}} \quad \text{and} \quad n = 1, 2, 3, \ldots$$

In the given problem $b = 6$ m $= 600$ cm, $a = 3$ m $= 300$ cm

and

$$\lambda = 4900 \text{ Å} = 6 \times 10^{-8} \text{ cm}$$

$$k = \left[\frac{600\,(300 + 600) \times 4900 \times 10^{-8}}{300}\right]^{1/2} = 0.297 \text{ cm}$$

or

$$x_n = 0.297\sqrt{2n}$$

For the first minimum, $n = 1$ so $x_1 = 0.297\sqrt{2} = 0.420$ cm

For the fourth minimum, $n = 4$ so $x_4 = 0.297\sqrt{8} = 0.840$ cm.

The distance between the first and the fourth band

$$= x_4 - x_1 = 0.840 - 0.420 = \mathbf{0.42\ cm.}$$

3.6 FRESNEL AND FRAUNHOFFER DIFFRACTION

The diffraction phenomenon is usually divided into two categories: (*i*) Fresnel diffraction and (*ii*) Fraunhoffer diffraction.

In the Fresnel class of diffraction the source of light or the screen or both, are in general at a finite distance from the diffracting element but no lenses are needed for rendering the rays parallel or convergent. Therefore the incident wavefront is spherical or cylindrical instead of being plane. In the Fraunhoffer class of diffraction, the source and the screen are effectively at infinite distance from the diffracting element. It is observed by employing two convergent lenses; one to render the incoming light parallel and the other to focus the parallel diffracted rays on the screen. Therefore the incident wavefront is plane.

In Fresnel diffraction lateral distance is important, where as in Fraunhoffer diffraction the angular inclinations are important. In Fresnel diffraction the observed pattern is a projection of a diffracting element modified by diffracting effects, and the geometry of the source, whereas in Fraunhoffer diffraction, the observed diffraction pattern is an image of the source modified by the diffraction at the diffracting element.

In Fresnel diffraction, the center of the diffraction pattern may be bright or dark depending upon the number of Fresnel zones, whereas in Fraunhoffer diffraction, the center of the diffraction pattern is always bright for all paths parallel to the axis of the lens.

Resultant of *n* Simple Harmonic Motions of Equal Amplitudes and Periods; Phases Increasing in Arithmetic Progression

Let a particle be affected simultaneously by n simple harmonic vibrations of equal amplitude 'a' and having the common phase difference δ between successive vibrations. To find the resultant amplitude R and phase ϕ of the particle, we construct a polygon of vector amplitudes having each side equal to a as shown in Figure 5. The closing side OP of the polygon gives the resultant amplitude R and the angle which this closing side subtends with the first vibration gives the resultant phase ϕ.

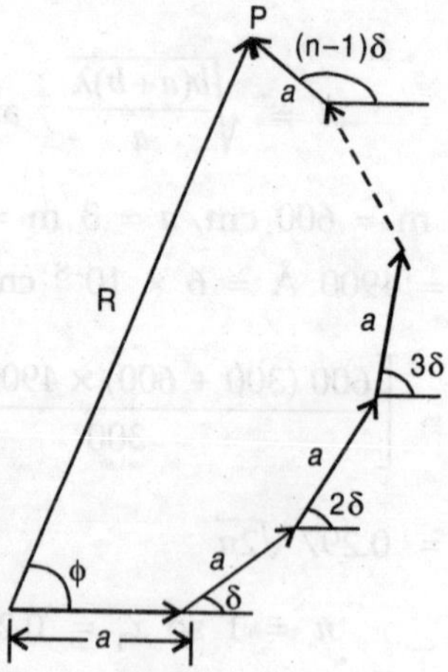

FIGURE 5

Resolving the amplitudes parallel and perpendicular to OA, we get

$$R\cos\phi = a + a\cos\delta + a\cos 2\delta + \ldots + a\cos(n-1)\delta \qquad (1)$$

and $R \sin \phi = 0 + a \sin \delta + a \sin 2\delta + \ldots + a \sin (n-1)\delta$ (2)

Multiplying equation (1) by $2 \sin \delta/2$, we get

$$2R \cos \phi \sin (\delta/2) = a\,[2\sin \delta/2 + 2\sin (\delta/2) \cos\delta + 2\sin (\delta/2) \cos 2\delta + \ldots + 2\sin(\delta/2)\cos (n-1)\delta]$$

Applying trigonometrical formula,

$2 \sin A \cos B = \sin (A + B) - \sin (A - B)$, we get

$$2R \cos \phi \sin \frac{\delta}{2} = a\left[2\sin\frac{\delta}{2} + \left(\sin\frac{3\delta}{2} - \sin\frac{\delta}{2}\right) + \left(\sin\frac{5\delta}{2} - \sin\frac{3\delta}{2}\right) + \ldots + \left\{\sin\left(n-\frac{1}{2}\right)\delta - \sin\left(n-\frac{3}{2}\right)\delta\right\}\right]$$

$$= a\left[\sin\frac{\delta}{2} + \sin\left(n-\frac{1}{2}\right)\delta\right] = 2a\sin\frac{n\delta}{2}\cos\frac{(n-1)\delta}{2}$$

or $$R \cos \phi = \frac{a \sin\left(\frac{n\delta}{2}\right)}{\sin\frac{\delta}{2}} \cos\left[\frac{(n-1)\delta}{2}\right] \quad (3)$$

Similarly the second series [equation (2)] after multiplication by $2 \sin (\delta/2)$ and on simplification gives

$$R \cos \phi = \frac{a \sin\left(\frac{n\delta}{2}\right)}{\sin\frac{\delta}{2}} \sin\left[\frac{(n-1)\delta}{2}\right] \quad (4)$$

Squaring equation (3) and (4), and adding, we get

$$R^2 = \frac{a^2 \sin^2 (n\delta/2)}{\sin^2 (\delta/2)} \quad (5)$$

or $$R = \frac{a \sin (n\delta/2)}{\sin (\delta/2)} \quad (6)$$

Dividing equation (4) by equation (3), we obtain

$$\tan \phi = \tan \frac{(n-1)\delta}{2} \text{ or } \phi = \frac{(n-1)\delta}{2} \quad (7)$$

Let us suppose that the number of vibrations n are infinitely large but the amplitude a and the common phase δ are infinitely small, so that na and $n\delta$ are finite.

Also suppose $n\delta = 2\alpha$; then equation (6) gives

$$R = \frac{a \sin \alpha}{\sin (\alpha/n)} = \frac{a \sin \alpha}{(\alpha/n)}, \text{ because } \alpha/n \text{ is very small.}$$

$$R = na\frac{\sin\alpha}{\alpha}$$

Now putting $n\alpha = A$, we obtain $R = A\dfrac{\sin\alpha}{\alpha}$ (8)

From equation (7), we have

$$\phi = \frac{(n-1)\delta}{2} = \frac{n\delta}{2} = \alpha$$

$$\left(\because \text{ For large } n, \frac{n\delta}{2} = \frac{(n-1)\delta}{2} \right) \quad (9)$$

Equations (8) and (9) give the resultant amplitude and phase of n simple harmonic vibrations of equal amplitude and periods and phases increasing in arithmetic progression.

3.7 FRAUNHOFFER DIFFRACTION AT A SINGLE SLIT

Let a collimated beam of monochromatic light of wavelength λ, produced by a point source S placed at the focus of a spherical lens L_1, incident normally on a narrow slit AB of width *'a'* (Figure 6). Let the diffracted light be focussed by another convex lens L_2 on the screen XY placed in the focal plane of lens L_2 and perpendicular to the plane of the paper. According to the laws of geometrical optics, we should get a sharp image of the slit (AB) on the screen. But this is not the case. Actually, the transmitted beam spreads out perpendicularly to the length of the slit. Therefore when it is focussed on the screen, a diffraction pattern having alternate dark and bright fringes of decreasing intensity, is obtained on both the sides of the point C.

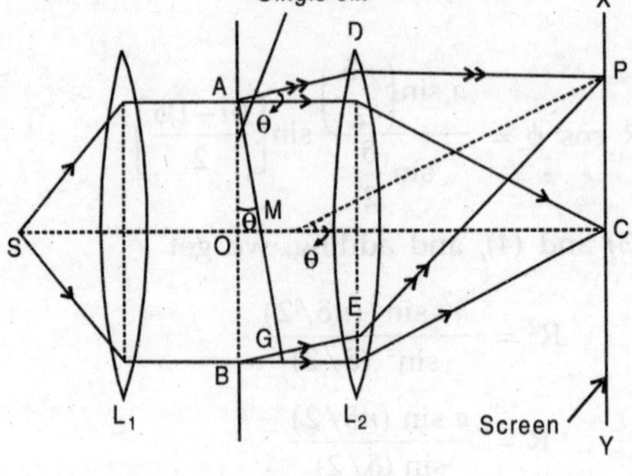

FIGURE 6

Explanation

Let a plane wavefront of monochromatic light of wavelength λ propagating normally to the slit AB. According to Huygens' theory every point within the slit is to be regarded as the origin of secondary wavelets, which speed out in all directions. These wavelets then interfere to produce the diffraction pattern. All the secondary wavelets diffracted at an angle θ with the normal are brought to focus at P. The intensity of the diffracted beams will be different in different directions and there will be some directions in which there will be no light. The point C is equidistance from all points on either side of the center of slit O. Therefore at the point C intensity is maximum.

In order to find the resultant intensity at P, draw a perpendicular AG on BE. The path difference between the rays emanating from extreme point A and B of the slit AB is given by

$$\Delta = BG = AB \sin\theta = a \sin\theta$$

where a is the width of slit AB and the corresponding phase difference is $\frac{2\pi}{\lambda}(a \sin\theta)$. Let the width AB of the slit be divided into a large number of n equal parts, each part being the source of secondary wavelets. The amplitude of the wave due to each part is equal to a but their phase will vary gradually from zero to $\frac{2\pi}{\lambda}(a\sin\theta)$, as the path difference between the rays originating from points in AB vary from O to $(a \sin\theta)$.

Thus, the phase difference between the waves from any two successive parts would be

$$\frac{1}{n}\left(\frac{2\pi}{\lambda}a\sin\theta\right) = \delta.$$

According to the theory of composition of n simple harmonic motions of equal amplitude (a) and common phase difference δ between successive vibrations, the resultant amplitude at P is given by

$$R = a\frac{\sin(n\delta/2)}{\sin(\delta/2)} = a\frac{\sin\{(\pi a\sin\theta/\lambda)\}}{\sin\{(\pi a\sin\theta/n\lambda)\}}$$

or

$$R = a\frac{\sin\alpha}{\sin(\alpha/n)} = \frac{a\sin\alpha}{(\alpha/n)}$$

where

$$\alpha = \frac{\pi a\sin\theta}{\lambda}$$

and for large value of n, $\sin\left(\frac{\alpha}{n}\right) = \frac{\alpha}{n}$

$$\therefore \quad R = na\frac{\sin\alpha}{\alpha} = \frac{A\sin\alpha}{\alpha} \tag{1}$$

where

$$A = na$$

The resultant intensity at P being proportional to the square of the resultant amplitude R, is given by

$$I = R^2 = A^2\left(\frac{\sin\alpha}{\alpha}\right)^2 \tag{2}$$

the constant of proportionality being taken as unity for simplicity.

Equation (2) gives the variation of intensity with θ (or α).

Positions of Maxima and Minima

Principal Maxima

The resultant amplitude given by equation (1) can be expanded as

$$R = \frac{A}{\alpha}\sin\alpha = \frac{A}{\alpha}\left[\alpha - \frac{\alpha^3}{3!} + \frac{\alpha^5}{5!} - \frac{\alpha^7}{7!} + \ldots\right]$$

$$= A\left[1 - \frac{\alpha^2}{3!} + \frac{\alpha^4}{5!} - \frac{\alpha^6}{7!} + \ldots\right]$$

R will be maximum if the negative terms vanish. This is possible only when

$$\alpha = \frac{\pi}{\lambda}(a \sin\theta) = 0 \text{ or } \theta = 0$$

Thus the maximum value of resultant amplitude R at C is A and corresponding maximum intensity I_{max} is proportional to A^2.

Position of Minima

From equation (2) it is clear that the intensity is minimum when $\sin\alpha = 0$ and $\alpha \neq 0$

i.e., $$\alpha = \pm\pi, \pm 2\pi, \pm 3\pi, \ldots, \pm n\pi \text{ or } \alpha = \pm n\pi$$

$$\therefore \quad \alpha = \frac{\pi a \sin\theta}{\lambda} = \pm n\pi$$

or $$a \sin\theta = \pm n\lambda \tag{3}$$

where n = 1, 2, 3, etc. gives the directions of first, second, third, ... minima. Here $n \neq 0$, because $n = 0$ or $\theta = 0$ corresponds to principal maximum.

Secondary Maxima

In addition to the principal maximum, there are less intense secondary maxima between equally spaced minima. In order to locate the positions of secondary maxima we apply the mathematical method for finding maxima and minima. Let us differentiate equation (2) with respect to α and equate it to zero, that is,

$$\frac{dI}{d\alpha} = \frac{d}{d\alpha}\left(A^2\frac{\sin^2\alpha}{\alpha^2}\right) = A^2 \times \frac{2\sin\alpha}{\alpha}\left[\frac{\alpha\cos\alpha - \sin\alpha}{\alpha^2}\right] = 0$$

that is, either $\dfrac{\sin\alpha}{\alpha} = 0$ or $\dfrac{\alpha\cos\alpha - \sin\alpha}{\alpha^2} = 0$

$\therefore \sin\alpha = 0$ or $\alpha\cos\alpha - \sin\alpha = 0$ or $\alpha = \tan\alpha$

The equation $\sin\alpha = 0$ gives the position of primary minima except when $\alpha = 0$. Thus the positions of secondary maxima are given by the equation

$$\alpha = \tan\alpha \tag{4}$$

Equation (4) can be solved graphically by plotting the curves

$$y = \alpha \text{ and } y = \tan\alpha$$

The equation $y = \alpha$, is an equation of straight lines passing through an origin and making a angle 45° with the axis as shown in Figure 7. The equation $y = \tan\alpha$ represents a discontinuous curve having a number of branches with asymptotes at intervals $\alpha = \pi$ (as shown in Figure 7). The point of intersection of these two curves gives the value of α satisfying the equation $\alpha = \tan\alpha$.

The first value $\alpha = 0$ gives principal maximum. The remaining values of α which give the secondary maxima are

$$\alpha = \frac{3\pi}{2}, \frac{5\pi}{2}, \frac{7\pi}{2}, \frac{9\pi}{2}, \ldots \tag{5}$$

The more exact values of α are

$$\alpha = 1.430\pi, 2.46\pi, 3.47\pi, 4.471\pi, \ldots$$

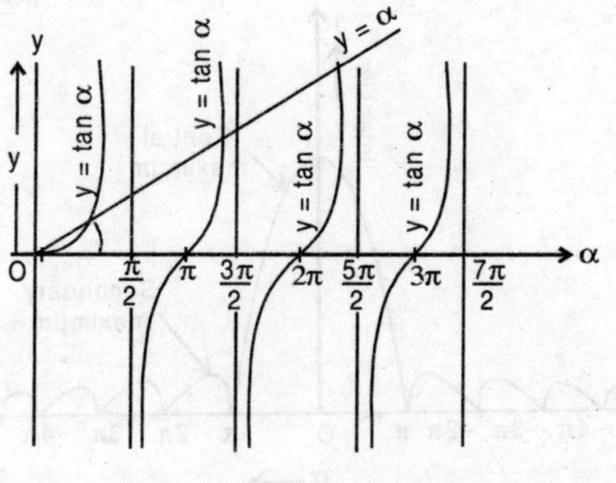

FIGURE 7 **Graphical representation of positions of secondary maxima's in the diffraction pattern.**

The direction of secondary maxima are approximately given by the relation

$$\alpha = \pm\frac{(2n+1)\pi}{2}, \text{ where } n = 1, 2, 3, ...$$

or
$$a \sin\theta = \pm\frac{(2n+1)\lambda}{2}$$

From $\left(\alpha = \frac{\pi}{\lambda}(a \sin\theta)\right)$ (6)

Substituting the approximate value of *a* from equation (5) in equation (2), we get intensities of various secondary maxima.

(*i*) The intensity of first secondary maximum

$$I_1 = A^2\frac{\left(\sin\frac{3\pi}{2}\right)^2}{\left(\frac{3\pi}{2}\right)^2} = \frac{A^2}{22} = \frac{I_0}{22}$$

where I_0 is the intensity of primary maximum.

(*ii*) The intensity of second secondary maximum

$$I_1 = A^2\frac{\left(\sin\frac{5\pi}{2}\right)^2}{\left(\frac{5\pi}{2}\right)^2} = \frac{A^2}{62} = \frac{I_0}{62}$$

similarly, the intensity of third secondary maximum $I_3 = \frac{I_0}{120}$ and so on.

From the expressions of intensities of secondary maxima, it is clear that most of the light is concentrated in the principal maximum, the first secondary maximum on either side of the principal maximum have only 4% of the light of the principal maximum. The minima are equally spaced and their intensity is zero.

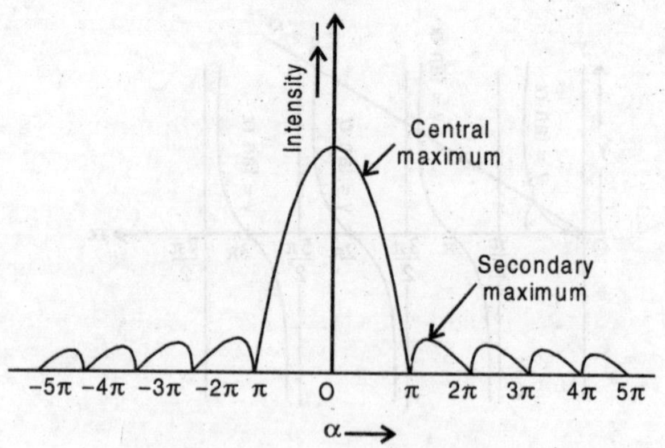

FIGURE 8 **Intensity distribution in the diffraction pattern of a single slit.**

The intensity distribution graph as depicted in Figure 8 indicates that the diffraction pattern consists of a bright central maximum surrounded alternatively by minima of zero intensity and less intensed secondary maxima of rapidly decreasing intensity.

Example 2. *The light of wavelength 5500 Å falls normally on a slit of a width 22.0 × 10^{-5} cm. Calculate the angular position of the first two minima on either side of the central maximum.*

Solution. The angular positions of minima in a single slit diffraction pattern are given by

$$a \sin \theta = \pm n\lambda \quad \text{or} \quad \sin \theta = n\lambda/a$$

where

$$n = 1, 2, 3,....$$

Here,

$$a = 22.0 \times 10^{-5} \text{ cm} \quad \text{and} \quad \lambda = 5500 \times 10^{-8} \text{ cm}$$

From the first order minimum, $n = 1$

$$\therefore \quad \sin \theta_1 = \frac{\lambda}{a} = \frac{5500 \times 10^{-8}}{22.0 \times 10^{-5}} = 0.25$$

or

$$\theta_1 = \sin^{-1}(0.25) = 14° \ 29'$$

For the second order minimum, $n = 2$

$$\therefore \quad \sin \theta_2 = 2\lambda/a = 2 \times 0.25 = 0.5$$

$$\therefore \quad \theta_2 = \sin^{-1} (0.5) = \mathbf{30°}$$

Therefore, the first two minima will occur at angle 14° 29' and 30° on either side of the central maximum.

Example 3. *A lens of focal length 100 cm forms a Fraunhoffer diffraction pattern of a single slit of width 0.04 cm in its focal plane. The incident light contains two wavelengths λ_1 and λ_2. It is found that the fourth minimum corresponding to λ_1 and the fifth minimum corresponding to λ_2 occur at the same point 0.5 cm from the central maximum. Calculate λ_1 and λ_2.*

Solution. In the Fraunhoffer diffraction pattern due to a single slit of width *a*, the directions of minima are given by

$$a \sin \theta = \pm n\lambda$$

For wavelength λ_1 the direction of fourth minimum ($n = 4$) is given by

$$a \sin \theta_1 = 4\lambda_1 \tag{1}$$

Similarly for wavelength λ_2, the direction of fifth minimum ($n = 5$) is give by

$$a \sin \theta_2 = 5\lambda_2 \qquad (2)$$

According to the problem, the direction of fourth minimum (θ_1) due to λ_1 and the fifth minimum (θ_2) due to λ_2 occur at the same point, that is,

$$\theta_1 = \theta_2 = \theta.$$

Therefore $$a \sin \theta = 4\lambda_1 = 5\lambda_2 \quad \text{[from equations (1) and (2)]} \qquad (3)$$

If x is the linear distance of fourth minimum for λ_1 or fifth minimum for λ_2, we have $\theta = x/f$ (as θ is small) where f is the local length of the lens here, $x = 0.5$ cm and $f = 100$ cm

$$\therefore \qquad \theta = 0.5/100 = 0.005 \text{ rad.}$$

From equation (3), we have $a \sin \theta = 4\lambda_1$ or $\lambda_1 = \frac{a \sin \theta}{4} = \frac{a\theta}{4}$ (as θ is small). Here $a = 0.04$ cm and $\theta = 0.005$ rad $\therefore$ $\lambda_1 = \frac{0.04 \times 0.005}{4} = 5 \times 10^{-5}$ cm.

Substituting value of λ_1 in equation (3), we have

$$\lambda_2 = \frac{4\lambda_1}{5} = \frac{4 \times 5 \times 10^{-5}}{5} = 4 \times 10^{-5} \text{ cm.}$$

3.8 PLANE DIFFRACTION GRATING (DIFFRACTION AT N PARALLEL SLITS)

A plane diffraction grating is an arrangement of a large number of close parallel straight transparent equidistant slits, each of equal width '*a*', with neighboring slits being separated by an opaque region of width '*b*'. A grating is made by drawing a series of very fine equidistant parallel lines on a glass plate by means of a fine diamond pen. The light cannot go through the lines drawn by the diamond; while the spacing between the lines is transparent to the light. There are about 15,000 lines per inch in such a grating to produce a diffraction of visible light. The spacing ($a + b$) between adjacent slits is called the diffraction element or grating element. If the lines are drawn on a silvered surface (plane or concave), then light is reflected from the positions of mirrors in between any two lines and it forms a plane or concave reflection grating. Since the original gratings are quite expensive, for practical purposes their photographic reproductions are generally used.

The commercial gratings are produced by taking the cast of an actual grating on a transparent film such as cellulose acetate. A thin layer of collodion solution (celluloid dissolved in a volatile solvent) is poured on the surface of the ruled grating and allowed to dry. Thin collodion film is stripped off from the grating surface. This film, which retains the impressions of the original grating, is preserved by mounting the film between two glass sheets. Now-a-days holographic gratings are also produced. Holographic gratings have a larger number of lines per cm than a ruled grating.

Theory

Suppose a plane diffraction grating, consisting of a large number of N parallel slits each of width a and separated by an opaque space b, is illuminated normally by a plane wavefront of monochromatic light of wavelength λ as in Figure 9. The light diffracted through N slits is focussed by a convex lens on the screen XY placed in the focal plane of the lens L. The diffraction pattern obtained on the screen with a very large number of slits consists of extremely sharp principal interference maximum; while the intensity of secondary maxima becomes negligibly small so that these are not visible in the diffraction pattern.

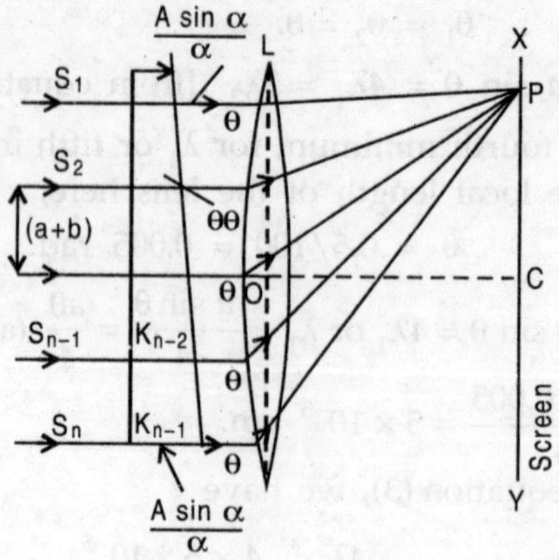

FIGURE 9

According to the Huygen-Fresnel principle, all points within each slit become the sources of secondary wavelets, which spread out in all directions. From the theory of Fraunhoffer diffraction at a single slit, the resultant amplitude R due to all waves diffracted from each slit in the direction θ is given by $R = \dfrac{A \sin \alpha}{\alpha}$, where A is the constant and $\alpha = \dfrac{\pi a \sin \theta}{\lambda}$. Thus the resultant of all the secondary waves may be treated as a single wave of amplitude R and phase α starting from the middle point of the slit and travelling at an angle θ with the normal. Thus the waves, diffracted from all the N slits in direction θ are equivalent to N parallel waves, each starting from the middle points S_1, S_2, S_3, ..., S_{n-1}, S_n, of the slits. Drawing perpendicular S_1K_{n-1} on the parallel paths of these waves, then the path difference between the waves from the slits S_1 and S_2 is

$$S_2K_1 = S_1S_2 \sin \theta = (a + b) \sin \theta$$

Thus the path difference between the waves from any two consecutive slits is $(a + b) \sin \theta$. Therefore the corresponding phase difference $= \dfrac{2\pi}{\lambda}(a+b) \sin\theta = 2\beta$. Thus as we pass from one vibration to another, the path goes on increasing by an amount $(a + b) \sin \theta$ and the corresponding phase goes on increasing by an amount $\dfrac{2\pi}{\lambda}(a+b)\sin\theta$. Thus the phase increases in arithmetical progression. Therefore, the problem reduces to find the resultant amplitude of N waves of equal amplitude $(A \sin \alpha/a)$ and period but the phase increasing in the arithmetical progression. In this case the resultant amplitude in the direction θ is given by

$$R' = R\frac{\sin N\beta}{\sin \beta} = \frac{A \sin \alpha}{\alpha} \cdot \frac{\sin N\beta}{\beta}$$

The corresponding intensity at P is given by

$$I = R'^2 = \frac{A^2 \sin^2 \alpha}{\alpha^2} \cdot \frac{\sin^2 N\beta}{\sin^2 \beta} \qquad (1)$$

The first factor of equation (1) $\frac{A^2 \sin^2 \alpha}{\alpha^2}$ represents intensity distribution due to diffraction at a single slit; whereas the second factor $\frac{\sin^2 N\beta}{\sin^2 \beta}$ gives the distribution of intensity in the diffraction pattern due to interference in the waves from all the slits.

The Principal Maxima

The condition for the principal maxima is

$$\sin \beta = 0, \text{ that is, } \beta = \pm n\pi, \text{ where } n = 0, 1, 2, 3, ...$$

we have $\sin N\beta = 0$, so $\frac{\sin N\beta}{\sin \beta} = \frac{0}{0}$ becomes indeterminate. But it may be evaluated by applying the usual method of differentiating the numerator and denominator and applying its limits as β tends to $\pm n\pi$.

Thus
$$\lim_{\beta \to \pm n\pi} \frac{\sin N\beta}{\sin \beta} = \lim_{\beta \to \pm n\pi} \frac{\frac{d}{d\beta}(\sin N\beta)}{\frac{d}{d\beta}(\sin \beta)}$$

$$= \lim_{\beta \to \pm n\pi} \frac{N \cos N\beta}{\cos \beta} = N$$

Therefore, the intensity of the principal maxima is proportional to N^2 and increases as the number of slits increase.

Substituting the value of $\frac{\sin N\beta}{\sin \beta} = N$ in equation (1),

we get
$$I = \frac{A^2 \sin^2 \alpha}{\alpha^2} N^2 \qquad (2)$$

Thus the actual intensity of the principal maxima is $\frac{A^2 \sin^2 \alpha}{\alpha^2} N^2$.

Thus the directions of the principal maxima are given by

$$\sin \beta = 0, \text{ that is, } \beta = \pm n\pi, \text{ where } n = 0, 1, 2, ...$$

or
$$\frac{\pi}{\lambda}(a+b)\sin\theta = \pm n\pi \quad \text{or} \quad (a + b) \sin \theta = \pm n\lambda \qquad (3)$$

If we put $n = 0$, we get $\theta = 0$ and equation (3) gives the direction of the zero order principal maximum. The first, second, third, ..., order principal maxima may be obtained by putting n = 1, 2, 3, ... in equation (3).

The Minima

The intensity is minimum when $\sin N\beta = 0$; but $\sin \beta \neq 0$

or
$$N\beta = \pm m\pi$$

or $$N\frac{\pi}{\lambda}(a+b)\sin\theta = \pm m\pi$$

or $$N(a + b) \sin \theta = \pm m\lambda \quad (4)$$

Here m can have all integral values except 0, N, 2N, 3N, ... because for these values of m, $\sin \beta = 0$ which gives the positions of the principal maxima. Positive and negative signs show that the minima lie symmetrically on both sides of the central principal maximum.

The Secondary Maxima

The above study reveals that there are (N – 1) minima between two successive principal maxima. Hence there are (N – 2) other maxima coming alternatively with the minima between two successive principal maxima. These maxima are called secondary maxima. To find the positions of the secondary maxima, we first differentiate equation (1) with respect to β and equating it to zero

$$\frac{dI}{d\beta} = \frac{A^2 \sin^2 \alpha}{\alpha^2}.2\left[\frac{\sin \beta N}{\sin \beta}\right]\left[\frac{N \cos N\beta \sin \beta - \sin N\beta \cos \beta}{\sin^2 \beta}\right] = 0 \quad (5)$$

or $$N \cos N\beta \sin \beta - \sin N\beta \cos \beta = 0$$

or $$\tan N\beta = N \tan \beta$$

To find the intensity of the secondary maxima, we make the use of the triangle shown in Figure 10.

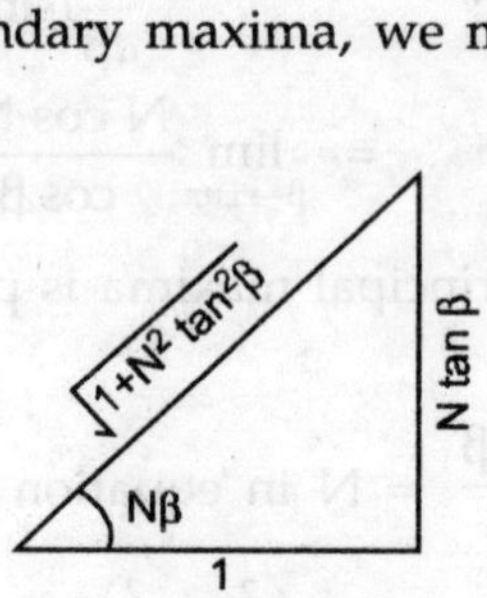

FIGURE 10

We have $$\sin N\beta = \frac{N \tan \beta}{\sqrt{(1+N^2 \tan^2 \beta)}}$$

Therefore, $$\frac{\sin^2 N\beta}{\sin^2 \beta} = \frac{(N^2 \tan^2 \beta)/(1+N^2 \tan^2 \beta)}{\sin^2 \beta}$$

or $$\frac{\sin^2 N\beta}{\sin^2 \beta} = \frac{N^2 \tan^2 \beta}{(1+N^2 \tan^2 \beta) \sin^2 \beta} = \frac{N^2}{1+(N^2-1)\sin^2 \beta}$$

Putting this value of $(\sin^2 N\beta)/\sin^2 \beta$ in equation (1), we get the intensity of the secondary maxima as

$$I' = \frac{A^2 \sin^2 \alpha}{\alpha^2}.\frac{N^2}{\left[1+(N^2-1)\sin^2 \beta\right]} \quad (6)$$

The ratio of the intensity of these secondary maxima to the intensity of the principal maxima is obtained by dividing equation (6) by equation (2). Thus

$$\frac{\text{Intensity of secondary maxima}}{\text{Intensity of primary maxima}} = \frac{I'}{I} = \frac{1}{1 + (N^2 - 1)\sin^2\beta}$$

Hence, the greater the value of N, the weaker are the secondary maxima. In an actual grating N is extremely large. Hence the secondary maxima are not visible in the grating spectrum.

The intensity distribution curve for the plane diffraction grating is depicted in Figure 11.

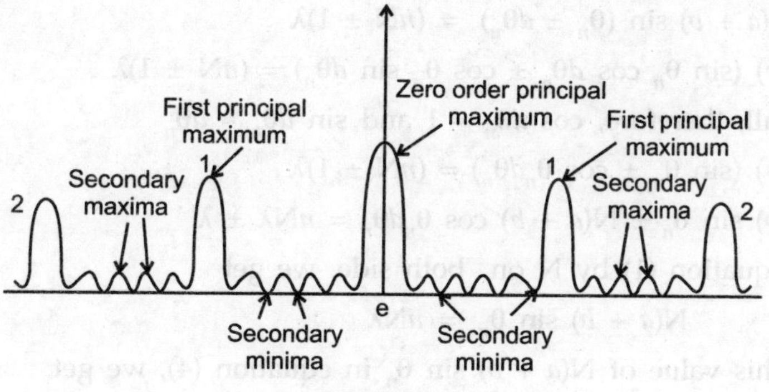

FIGURE 11

3.9 WIDTH OF THE PRINCIPAL MAXIMA

The angular separation between the first two minima lying on either side of the principal maximum is called the angular width of the principal maximum of any order. As shown in Figure 12, $2d\theta_n$ represents the angular width of nth order principal maximum.

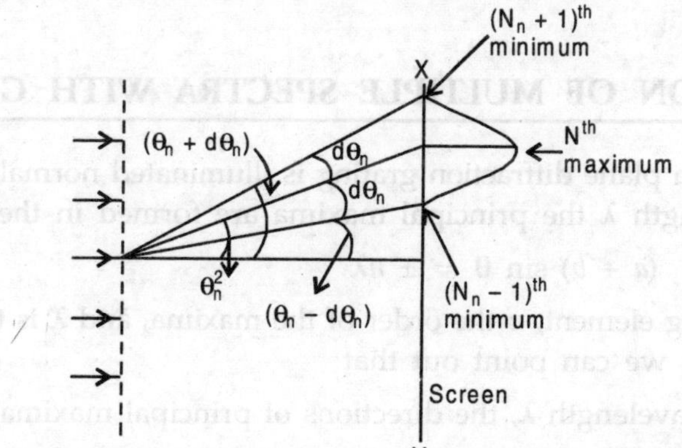

FIGURE 12

The nth order principal maximum lie in the direction given by the equation

$$(a + b)\sin\theta_n = n\lambda \tag{1}$$

If $(\theta_n + d\theta_n)$ and $(\theta_n - d\theta_n)$ represent the directions of first outer and inner sided minima adjacent to the nth maximum, then $d\theta_n$ will give the angular half width of nth principal maximum. The directions of minima are represented by

$$N(a + b) \sin \theta = m\lambda \tag{2}$$

where m is an integer and takes all integer values except 0, N, 2N, ...nN, because these values of equation (2) give the directions of zero, first, second, ...n^{th} order principal maxima. As the first order and inner sided minima adjacent to n^{th} order principal maxima lie in the direction $(\theta_n \pm d\theta_n)$, this corresponds to the value of $m = (nN \pm 1)$.

Hence equation (2) gives

$$N(a + b) \sin (\theta_n \pm d\theta_n) = (nN \pm 1)\lambda \tag{3}$$

or $$N(a + b) (\sin \theta_n \cos d\theta_n \pm \cos \theta_n \sin d\theta_n) = (nN \pm 1)\lambda$$

As $d\theta_n$ is small, therefore, $\cos d\theta_n \approx 1$ and $\sin d\theta_n \approx d\theta_n$

Thus, $N(a + b) (\sin \theta_n \pm \cos \theta_n d\theta_n) = (nN \pm 1)\lambda$

or $$N(a + b) \sin \theta_n \pm N(a + b) \cos \theta_n d\theta_n = nN\lambda \pm \lambda \tag{4}$$

Multiplying equation (1) by N on both side, we get

$$N(a + b) \sin \theta_n = nN\lambda \tag{5}$$

Substituting this value of $N(a + b) \sin \theta_n$ in equation (4), we get

$$N(a + b) \cos \theta_n d\theta_n = \lambda$$

or $$d\theta_n = \frac{\lambda}{N(a+b) \cos \theta_n}$$

The width of the nth order principal maximum $= 2d\theta_n = \dfrac{2\lambda}{N(a+b)\cos\theta_n}$

Thus, the angular width of principal maxima is inversely proportional to N, that is, as N increases, width of the maxima decreases.

3.10 THE FORMATION OF MULTIPLE SPECTRA WITH GRATING

We have seen that when a plane diffraction grating is illuminated normally by a monochromatic beam of light of wavelength λ the principal maxima are formed in the direction given by

$$(a + b) \sin \theta = \pm n\lambda \tag{1}$$

where $(a + b)$ is the grating element, n the order of the maxima, and λ is the wavelength of light. From the above equation we can point out that

- For a particular wavelength λ, the directions of principal maxima of different orders are different.
- For a particular order n, the light of different wavelengths will be diffracted in different directions.

The longer the wavelength, the greater is the angle of diffraction θ. Hence if white light is incident normally on a diffraction grating, each order of spectrum will contain principal maxima of different wavelengths in different directions.

Equation (1) shows that, for $n = 0$, $\theta = 0$ for all values of λ. Thus zero order principal maximum for all values of λ lie in the same direction. Hence the zero order ($n = 0$) principal maximum will be white. The first order ($n = 1$) principal maxima of all visible wavelengths form the first order spectrum and the second order principal maxima of all wavelengths form the second order spectrum and so on. As $\lambda_r > \lambda_v$, the angle of diffraction for red is greater than that for violet and hence in every spectrum (as shown in Figure 13), violet being in the innermost position and red in the outermost position except the central maximum. Thus the spectrum consists of the white principal maximum of zero order having on either side of it the first order spectra, the second order spectra, and so on. Most of the light is concentrated in the zero order principal maximum and the intensity goes on decreasing gradually as we move to the higher order spectrum.

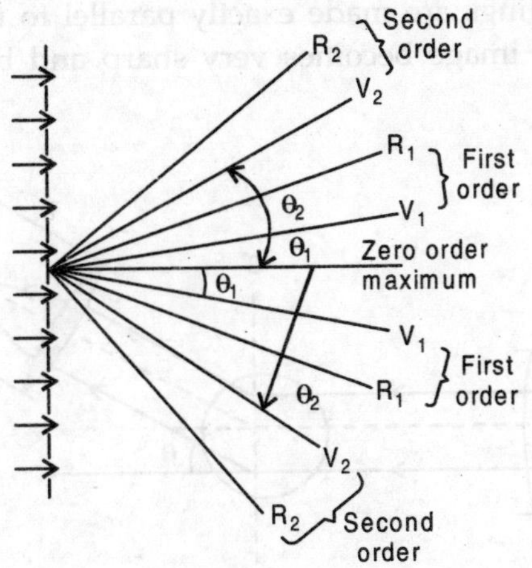

FIGURE 13

Determination of the Wavelength of Light with a Plane Transmission Diffraction Grating

The plane transmission diffraction grating is the most important tool for the accurate determination of the wavelength of light. It is evident from the grating equation

$$(a + b) \sin \theta = n\lambda \tag{2}$$

that if the angle of diffraction θ for a particular order and for a particular value of λ is measured accurately the wavelength λ can be determined with a great accuracy.

The grating element $(a + b)$ is measured directly from the number of rulings per inch written on the grating by the manufacturer. If this number is N′, then

$$N'(a + b) = 1 \text{ inch} = 2.54 \text{ cm or } (a + b) = \frac{2.54}{N'} \text{ cm}$$

For the accurate measurement of the angle of diffraction θ by this method an ordinary spectrometer is employed. Figure 14 shows the basic elements of a grating spectrometer. The slit of the spectrometer is illuminated by the given source of sodium light and the telescope and the collimator are adjusted for parallel rays by Schuster's method. For the adjustment of the grating surface normal to the incident light, the collimator and the telescope are placed in a line so that the direct image of the slit falls on the vertical cross-wire of the eye-piece. This position of the telescope is noted and then the telescope is turned through 90° and clamped. In this situation the axis of the telescope is perpendicular to the collimator axis. Now the grating is mounted vertically on the prism table such that its ruled surface lies over the center of the table. The prism table is rotated slowly until the image of the slit obtained by reflection from the grating surface is formed on the vertical cross-wire of the eye-piece of the telescope. In this situation the grating surface is inclined at 45° to the incident light. The prism table is now rotated 45° or 135° from this position so that the ruled surface of the grating becomes exactly normal to the incident light. In this position the grating surface is normal to the incident light. The prism table is now clamped and the telescope is unclamped and rotated until the positions of the diffracted images, on either side of the central direct image, are observed in the field of view of the telescope. Now rulings are made exactly parallel to the slit by tilting the slit in its own plane until the diffracted image becomes very sharp and bright.

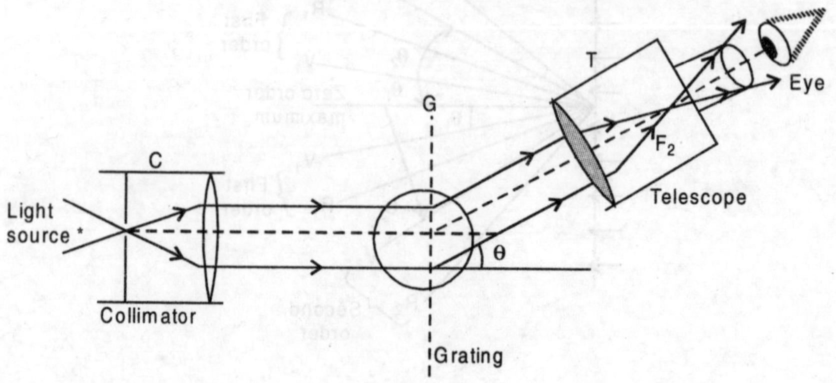

FIGURE 14

Measurement of θ

Now the sodium light is replaced by a polychromatic source of white light so that a sharp and bright grating spectrum is obtained in the field of view of the telescope as shown in Figure 13.

To measure θ, the telescope is rotated to the left of the direct image and the vertical cross-wire is adjusted to a spectral line of a particular color for the first order and its position is noted on both the verniers. The telescope is then turned to the right of the direct image and the positions of both the circular scales for the same line in the first order are noted. The difference between the two readings of the same vernier gives 2θ for that particular spectral line. The same procedure is followed for the second order spectrum. Thus, knowing the grating element $(a + b)$, angle of diffraction θ and order of the spectrum n, the wavelength λ of any spectral line can be determined by equation (2).

The Dispersive Power of a Grating

The dispersive power of a diffraction grating is defined as the rate of change of the angle of diffraction with the change in the wavelength of light used. Thus, if the wavelength changes from λ to $(\lambda + d\lambda)$ and the corresponding angle of diffraction changes from θ to $(\theta + d\theta)$, then the ratio $d\theta/d\lambda$ is called the dispersive power of the grating.

For the plane diffraction grating, we have the grating equation for normal incidence as

$$(a + b)\sin\theta = n\lambda \quad (3)$$

where $(a + b)$ is the grating element and θ the angle of diffraction for n^{th} order spectrum.

Differentiating equation (3) with respect to λ, we get $(a + b)\cos\theta \dfrac{d\theta}{d\lambda} = n$

Therefore, the dispersive power, $$\frac{d\theta}{d\lambda} = \frac{n}{(a+b)\cos\theta} \quad (4)$$

or $$\frac{d\theta}{d\lambda} = \frac{n}{(a+b)(1-\sin^2\theta)^{1/2}} \quad (5)$$

From equation (3), we have, $\sin\theta = \dfrac{n\lambda}{(a+b)}$

Therefore, equation (5) can be written as

$$\frac{d\theta}{d\lambda} = \frac{n}{(a+b)\sqrt{1-\left(\dfrac{n\lambda}{a+b}\right)^2}} \quad \text{or} \quad \frac{d\theta}{d\lambda} = \frac{1}{\sqrt{\left(\dfrac{a+b}{n}\right)^2 - \lambda^2}}$$

In the above equation $d\theta$ is the angular separation between two spectral lines with the difference in wavelengths as $d\lambda$.

The above equation indicates that

- The dispersive power is directly proportional to the order of the spectrum n, that is, the higher the order, the greater is the dispersive power.
- The dispersive power is inversely proportional to the grating element, that is, the dispersive power of two given lines is greater for a grating having larger number of lines per cm.
- The dispersive power is inversely proportional to cos θ, that is, the larger the value of θ, the higher is the dispersive power.

If θ is small, $\cos\theta \simeq 1$ and the influence of the factor cos θ on the dispersive power may be taken as negligible, then ($d\theta \propto d\lambda$) the angular dispersion of two spectral lines in a particular order is directly proportional to the difference in wavelength between the two spectral lines. Such a spectrum is called normal spectrum.

Example 4. *A parallel beam of monochromatic light is allowed to be incident normally on a plane grating having 1250 lines per cm and a second order spectral line is observed to be deviated through 30°. Calculate the wavelength of the spectral line.*

Solution. The grating equation for normal incidence is given by

$$(a + b)\sin\theta = n\lambda \quad (1)$$

where $(a + b)$ is the grating element, θ is the direction of the principal maxima and n is the order of maxima.

In the given problem, there are 1250 lines per cm, therefore the grating element,

$$(a+b) = \frac{1}{1250}\text{ cm}$$

$$\theta = 30° \text{ and } n = 2$$

From equation (1), we have

$$\lambda = \frac{(a+b)\sin\theta}{n} = \frac{(1/1250)\times\sin 30°}{2}$$

$$= \frac{1\times1}{125\times2\times2} = 2\times10^{-4}\text{ cm} = \mathbf{20000\ Å.}$$

Example 5. *In a plane transmission grating the angle of diffraction for the second order principal maxima for* $\lambda = 5\times10^{-5}$ *cm is 30°. Calculate the number of lines in 1 cm of the grating surface.*

Solution. The grating equation for normal incidence is given by

$$(a+b)\sin\theta = n\lambda$$

where $(a+b)$ is the grating element, n is the order of maxima and θ the angle of diffraction

$$\therefore \qquad (a+b) = \frac{n\lambda}{\sin\theta}$$

The number of lines in 1 cm = $\frac{1}{(a+b)}$ $\therefore$ $\frac{1}{(a+b)} = \frac{\sin\theta}{n\lambda}$

Here, $n = 2$, $\theta = 30°$ and $\lambda = 5\times10^{-5}$ cm

$$\therefore \qquad \frac{1}{(a+b)} = \frac{\sin 30°}{2\times5\times10^{-5}} = \frac{10^5}{2\times2\times5} = \mathbf{5000}$$

Therefore, the number of lines in 1 cm of the grating surface is 5000.

Example 6. *Calculate the angle between the central image of a lamp filament and its first diffracted image produced by a fabric with 160 threads per cm* ($\lambda = 6\times10^{-5}$ *cm).*

Solution. The grating equation is

$$(a+b)\sin\theta = n\lambda$$

$\therefore \sin\theta = \frac{n\lambda}{(a+b)}$. Here $(a+b) = \frac{1}{160}$ cm, $n = 1$ and $\lambda = 6\times10^{-5}$ cm

$$\therefore \qquad \sin\theta = \frac{1\times6\times10^{-5}}{(1/160)} = 160\times6\times10^{-5} = 0.0096$$

Hence $\qquad \theta = \sin^{-1}(0.0096) = \mathbf{33\ min.}$

Example 7. *Light of wavelength* 5×10^{-5} *cm is incident normally on a plane transmission grating of width 3 cm and 15000 lines. Find the angle of diffraction in first order.*

Solution. At normal incidence, the direction (θ) of principal maxima are given by

$$(a+b)\sin\theta = n\lambda$$

$$\sin\theta = \frac{n\lambda}{(a+b)}$$

where $(a + b)$ is the grating element and n is the order of maxima.

$$\text{Here } (a+b) = \frac{\text{Width of grating}}{\text{Total number of lines on the grating}} = \frac{3}{15000} \text{ cm}$$

$$n = 1 \text{ and } \lambda = 5 \times 10^{-5} \text{ cm}$$

$$\therefore \quad \sin\theta = \frac{1 \times 5 \times 10^{-5}}{(3/15000)} = \frac{1 \times 5 \times 10^{-5} \times 15000}{3} = 0.25$$

or
$$\theta = \sin^{-1}(0.25) = \mathbf{14° \, 29'}.$$

Example 8. *For a grating with a grating element (a + b) = 18000 Å, calculate the dispersive powers in the first and third order spectra around λ = 5000 Å, assuming normal incidence.*

Solution. The dispersive power or the angular dispersion produced by a grating around a mean wavelength λ is given by

$$\frac{d\theta}{d\lambda} = \frac{n\lambda}{(a+b)\cos\theta} \quad \text{or} \quad \frac{d\theta}{d\lambda} = \frac{1}{\sqrt{\left(\frac{(a+b)}{n}\right)^2 - \lambda^2}}$$

where $(a + b)$ is the grating element and n is the order of spectra.

Here, for the first order, that is, for $n = 1$, $\lambda = 5000$ Å and $(a + b) = 18000$ Å

$$\therefore \quad \frac{d\theta}{d\lambda} = \frac{1}{\sqrt{(18000)^2 - (5000)^2}} = \frac{1}{10^3\sqrt{\left[(18)^2 - (5)^2\right]}}$$

$$= \frac{1}{10^3\sqrt{299}} = 5.78 \times 10^{-5} \text{ radian/Å}$$

Similarly for the third order, that is, for $n = 3$

$$\frac{d\theta}{d\lambda} = \frac{1}{\sqrt{\left(\frac{18000}{3}\right)^2 - (5000)^2}}$$

$$= \frac{1}{\sqrt{(6000)^2 - (5000)^2}} = \frac{1}{10^3\sqrt{11}}$$

$$= \mathbf{3.0 \times 10^{-4} \text{ radian/Å}}.$$

Example 9. *A plane transmission grating produces an angular separation of 0.01 radian between two wavelengths observed at an angle of 30°. If the mean value of the wavelength is 5000 Å and the spectrum is observed in the second order, calculate the difference in two wavelengths.*

Solution. The grating equation is $(a + b)\sin\theta = n\lambda$ (1)

Differentiating it with respect to λ, we get $(a+b)\cos\theta \dfrac{d\theta}{d\lambda} = n$

Dividing it by equation (1), we get, $\cot\theta \dfrac{d\theta}{d\lambda} = \dfrac{1}{\lambda}$ or $d\lambda = \lambda \cot\theta \, d\theta$

In the given problem, $\theta = 30°$, $d\theta = 0.01$ rad. and $\lambda = 5000$ Å $= 5000 \times 10^{-8}$ cm

Therefore, $$d\lambda = 5000 \times 10^{-8} \times \cot 30° \times 0.01$$

$$= 5000 \times 10^{-8} \times \sqrt{3} \times 0.01 = \mathbf{86.6\ Å.}$$

Example 10. *A diffraction grating having 15000 lines to an inch is used to photograph a spectrum. Calculate the angular dispersion in the second order spectrum of wavelength region 5.9 × 10⁻⁵ cm. If the focal length of the convex lens is 25 cm, calculate the linear dispersion in the spectrograph and also the separation between the spectral lines 5890 Å and 5896 Å in the second order.*

Solution. The grating equation is $(a + b) \sin \theta = n\lambda$

or $$\sin \theta = \frac{n\lambda}{(a+b)}$$

Here $$(a + b) = \frac{2.54}{15000} \text{ cm} = 17 \times 10^{-5} \text{ cm}, n = 2 \text{ and } \lambda = 5.9 \times 10^{-5} \text{ cm}$$

∴ $$\sin \theta = \frac{n\lambda}{(a+b)} = \frac{2 \times 5.9 \times 10^{-5}}{(2.54/15000)} = 0.69$$

Hence $$\cos \theta = \sqrt{(1 - \sin^2 \theta)} = \sqrt{\left[1 - (0.69)^2\right]} = 0.72$$

The angular dispersion produced by the grating is given by

$$\frac{d\theta}{d\lambda} = \frac{n}{(a+b)} = \frac{2}{17 \times 10^{-5} \times 0.72} = 1.6 \times 10^4 \text{ rad/cm}$$

or $$= 1.6 \times 10^{-4} \text{ rad/Å}$$

If f is the focal length of the camera lens, then the linear dispersion is given by

$$\frac{dx}{d\lambda} = f\frac{d\theta}{d\lambda}$$

∴ $$\frac{dx}{d\lambda} = 25 \times 1.6 \times 10^{-4} \text{ rad/Å}$$

$$= 4 \times 10^{-3} \text{ cm/Å} \qquad [f = 25 \text{ cm (given)}]$$

The separation between the spectral lines 5896 Å and 5890 Å is ($d\lambda$ = 5896 – 5890 = 6 Å)

∴ $$dx = 4 \times 10^{-3} \times d\lambda = 4 \times 10^{-3} \times 6 = 2.4 \times 10^{-2} \text{ cm.}$$

Example 11. *Light which is a mixture of two wavelengths 5000 Å and 5200 Å is incident normally on a plane transmission grating having 10000 lines per cm. A lens of focal length 150 cm is used to observe the spectrum on a screen. Calculate the separation in cm of the two lines in the first order spectrum.*

Solution. The grating equation is $(a + b) \sin \theta = n\lambda$

or $$\sin \theta = \frac{n\lambda}{(a+b)}$$

For the first wavelength $$\lambda_1 = 5000 \text{ Å} = 5000 \times 10^{-8} \text{ cm}$$

$$(a + b) = \frac{1}{10000} = 10^{-4} \text{ cm}$$

$$\sin \theta_1 = \frac{n\lambda_1}{(a+b)} = \frac{1 \times 5.0 \times 10^{-5}}{10^{-4}} = 0.5$$

$$\therefore \quad \theta_1 = \sin^{-1} (0.5) = 30°$$

Similarly for wavelength $\lambda_2 = 5200$ Å $= 5.2 \times 10^{-5}$ cm

$$\sin \theta_2 = \frac{n\lambda_2}{(a+b)} = \frac{1 \times 5.2 \times 10^{-5}}{10^{-4}} = 0.52$$

$$\therefore \quad \theta_2 = \sin^{-1} (0.52) = 31.3°$$

Further $\quad \tan \theta_1 = \frac{x_1}{f}$ and $\tan \theta_2 = \frac{x_2}{f}$

Here $\quad \theta_1 = 30°, \theta_2 = 31.3°$ and $f = 150$ cm

$$\therefore \quad x_2 - x_1 = f(\tan \theta_2 - \tan \theta_1) = 150\ (0.6080 - 0.5774)$$

$$= 150 \times 0.0306 = \mathbf{4.59\ cm.}$$

3.11 THE DIFFERENCE BETWEEN PRISM AND GRATING SPECTRA

The following are the points of difference between the spectra obtained with a grating and a prism:

- A prism produces only one spectrum, whereas the grating gives a number of spectra on either side of the central maximum.
- The prism spectrum is produced by the dispersion of light; whereas the grating spectrum is produced by the diffraction of light.
- The prism spectrum is brighter than the grating spectra.
- In the case of a prism, the deviation is the least for red and the greatest for violet; while in case of *a*, grating, the deviation is the least for violet and the greatest for the red. Hence in prism spectrum, the spectral colors are in the order from red to violet; whereas in grating spectra, the spectral colors are in the order from violet to red.
- The prism spectrum depends on the material of the prism; whereas the grating spectra are independent of the material of the grating.
- The dispersive power of a grating is given by $\frac{d\theta}{d\lambda} = \frac{n}{(a+b)\cos\theta}$, which is constant for a particular order; whereas the dispersive power of a prism is given by $\frac{d\mu}{(\mu - 1)}$.
- The resolving power of a grating (nN) is very large; whereas the resolving power of a prism $\left(t.\frac{d\mu}{d\lambda}\right)$ is quite small.
- **In the grating spectra the spectral lines are almost straight; whereas in the prism spectrum the spectral lines are slightly curved.**

3.12 THE DIFFERENCE BETWEEN INTERFERENCE AND DIFFRACTION

The following are the points of difference between the interference and diffraction phenomena:

- In the phenomenon of interference, the interference occurs between the two separate wavefronts emanating from two coherent sources; whereas in the phenomenon of diffraction, the interference occurs between the innumerable secondary wavelets produced by the unobstructed portion of the same wavefront.
- In an interference pattern the regions of minimum intensities are usually almost perfectly dark; whereas it is not so in the diffraction pattern.
- The interference fringes are usually equally spaced; whereas the diffraction fringes are never equally spaced.
- In an interference pattern all the bright fringes are of equal intensity; while in the diffraction pattern the intensity of central maximum is maximum and goes on decreasing as the order of maxima increases on either side of the central maximum.

3.13 THE RESOLVING POWER OF AN OPTICAL INSTRUMENT

When two objects are very close together they may appear as one object and it may be difficult for the naked eye to see them as separate. Similarly, if there are two point sources very close together, the two diffraction patterns produced by each of them may overlap and hence it may be difficult to distinguish them as separate. To see the two objects or two spectral lines which are very close together, optical instruments like telescopes, microscopes, prisms, grating etc., are employed. The capacity of an optical instrument to show two close objects separately is called resolution, and the ability of an optical instrument to resolve the images of two close point objects is called its "resolving power". The eye can see two objects as separate only if the angle subtended by them at the eye is greater than one minute, which is the resolving limit of the normal eye.

3.14 THE RAYLEIGH CRITERION FOR THE LIMIT OF RESOLUTION

To get a definite measure of the resolving power, Lord Rayleigh, on the basis of a detailed study of the resultant intensity distribution in the diffraction pattern of closely spaced point sources, laid down the following criterion for resolution.

According to Rayleigh, the two point sources or two equally intense spectral lines, are just resolved by an optical instrument when the central maximum of the diffraction pattern, due to one source, falls exactly on the first minimum of the diffraction pattern of the other and vice-versa. This criterion is applicable for the evaluation of resolving powers of a telescope, microscope, grating, prism, etc.

To illustrate the criterion, consider the intensity distribution curves of two wavelength λ and $\lambda + d\lambda$. The separation between their central maximum will depend upon the wavelength difference $d\lambda$. As depicted in Figure 15, $d\lambda$ is sufficiently large so that the central maximum due to both wavelengths are quite separated and the two spectral lines are distinctly resolved. If, however, the difference in the wavelengths is smaller and has a limiting value for which the angular separation between their principal maxima is such that the principal central maximum of one coincides with the first minimum of the other and vice-versa (as in Figure 16). In this case, the curve shows a distinct dip in the middle, indicating the presence of two spectral lines corresponding to wavelengths λ and $\lambda + d\lambda$. The two spectral lines under this condition are said to be just resolved. If the wavelength difference ($d\lambda$) is smaller than the limiting value, the two principal maxima show considerable overlapping. The resultant intensity curve shown in Figure 17 indicates no dip and appears as if it is a diffraction pattern of only one spectral line. Hence under this condition, the two spectral lines are not resolved.

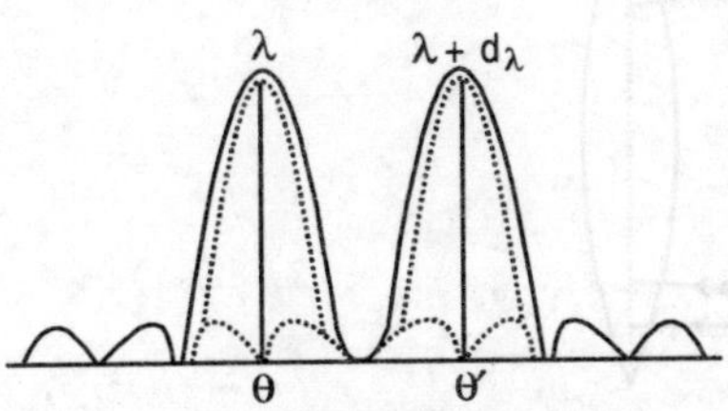

FIGURE 15

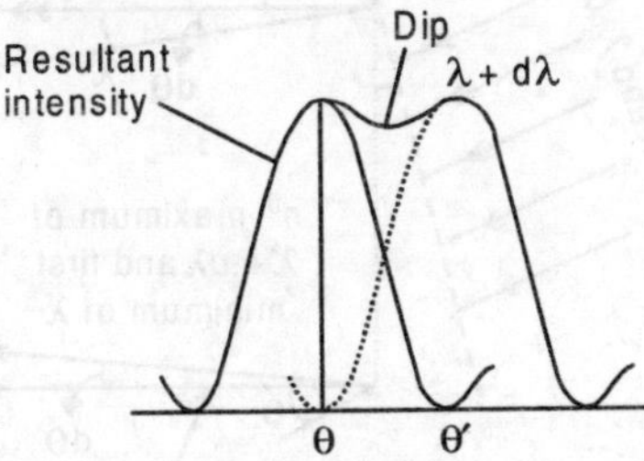

FIGURE 16

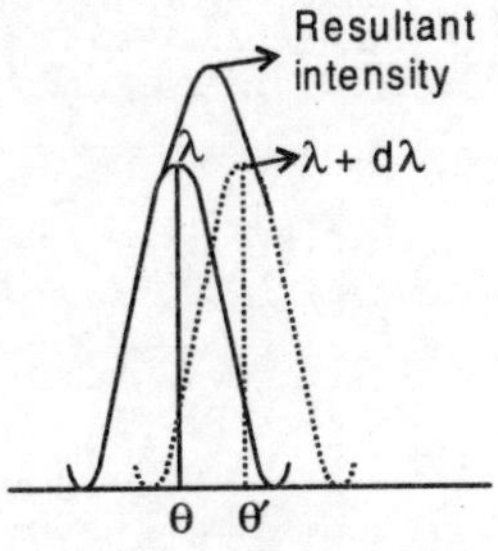

FIGURE 17

The distribution of intensity in the grating or prism spectra is of the form

$$I = I_{max} \frac{\sin^2 x}{x^2} \quad (1)$$

For the first minimum, I is minimum if $x = \pi$. At the mid point of the two maxima the intensity due to each wavelength is obtained by substituting $x = \frac{\pi}{2}$ in equation (1) (Figure 16). Thus the total intensity at the mid point of the maxima is given by

$$I_{min} = 2I_{max} \frac{\sin^2(\pi/2)}{(\pi/2)^2} = \frac{8}{\pi^2} I_{max}$$

$$\therefore \quad \frac{I_{min}}{I_{max}} = \frac{8}{\pi^2}$$

Thus the Rayleigh criterion may also be stated such that the two spectral lines are just resolved if the intensity at the dip in the middle is $\frac{8}{\pi^2}$ times the intensity at either of the maxima.

The Resolving Power of a Diffraction Grating

The resolving power of a grating is defined as its ability to show the two neighboring spectral lines in a spectrum as separate. It can also be defined as the ratio of the wavelength of any spectral line of the smallest wavelength difference between neighboring lines for which the spectral lines can be just resolved at the wavelength λ. It can be expressed mathematically as $\lambda/d\lambda$. We shall apply Rayleigh's criterion for resolution to derive an expression for the resolving power of the plane diffraction grating. Let a plane diffraction grating be illuminated normally by a parallel beam of light of two nearby wavelengths λ and $\lambda + d\lambda$ as shown in Figure 18. The light of each wavelength would form a separate diffraction pattern of the slit. Let n^{th} principal maximum of wavelength λ be formed in the direction θ, then we have

$$(a + b) \sin\theta = n\lambda \quad \text{or} \quad N(a + b) \sin\theta = Nn\lambda \tag{2}$$

where N is the total number of lines on the grating surface and $(a + b)$ is the grating element.

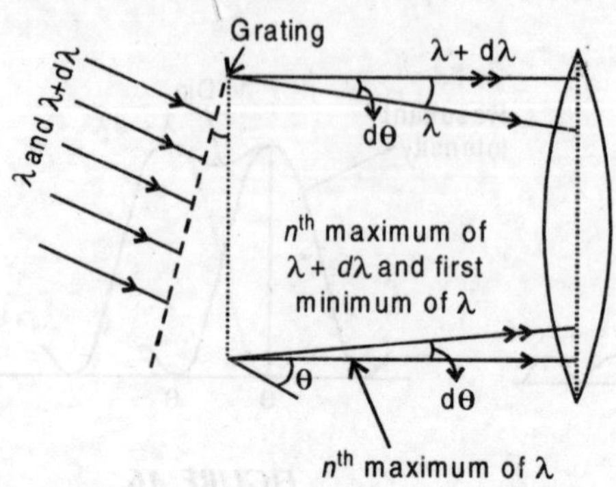

FIGURE 18

Let the first minimum adjacent to the nth maximum be formed in the direction $(\theta + d\theta)$. Then from the grating equation for the directions of minima, we have

$$N(a + b) \sin(\theta + d\theta) = (Nn + 1)\lambda \tag{3}$$

According to Rayleigh the two spectral lines of wavelength λ and $\lambda + d\lambda$ will appear just resolved when n^{th} maximum of wavelength $\lambda + d\lambda$ falls over the first minimum of wavelength λ adjacent to its nth maximum. Hence for just resolution of spectral lines n^{th} maximum of $\lambda + d\lambda$ and first minimum of wavelength λ adjacent to its nth maximum should be formed in the direction $\theta + d\theta$. Hence for n^{th} maximum of wavelength $\lambda + d\lambda$, we have

$$(a + b) \sin(\theta + d\theta) = n(\lambda + d\lambda)$$

or

$$N(a + b) \sin(\theta + d\theta) = Nn(\lambda + d\lambda) \tag{4}$$

Equating equations (3) and (4), we get

$$(Nn + 1)\lambda = Nn(\lambda + d\lambda) \text{ or } \lambda = Nnd\lambda$$

$\therefore$ Resolving power, $$\frac{\lambda}{d\lambda} = Nn$$

Thus, the resolving power of a grating is equal to the product of the order of the spectrum and the total number of lines on the grating

From equation (2), $$n = \frac{(a+b)\sin\theta}{\lambda}$$

Therefore, the resolving power of a grating $= nN = \dfrac{N\,(a+b)\sin\theta}{\lambda}$.

3.15 THE DIFFERENCE BETWEEN THE DISPERSIVE POWER AND THE RESOLVING POWER OF A GRATING

The following are the points of difference between the dispersive power and the resolving power of a grating:

- The dispersive power of a grating is defined as the rate of change of angle diffraction with the wavelength of light used ($d\theta/d\lambda$); whereas the resolving power of a grating is defined as the ratio of the wavelength of any spectral line to the smallest wavelength difference between neighboring lines for which the spectral lines can be just resolved ($\lambda/d\lambda$).
- The dispersive power is given by, $\dfrac{d\theta}{d\lambda} = \dfrac{n}{(a+b)\cos\theta}$; whereas the resolving power of a grating is given by $\lambda/d\lambda = n\text{N}$.
- The resolving power increases as the number of lines on the grating surface increases; whereas the dispersive power is independent of the number of lines on the grating surface.
- The dispersive power depends upon the grating element; while the resolving power is independent of the grating element.
- The dispersive power gives an idea of the angular separation of the two spectral lines; whereas the resolving power tells us about the closeness of the two spectral lines.
- The dispersive power depends upon the number of lines per centimeter on the grating; whereas the resolving power depends on the total number of lines on the ruled surface.

Example 12. *A plane transmission diffraction grating has 40,000 lines. Determine its resolving power in the second order (n = 2) for a wavelength of 5000 Å.*

Solution. The resolving power of a grating is given by $\lambda/d\lambda = n\text{N}$

Here $n = 2$ and $\text{N} = 40{,}000$

$$\therefore \quad \frac{\lambda}{d\lambda} = 2 \times 40{,}000 = \mathbf{80{,}000.}$$

Example 13. *Find the minimum number of lines in a plane diffraction grating required to just resolve the sodium doublet (5890 Å and 5896 Å) in the (i) first order, (ii) second order.*

Solution. The resolving power of a grating is given by

$$\frac{\lambda}{d\lambda} = n\text{N}$$

where n is the order and N is the total number of lines on the grating

$$\therefore \quad \text{N} = \frac{1}{n}\left(\frac{\lambda}{d\lambda}\right)$$

(*i*) For the first order, $n = 1$

Mean wavelength $\lambda = \dfrac{\lambda_1 + \lambda_2}{2} = \dfrac{5890 + 5896}{2} = 5893$ Å

and $d\lambda = 5896 - 5890 = 6$ Å

$$\therefore \qquad N = \frac{5893}{1 \times 6} = \mathbf{982}$$

(*ii*) For the second order, $n = 2$, the minimum number of lines,

$$N = \frac{5893}{2 \times 6} = \mathbf{491}$$

Example 14. *A grating has 1000 lines ruled on it. What is the difference between the two wavelengths that just resolved in the first order spectrum in the region of wavelength $\lambda = 6 \times 10^{-5}$ cm?*

Solution. Let λ and $\lambda + d\lambda$ be the two wavelengths for the grating in the first order spectrum. The resolving power of a grating is given by

$$\frac{\lambda}{d\lambda} = nN \quad \text{or} \quad d\lambda = \frac{\lambda}{nN}$$

In the given problem, $N = 1000$, $n = 1$ and $\lambda = 6 \times 10^{-5}$ cm

$$\therefore \qquad d\lambda = \frac{6 \times 10^{-5}}{1 \times 1000} = 6 \times 10^{-8} \text{ cm} = \mathbf{6\ \text{Å}}.$$

Example 15. *Show that for a transmission grating with one inch ruled space, the resolving power can not exceed 5×10^4 at normal incidence for $\lambda = 5080$ Å.*

Solution. The resolving power of a grating is given by $\dfrac{\lambda}{d\lambda} = nN$

For normal incidence, the grating equation is $(a + b) \sin \theta = n\lambda$

or
$$n = \frac{(a+b)\sin\theta}{\lambda}$$

Therefore, in terms of the angle of diffraction, the resolving power is

$$\frac{\lambda}{d\lambda} = \frac{(a+b)\sin\theta . N}{\lambda}$$

The resolving power will be maximum if $\theta = 90°$ or $\sin \theta = 1$

Thus
$$\left(\frac{\lambda}{d\lambda}\right)_{max} = \frac{N(a+b)}{\lambda}$$

where $N(a + b)$ is the total ruled width of the grating. In the given problem, $N(a + b) = 1$ inch $= 2.54$ cm and $\lambda = 5080$ Å $= 5080 \times 10^{-8}$ cm

$$\therefore \qquad \left(\frac{\lambda}{d\lambda}\right)_{max} = \frac{2.54}{5080 \times 10^{-8}} = \mathbf{5 \times 10^4}.$$

Example 16. *A diffraction grating is just able to resolve two lines of wavelengths 5140.34 Å and 5140.85 Å in the first order. Will it resolve the lines 8037.20 Å and 8037.50 Å in the second order?*

Solution. The resolving power of a grating is given by

$$\frac{\lambda}{d\lambda} = nN$$

Therefore
$$N = \frac{1}{n}\left(\frac{\lambda}{d\lambda}\right)$$

where $d\lambda$ is the smallest wavelength difference at the mean wavelength λ, N is the number of the lines on the grating and n is the order of the spectrum.

In the given problem, $\lambda = \dfrac{5140.34 + 5140.85}{2} = 5140.595$ Å

$$d\lambda = 5140.85 - 5140.34 = 0.51 \text{ Å} \quad \text{and} \quad n = 1$$

$$\therefore \quad N = \frac{1}{1}\left(\frac{5140.595}{0.51}\right) = 10080$$

Hence the resolving power ($\lambda/d\lambda$) of a grating in the second order should be = nN = 2 × 10080 = 20160. The resolving power required to resolve the lines 8037.20 Å and 8037.50 Å in the second order is $\lambda/d\lambda$.

In this case $\lambda = \dfrac{8037.20 + 8037.50}{2} = 8037.35$ Å

and $d\lambda = 8037.50 - 8037.20 = 0.30$ Å

$$\therefore \quad \text{Resolving power} = \frac{8037.35}{0.30} = 26791.17$$

Thus, the grating will not be able to resolve the lines 8037.20 Å and 8037.50 Å in the second order because the required resolving power (26791.17) is greater than the actual resolving power (20160).

Example 17. *In the second order spectrum of a plane diffraction grating, a certain spectral line appears at an angle of 10°, while another line of wavelength 5 × 10^{-9} cm greater appears at angle 3" greater. Find the wavelength of the lines and the minimum grating width required to resolve them.*

(Given: sin 10° = 0.1736 and cos 10° = 0.9848).

Solution. The grating equation is $(a + b) \sin\theta = n\lambda$ (1)

Differentiating it with respect to λ, we get $(a + b)\cos\theta \, d\theta = n d\lambda$

Dividing it by equation (1), we get $\dfrac{\cos\theta}{\sin\theta} d\theta = \dfrac{d\lambda}{\lambda}$ or $\lambda = \dfrac{\sin\theta \, d\lambda}{\cos\theta \, d\theta}$

In the given problem, $\sin\theta = \sin 10° = 0.1736$, $\cos\theta = \cos 10° = 0.9848$

and $d\lambda = 5 \times 10^{-9}$ cm $= 0.5 \times 10^{-8}$ cm

and $d\theta = 3'' = \left(\dfrac{3}{3600}\right)^{\circ} = \dfrac{3}{3600}\left(\dfrac{\pi}{180}\right)$ in radian.

$$\therefore \quad \lambda = \frac{0.1736 \times 0.5 \times 10^{-8}}{0.9848 \times (3\pi/3600 \times 180)}$$

$$= \frac{0.1736 \times 0.5 \times 10^{-8} \times 3600 \times 180}{0.9848 \times 3\pi}$$

or $\lambda = 6063 \times 10^{-8}$ cm or 6063 Å

Thus the wavelength of lines are λ and $\lambda + d\lambda$, that is,

6063×10^{-8} and $6063 \times 10^{-8} + 0.5 \times 10^{-8} = 6063.5$ Å

The resolving power of the grating required to resolve the lines in the second order is given by

$$\frac{\lambda}{d\lambda} = N.n \quad \text{or} \quad \frac{6063 \times 10^{-8}}{0.5 \times 10^{-8}} = N \times 2$$

or
$$N = \frac{6063 \times 10^{-8}}{2 \times 0.5 \times 10^{-8}} = 6063$$

Hence the minimum grating width required

$$= N(a + b) = N\frac{n\lambda}{\sin\theta} \qquad \text{[From equation (1)]}$$

Here
$$n = 2$$

$$\therefore \qquad \text{Grating width} = \frac{6063 \times 2 \times 6063 \times 10^{-8}}{0.1736} = \mathbf{4.2\ cm.}$$

Example 18. *A plane transmission grating has 16,000 lines to an inch over a length of 5 inches. Find in the wavelength region of 6000 Å, in the second order (a) the resolving power of the grating and (b) the smallest wavelength difference that can be resolved.*

Solution. The smallest resolvable wavelength difference $d\lambda$ in given by $\frac{\lambda}{d\lambda} = nN$ or $d\lambda = \frac{\lambda}{n.N}$

Number of lines per inch, $\frac{1}{(a+b)} = 16000$, length of the ruled grating = 5 inches

Therefore, the total number of lines on the grating is,

$$N = 5 \times 16000 = 80000$$

In the given problem, $\lambda = 6000$ Å $= 6000 \times 10^{-8}$ cm, $n = 2$

$$\therefore \qquad d\lambda = \frac{6000 \times 10^{-8}}{2 \times 80000} = 375 \times 10^{-4} = \mathbf{0.0375\ Å.}$$

3.16 THE RESOLVING POWER OF A PRISM

The resolving power of a prism is defined as its ability to just resolve the spectral lines of two wavelengths very close together. If λ and $(\lambda + d\lambda)$ are the wavelength of two close lines which are just resolved, then $(\lambda / d\lambda)$ measures the resolving power of the prism. Figure 19 illustrates essential parts of a prism spectrograph. Let a prism of transparent material be illuminated by light consisting of wavelength λ and $\lambda + d\lambda$. After refraction by the prism, the light is brought to the focal plane of a telescope objective L, where it is viewed by the telescope eye-piece. Let the prism ABC be placed in the position of minimum deviation and BP be the incident plane wavefront. CQ and CQ' are the emergent wavefronts corresponding to wavelengths λ and $\lambda + d\lambda$ and P_1 and P_2 are the positions of the central principal maxima of diffraction patterns corresponding to the two spectral lines of wavelengths λ and $\lambda + d\lambda$ respectively.

According to Fermat's principle, the optical path between the incident and the emergent wavefronts for any particular wavelength must be equal. Hence, for the wavelength λ, we have

$$PA + AQ = \mu.BC \qquad (1)$$

and for the wavelength $\lambda + d\lambda$, we have

$$PA + AQ' = (\mu - d\mu)BC \tag{2}$$

where μ and $(\mu - d\mu)$ are the refractive indices for the material of a prism for wavelength λ and $\lambda + d\lambda$

Subtracting equation (2) from equation (1), we get

$$AQ - AQ' = d\mu.BC = td\mu \tag{3}$$

where t (=BC) is the width of the base of the prism. If θ and $\theta + d\theta$ are the angles of deviation for wavelength λ and $\lambda + d\lambda$ respectively then the angle between the refracted wavefronts CQ and CQ' is $d\theta$. Hence from the geometry of Figure 19

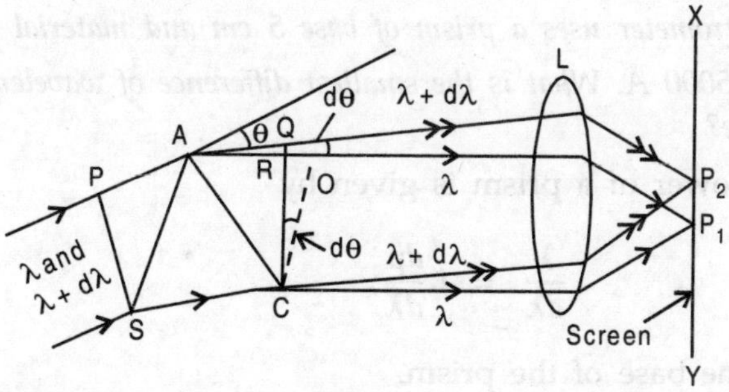

FIGURE 19

$$AQ' = AR$$

$\therefore$ $$AQ - AQ' = AQ - AR = RQ \tag{4}$$

Equating equations (3) and (4), we get

$$RQ = td\mu$$

But $$RQ = CQ.d\theta = ad\theta, \text{ where } CQ = a$$

$\therefore$ $$td\mu = ad\theta$$

or $$d\theta = (t/a)\, d\mu \tag{5}$$

The face AC of the prism behaves like a rectangular aperture. We can consider the rays of light corresponding to wavelengths λ and $\lambda + d\lambda$ to be diffracted in all directions for every point of the face AC. According to Rayleigh's criterion for just resolution, the two spectral lines are just resolved when the central maximum of one falls over the first minimum of the other and vice-versa. Hence the Rayleigh criterion can be applied to the diffraction pattern due to the rectangular aperture. The principal maximum of λ is formed at P_1 and that due to $\lambda + d\lambda$ is formed at P_2. Hence the angular separation between the two maxima is the angle between CQ and CQ'.

The angle between the central maximum and first minimum of λ due to rectangular aperture is given by

$$a \sin d\theta_1 = \lambda$$

or $$d\theta_1 = \lambda/a \tag{6}$$

According to the Rayleigh criterion for just resolution, the angle between the central maximum and first minimum ($d\theta_1$) must be equal to the angular separation ($d\theta$) of the two wavelengths, that is, $d\theta = d\theta_1$

From equations (5) and (6), we have

$$(t/a)\,d\mu = \lambda/a \text{ or } \lambda = t d\mu$$

Therefore, the resolving power of the prism $\frac{\lambda}{d\lambda} = t\frac{d\mu}{d\lambda}$

This expression holds good only when the prism is at a position of minimum deviation. From the above equation for resolving power, it is clear that the resolving power of a prism is directly proportional to the length of the base of the prism and also to the rate of change of the refractive index with respect to wavelength ($d\mu/d\lambda$).

Example 19. *A prism spectrometer uses a prism of base 5 cm and material whose dispersion* $\frac{d\mu}{d\lambda}$ *is 200* cm^{-1} *in the range* λ *= 5000 Å. What is the smallest difference of wavelength in this range which this spectrometer can resolve?*

Solution. The resolving power of a prism is given by

$$\frac{\lambda}{d\lambda} = t.\frac{d\mu}{d\lambda},$$

where t is the width of the base of the prism.

The smallest wavelength difference $d\lambda$ that can be resolved is $d\lambda = \lambda/t\left(\frac{d\mu}{d\lambda}\right)$

Here, λ = 5000 Å, t = 5 cm and $\frac{d\mu}{d\lambda} = 200 \text{ cm}^{-1}$

$$\therefore \quad d\lambda = \frac{5000}{5 \times 200} = \mathbf{5\ Å.}$$

Example 20. *What must be the length of the base of a flint glass prism* ($d\mu/d\lambda$ *= – 982* cm^{-1}*) necessary to resolve the sodium D lines* (D_1 *= 5896 Å and* D_2 *= 5890 Å)?*

Solution. The resolving power of a prism necessary to resolve the sodium D lines is $\lambda/d\lambda$.

Here $\lambda = \frac{5896\text{ Å} + 5890\text{ Å}}{2} = 5893$ Å

and $d\lambda = 5896\text{ Å} - 5890\text{ Å} = 6$ Å

$$\therefore \quad \frac{\lambda}{d\lambda} = \frac{5893}{6} = 982.16 \approx 982$$

In terms of the length of base t

$$\frac{\lambda}{d\lambda} = t.\frac{d\mu}{d\lambda} \text{ or } t = \frac{1}{(d\mu/d\lambda)}\frac{\lambda}{d\lambda}$$

Here $\frac{d\mu}{d\lambda} = 982 \text{ cm}^{-1}$ $\quad \therefore\ t = \frac{982}{982} = \mathbf{1.0\ cm.}$

Example 21. *For flint glass* $\mu_C = 1.6389$ *and* $\mu_F = 1.7168$ *where the wavelengths of C and F lines are 6563 Å and 4861 Å respectively. What is the smallest difference of wavelengths in this region 5.0 ×* 10^{-3} *Å resolved by a prism of flint glass of base 3.0 cm?*

Solution. The resolving power of a prism is given by

$$\frac{\lambda}{d\lambda} = t.\frac{d\mu}{d\lambda}$$

Here $\mu_C = 1.6389$, $\mu_F = 1.7168$, $\lambda_C = 6563$ Å and $\lambda_F = 4861$ Å

$$\therefore \quad \frac{d\mu}{d\lambda} = \frac{\mu_F - \mu_C}{\lambda_C - \lambda_F} = \frac{(1.7168-1.6389)}{(6563-4861)\times 10^{-8}}$$

or
$$\frac{d\mu}{d\lambda} = \frac{0.0779}{1702\times 10^{-8}} = 457.69$$

Given, $\lambda = 5.0 \times 10^{-3}$ Å $= 5.0 \times 10^{-5}$ cm and $t = 3$ cm

$$\therefore \quad d\lambda = \lambda \Big/ \left(t\frac{d\mu}{d\lambda}\right) = \frac{5.0\times 10^{-5}}{3\times 457.69} = 3.64\times 10^{-8} \text{ cm}$$

$$d\lambda = \mathbf{3.64\ Å.}$$

Example 22. *The ruled space on a grating is as wide as the base of a 60° rocksalt prism. It has the same resolving power in the second order as the prism has at 6000 Å. Calculate the size of the grating interval, given the following data. λ = 6708 Å and 6438 Å, μ = 1.5400 and 1.5412.*

Solution. The resolving power of a grating is, $\frac{d}{d\lambda} = n.\text{N}$ and that of a prism is, $\frac{\lambda}{d\lambda} = t\left(\frac{d\mu}{d\lambda}\right)$

The ruled space on the grating is as wide as the base of the prism

$$\text{N}(a + b) = t \qquad (1)$$

According to the problem two resolving powers are equal

$$\therefore \quad \frac{\lambda}{d\lambda} = n\text{N} = t\left(\frac{d\mu}{d\lambda}\right) \text{ or } n\text{N} = t\left(\frac{d\mu}{d\lambda}\right) \qquad (2)$$

Substituting the value of t in equation (2) from equation (1), we have

$$n\text{N} = \text{N}(a+b)\frac{d\mu}{d\lambda} \text{ or } (a+b) = n\Big/\left(\frac{d\mu}{d\lambda}\right)$$

In the second order $n = 2$

and
$$\frac{d\mu}{d\lambda} = \frac{1.5412-1.5400}{(6708-6438)\times 10^{-8}} = \frac{0.0012}{270\times 10^{-8}}$$

$$\therefore \quad (a + b) = 2\Big/\left(\frac{0.0012}{270\times 10^{-8}}\right) = \frac{2\times 270\times 10^{-8}}{0.0012} = \mathbf{4.5 \times 10^{-8}\ cm.}$$

3.17 THE RESOLVING POWER OF A TELESCOPE

A telescope is used to see a clear and distinct image of an object which is at a very large distance from the observer. The resolving power of a telescope is defined as its ability to form separate images of two distant point objects situated close to each other. It depends on the angle the two point objects subtend at the objective. The resolving power of the telescope can also be defined as the inverse of the smallest angle subtended at the objective by two nearby distant objects whose images are just resolved or, in other words, the reciprocal of the limit of resolution is known as the geometrical resolving power of a telescope.

Let two point objects, such as two stars sending a monochromatic light of wavelength λ be situated at a far distance from the objective of the telescope AB. As the objects are distant, the wavefronts incident on the objective are plane. The objective of the telescope acts like a circular aperture and therefore Fraunhoffer diffraction patterns are produced in the focal plane of the lens. The diffraction pattern of S consists of a central bright disc with the center at P surrounded by alternate bright and dark rings. The image of another nearby object S' will also be a diffraction pattern consisting a central maximum at P′ (as shown in Figure 20). These diffraction patterns overlap. The central maximum of the diffraction pattern due to distant point object S lies at P. As the aperture is rectangular, the first minimum adjacent to central maximum of S will be at P' in the direction given by

$$a \sin \theta = \lambda \text{ or } \sin \theta = \frac{\lambda}{a} = \theta$$

where '*a*' is the width of the aperture.

This condition holds good for a rectangular aperture. The circular aperture treatment shows that

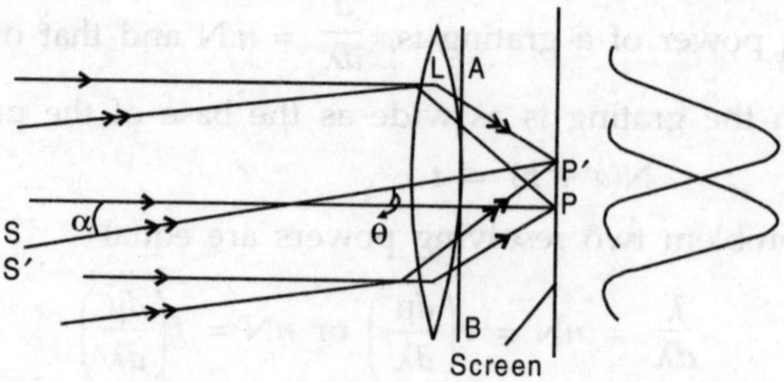

FIGURE 20

$$\theta = \frac{1.22\,\lambda}{a}$$

According to Rayleigh's criterion, the two distant object S and S′ are said to be just resolved if the position of the central maximum of the diffraction pattern of one exactly coincides with the adjacent minima of the other and vice-versa.

If α be the angle subtended at the objective by objects S and S′ for just resolution, we have $\alpha = \theta = \dfrac{1.22\,\lambda}{a}$ and the limit of resolution of the telescope,

$$\theta = \frac{1.22\,\lambda}{a}$$

Hence the resolving power of the telescope is $\dfrac{1}{\theta} = \dfrac{a}{1.22\,\lambda}$.

The Difference Between the Magnifying Power and the Resolving Power of a Telescope

The following are the points of difference between the magnifying power and the resolving power of a telescope:

- The magnifying power of a telescope represents its ability to show a magnified object to the eye; whereas the resolving power represents its ability to show the fine details in the object of the eye-piece.
- The magnifying power is measured by the ratio of the angle subtended by the final image at the eye to the angle subtended by the object in its actual position at the naked eye; whereas resolving power is measured by the angle subtended at the objective by two nearby point objects which are just resolved.
- The magnifying power is independent of the wavelength of light used; while the resolving power decreases as the wavelength increases.
- The magnifying powers of the two telescopes of the same focal lengths are equal; while the resolving powers of the two telescopes of the same focal lengths are different. The resolving power of the telescope is directly proportional to the size of the aperture.

Example 23. *The objective of a telescope has a diameter of 40 inches. Calculate the smallest angular separation of two stars, measured in seconds, that may be resolved by it. (Mean wavelength of light = 5600 Å).*

Solution. The smallest angular separation between the point stars that can be just resolved by the telescope is given by

$$\theta = \frac{1.22\,\lambda}{a},$$

where a is the diameter of the objective of the telescope.

Here $\lambda = 5600 \text{ Å} = 5600 \times 10^{-8}$ cm

and $a = 40$ inches $= 40 \times 2.54 = 101.6$ cm

$$\therefore \quad \theta = \frac{1.22\,\lambda}{a} = \frac{1.22 \times 5.6 \times 10^{-5}}{101.6} = 6.72 \times 10^{-7} \text{ rad}$$

In second, $\theta = 6.72 \times 10^{-7} \times \left(\frac{180}{\pi}\right) \times 60 \times 60 = \mathbf{0.13\ second.}$

Example 24. *What will be the diameter of a telescope objective required to resolve two stars separated by an angle of 10^{-3} degrees? Assuming $\lambda = 5000$ Å.*

Solution. The smallest angular separation θ resolvable by a telescope objective of diameter a is

$$\theta = \frac{1.22\,\lambda}{a} \text{ rad or } a = \frac{1.22\,\lambda}{\theta}$$

Here, $\lambda = 5000 \text{ Å} = 5000 \times 10^{-8}$ cm

and $\theta = 10^{-3}$ degree $= \left(\frac{\mu}{180} \times \frac{1}{1000}\right)$ rad

$$\therefore \quad a = \frac{1.22 \times 5000 \times 10^{-8}}{\left(\frac{3.14}{180} \times \frac{1}{1000}\right)} = \mathbf{3.5\ cm.}$$

Example 25. *A telescope of 20 cm objective is pointed towards the moon distant 4×10^5 km when two objects on its surface are just resolved. Calculate (i) the distance between them and (ii) the magnifying power of the telescope required for the purpose $\lambda = 5800$ Å and resolving power of eye = 1.5 min.*

Solution. The smallest angular separation that can be resolved by a telescope objective of a diameter is given by

$$\theta = \frac{1.22\,\lambda}{a}$$

Here $\lambda = 5800\ \text{Å} = 5800 \times 10^{-8}$ cm, $a = 20$ cm

$$\therefore \quad \theta = \frac{1.22 \times 5800 \times 10^{-8}}{20}\ \text{rad}$$

(*i*) Let x be the distance between the two objects on the moon. The distance of moon from the earth $= l = 4 \times 10^5$ km.

Therefore, $$\theta = \frac{x}{l} = \frac{x}{4 \times 10^5} = \frac{1.22 \times 5800 \times 10^{-8}}{20}$$

Hence $$x = 1.3 \text{ km}$$

Thus, the distance between two objects on the moon is **1.3 km.**

(*ii*) The smallest angle that can be resolved by the eye is $\theta' = \frac{1.22\,\lambda}{a'}$, where a' is the aperture of the eye.

The magnifying power of a telescope $= \frac{\theta'}{\theta} = \frac{a}{a'}$

Here $$\theta' = 1.5 \text{ min} = 1.5 \frac{\pi}{180°}\left(\frac{1}{60}\right)^{\circ}$$

$$\therefore \quad \text{M.P.} = \frac{\dfrac{1.5 \times 3.14}{(180 \times 60)}}{\dfrac{1.22 \times 5800 \times 10^{-8}}{20}} = \mathbf{120.}$$

Example 26. *Examine the validity of a claim by someone who saw a man 50 m away wearing a shirt with stripes about 1 cm wide. (The human eye is most sensitive for 5500 Å and has a pupil diameter of 2 mm).*

Solution. The smallest angular separation that can be resolved by the human eye is

$$\theta = \frac{1.22\,\lambda}{a}, \text{ where } a \text{ is the diameter of pupil.}$$

Here $\lambda = 5500\ \text{Å} = 5500 \times 10^{-8}$ cm $= 5.5 \times 10^{-5}$ cm

and $a = 0.2$ cm

$$\therefore \quad \theta = \frac{1.22 \times 5.5 \times 10^{-5}}{0.2} = 3.35 \times 10^{-4} \text{ rad}$$

The minimum linear resolvable separation x at a distance l is $x = l\theta$

Here $l = 50$ m $= 5000$ cm

$\therefore$ $x = 5000 \times 3.35 \times 10^{-4} = 1.675$ cm

Thus the distance of 1 cm cannot be resolved and hence the claim is false.

Example 27. *Calculate the least value of the angular separation of two stars which can be resolved by a telescope with a 2 meter aperture. If the aperture of the human eye is 2 mm and the focal length of*

the eye-piece is 2.54 cm, what must be the minimum focal length of the objective if the full resolving power of the telescope is to be utilized? (Assuming λ = 6000 Å).

Solution. The smallest angular separation that can be resolved by a telescope is given by

$$\theta = \frac{1.22\,\lambda}{a}$$

Here $a = 2\text{ m} = 200\text{ cm}$ and $\lambda = 6000\text{ Å} = 6000 \times 10^{-8}\text{ cm}$

$$\therefore \quad \theta = \frac{1.22 \times 6000 \times 10^{-8}}{200}\text{ rad} = 36.6 \times 10^{-8}\text{ rad}$$

The smallest angle that can be resolved by the eye is given by

$$\theta' = \frac{1.22\,\lambda}{a'},$$ where a' is the aperture of the eye.

Thus, the magnifying power $= \dfrac{\theta'}{\theta} = \dfrac{a}{a'}$

Here $a' = 2\text{ mm} = 0.2\text{ cm}$ and $a = 200\text{ cm}$

$$\therefore \quad \text{M.P.} = \frac{200}{0.2} = 1000 \qquad (1)$$

In terms of the focal lengths of the objective F and the eye-piece f, the magnifying power

$$= F/f \qquad (2)$$

Therefore equating equations (1) and (2), we get

$$1000 = F/f$$

or $F = 1000\,f$

Here $f = 2.54\text{ cm}$

$\therefore$ $F = 1000 \times 2.54 =$ **2540 cm.**

3.18 THE RESOLVING POWER OF A MICROSCOPE

A microscope is used to produce an enlarged distinct image of a small nearby object at the retina of an eye. The resolving power of a microscope is defined as its ability to form separate images of two nearby objects. It is measured by the smallest linear separation which two nearby object points may have so that their images appear just resolved by the objective of the microscope. The smallest linear separation is called the limit of resolution of the microscope. The reciprocal of the limit of resolution is called geometrical resolving power of the microscope. Let AB be the objective of the microscope and PQ be the two small objects placed very close to each other and perpendicular to its axis (as shown in Figure 21). The periphery of the objective acts like a circular aperture and hence a diffraction pattern corresponding to each object is formed in the focal plane of the objective. The center of the bright disc in the diffraction pattern of P is at P_1 while that of Q is at P_2. According to Rayleigh's criterion of resolution points P and Q will be just resolved if the first minimum of the diffraction pattern of Q coincides through P_1, the center of the bright disc of diffraction pattern due to P.

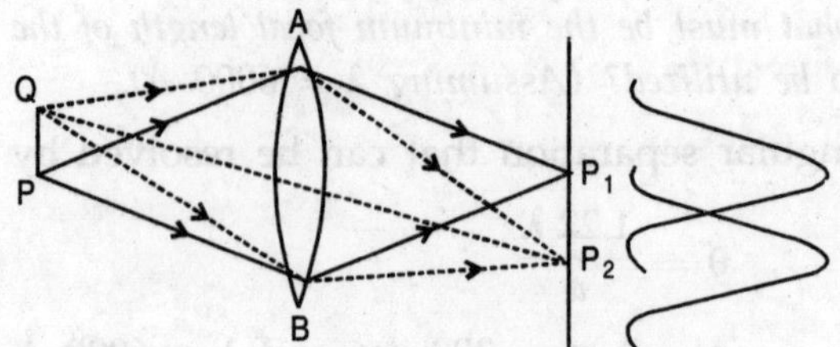

FIGURE 21

The path difference between extreme rays from point object Q and reaching P_1 is given by

$$QBP_1 - QAP_1 = (QB + BP_1) - (QA + AP_1)$$

$$= QB - QA \qquad (\because BP_1 = AP_1)$$

In Figure 22, PC is perpendicular to CA and PD to QB

$$\therefore \quad QB - QA = (QD + DB) - (CA - QC)$$

$$= (QD + PB) - (PA - QC)$$

$$\because \quad DB = PB \text{ and } CA = PA$$

$$\therefore \quad QB - QA = QD + QC \qquad (\because PB = PA)$$

Now $\angle QPC = \angle QPD = \angle APO = \theta$

From triangles QPC and QPD

$$\sin\theta = \frac{QC}{PQ} = \frac{QD}{PQ}$$

$$\therefore \quad QC = PQ\sin\theta = QD$$

Therefore the path difference, $QB - QA = PQ\sin\theta + PQ\sin\theta = 2PQ\sin\theta$

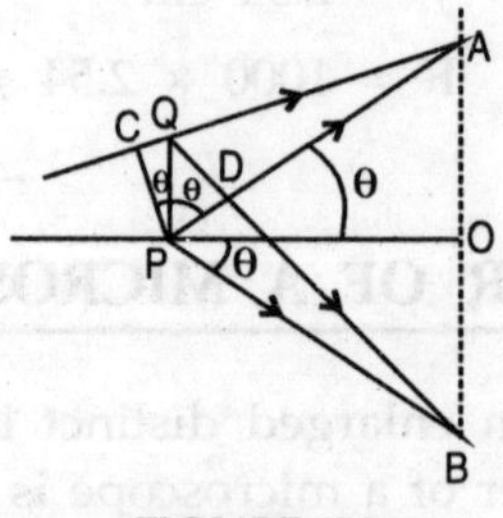

FIGURE 22

According to the theory of Fraunhoffer diffraction due to the circular aperture, for just resolution of two images this path difference must be equal to 1.22 λ, that is, $2PQ\sin\theta = 1.22\lambda$

or
$$PQ = \frac{1.22\lambda}{2\sin\theta}$$

Thus the linear distance x, between the two just resolvable objects is

$$x = \frac{1.22\lambda}{2\sin\theta}$$

If the medium between the object and objective is other than air, but one having a refractive index μ, then

$$x = \frac{1.22\,\lambda}{2\mu \sin\theta}$$

This equation gives the limit of resolution and is a measure of the resolving power of a microscope. The quantity 2μ sin θ is called the numerical aperture of the objective.

Example 28. *A microscope objective gathers light over a cone of semi-angle 30° and uses visible light (λ = 5500 Å). Estimate its resolving limit.*

Solution. The distance between the two nearby objects just resolved by the microscope is given by

$$x = \frac{1.22\,\lambda}{2\sin\theta},$$

where θ is the semi-angle of the cone of light.

Here $\lambda = 5500\text{ Å} = 5500 \times 10^{-8}$ cm and $\theta = 30°$

$$\therefore \quad x = \frac{1.22 \times 5500 \times 10^{-8}}{2 \sin 30°} = \mathbf{6.7 \times 10^{-5}\ cm.}$$

Example 29. *Calculate the resolving power of a microscope if N.A. given on the objective = 0.12 and the wavelength of light used = 6 × 10⁻⁵ cm.*

Solution. The minimum resolvable linear distance between the two nearby objects is given by

$$x = \frac{\lambda}{2\text{ N.A.}}$$

Hence the resolving power of a microscope $= \frac{1}{x} = \frac{2\text{ N.A.}}{\lambda}$, where N.A. is the numerical aperture

Here $\lambda = 6 \times 10^{-5}$ cm and N.A. = 0.12

$$\therefore \quad \text{R.P. of microscope} = \frac{2 \times 0.12}{6 \times 10^{-5}} = \mathbf{4000.}$$

Example 30. *A microscope is used to resolve two self luminous objects separated by a distance of 5 × 10⁻⁵ cm and placed in air. What should be the minimum overall magnifying power of the microscope in order to see the resolved images of the objects? The resolving power of the eye is 1.5′ and the least distance of distinct vision is 25 cm.*

Solution. The minimum overall magnifying power of the microscope in order to see the resolved images is

$$M = \frac{\text{Minimum linear separation resolved by the eye}}{\text{Minimum linear separation resolved by the microscope}}$$

or

$$M = \frac{\text{The linear limit of resolution of the eye}}{\text{The linear limit of resolution of the microscope}}$$

Here the least distance of distinct vision = 25 cm

∴ The angular limit of the resolution of the eye

$$= 1.5' = \left(\frac{1.5}{60}\right)^{\circ} = \frac{1.5}{60} \times \frac{\pi}{180}\text{ rad.}$$

∴ The linear limit of the resolution of eye

$$= \text{Distance of distinct vision} \times \text{angular limit}$$

$$= 25 \times \frac{1.5}{60} \times \frac{3.14}{180} = 1.09 \times 10^{-2} \text{ cm}$$

The limit of resolution of microscope $= 5 \times 10^{-5}$ cm

$$\text{The magnifying power M} = \frac{1.09 \times 10^{-2}}{5 \times 10^{-5}} = \mathbf{218.}$$

EXERCISES

Objective Questions

1. Large aperture telescopes are used for
   (*a*) greater magnification (*b*) greater resolution
   (*c*) reducing lens aberrations (*d*) ease of manufacture
2. The length of a telescope is 52 cm and its magnification is 25 for the relaxed eye. The focal lengths of the objective and eye-piece are respectively
   (*a*) 50 cm, 2 cm (*b*) 2 cm, 50 cm
   (*c*) (25 × 13)/6 cm, 13/6 cm (*d*) 26 cm, 2 cm
3. The magnifying power of a telescope can be increased by
   (*a*) increasing the focal length (*b*) using an eye-piece of a high power
   (*c*) using an eye-piece of a low power (*d*) increasing the distance of the object
4. In a plane transmission grating, the intensity of principal maximum
   (*a*) increases as the number of slits increases
   (*b*) decreases as the number of slits increases
   (*c*) remains constant as the number of slits increases
   (*d*) none of these
5. In a plane transmission grating, the intensity of the secondary maxima
   (*a*) increases as the number of slits increases
   (*b*) decreases as the number of slits increases
   (*c*) remains constant as the number of slits increases
   (*d*) none of these
6. In a plane diffraction grating, the width of the principal maximum
   (*a*) increases as the number of slits increases
   (*b*) decreases as the number of slits increases
   (*c*) remains constant as the number of slits increases
   (*d*) none of these
7. In a plane diffraction grating, the directions of mining are given by
   (*a*) $(a + b) \sin \theta = \pm n\lambda$ (*b*) $N (a + b) \sin \theta = \pm m\lambda$
   (*c*) $a \sin \theta = \pm m\lambda$ (*d*) none of these

8. In a plane diffraction grating, the angle of diffraction is

(*a*) directly proportional to the wavelength

(*b*) inversely proportional to the wavelength

(*c*) directly proportional to the square root of the wavelength

(*d*) inversely proportional to the square root of the wavelength

9. In a plane diffraction grating, the third, the sixth, the ninth etc., orders of spectrum will be missing if

(*a*) the width of the opacity is thrice the width of transparency

(*b*) the width of the opacity is double the width of transparency

(*c*) the width of the opacity is equal to the width of transparency

(*d*) none of these

10. The maximum number of orders available with a grating is

(*a*) directly proportional to the wavelength

(*b*) inversely proportional to the grating element

(*c*) independent of the grating element

(*d*) directly proportional to the grating element

11. The dispersive power of a grating can be defined as

(*a*) the increase in the angle of refraction corresponding to a change in wavelength

(*b*) the increase in the angle of incidence corresponding to a change in wavelength

(*c*) the increase in the angle of diffraction corresponding to a change in wavelength

(*d*) none of these

12. Which of the following statements is incorrect? The dispersive power of a grating

(*a*) is directly proportional to the order of the spectrum

(*b*) is inversely proportional to the grating element

(*c*) is directly proportional to the wavelength

(*d*) is inversely proportional to the cosine of the angle of diffraction

13. Which of the following statements is incorrect? In a grating spectroscope, when

(*a*) the grating is moved forward or backward between the collimating and focussing lenses, there will be no change on the diffraction pattern

(*b*) the grating is shifted in its own plane and perpendicular to the slit, there will be no change on the diffraction pattern

(*c*) the grating plane is rotated round the axis parallel to the slit, the position of maxima will change in the diffraction pattern

(*d*) none of the above statements are correct

14. The resolving power of a prism can be expressed as

(*a*) $\frac{\lambda}{d\lambda}=n\mathrm{N}$ (*b*) $\frac{\lambda}{d\lambda}=t\frac{d\mu}{d\lambda}$

(*c*) $\frac{1}{\theta}=\frac{2\mu\sin\theta}{1.22\lambda}$ (*d*) $\frac{1}{\theta}=\frac{a}{1.22\lambda}$

15. The resolving power of a grating

(*a*) increases as the total number of lines ruled on the grating increases

(*b*) increases as the total number of lines ruled on the grating decreases
(*c*) increases as the order of the spectrum decreases
(*d*) increases as the grating element decreases

16. The resolving power of a prism is
(*a*) inversely proportional to the thickness of the base of the prism
(*b*) inversely proportional to the rate of change of the refractive index with the wavelength
(*c*) directly proportional to the thickness of the base of the prism
(*d*) independent of the thickness of the base of the prism

17. According to Rayleigh's criterion of resolution, the two spectral lines are just resolved if the intensity at the dip in the middle is
(*a*) $4/\pi^2$ times the intensity at either of the maxima
(*b*) $6/\pi^2$ times the intensity at either of the maxima
(*c*) $8/\pi^2$ times the intensity at either of the maxima
(*d*) $10/\pi^2$ times the intensity at either of the maxima

18. According to Rayleigh's criterion of resolution, the two spectral lines of equal intensity are just resolved when the central maximum of the diffraction pattern due to one falls
(*a*) on the central maximum of the diffraction pattern of the other
(*b*) on the secondary maximum of the diffraction pattern of the other
(*c*) on the first minimum of the diffraction pattern of the other
(*d*) none of these

19. A telescope having a larger diameter will have
(*a*) a lower resolving power (*b*) a medium resolving power
(*c*) a higher resolving power (*d*) none of these

20. An observer looks at a tree of a height of 15 meters with a telescope of magnifying power 10. To him the tree appears greater by:
(*a*) 15 times (*b*) 10 times
(*c*) 150 times (*d*) 5 times

21. The limit of resolution of the telescope is equal to
(*a*) $\frac{1.22\lambda}{a}$ (*b*) $\frac{a}{1.22\lambda}$
(*c*) $\frac{1.22\lambda}{2\mu \sin\theta}$ (*d*) $\frac{1.22\lambda}{2\sin\theta}$

22. The resolving power of a microscope is
(*a*) larger for a smaller linear separation
(*b*) smaller for a smaller linear separation
(*c*) larger for a larger value of wavelength
(*d*) smaller for a larger value of numerical aperture

Short Questions for Discussion

1. What is diffraction of light?
2. Distinguish between Fresnel and Fraunhoffer diffraction.
3. Explain the difference between interference and diffraction of light.

4. What is the condition of obtaining a diffraction pattern?
5. What is a plane transmission grating?
6. What is a grating element? Discuss the role of a grating element in the function of grating.
7. Explain why white light gets dispersed after diffraction through grating.
8. Compare the grating spectrum with the prism spectrum.
9. Define the resolving power and the dispersive power of a grating and distinguish between them.
10. Derive an expression for the resolving power of a plane transmission grating.
11. Explain the Rayleigh criterion of resolution.

Long Questions for Discussion

1. Define the diffraction of light? Explain the difference between interference and diffraction of light.
2. A sharp razor blade is held parallel to a narrow slit illuminated by a monochromatic source of light. Discuss the intensity pattern obtained on the screen placed parallel to the blade.
3. Give the theory of the plane transmission grating and show how it can be used to determine the wavelength of light.
4. Explain the formation of a spectrum by grating and compare the spectrum of grating with that of a prism.
5. Define dispersive power of a grating and obtain an expression for it and discuss.
6. What is meant by the resolving power of an optical instrument? Explain Rayleigh's criterian for just resolution.
7. Define resolving power. Discuss and derive an expression for the resolving power of a
   - *(i)* Plane Transmission Grating
   - *(ii)* Prism
   - *(iii)* Telescope
   - *(iv)* Microscope

Chapter 4

Polarization of Light

4.1 OVERVIEW

The phenomena of interference and diffraction establish the wave nature of light. However, they fail to throw any light on the nature of light waves. Naturally the question arises, are the light waves longitudinal or transverse, or are the vibrations linear, circular, elliptical or torsional? From the study of sound we know that a longitudinal wave has the same properties with respect to any plane through its line of propagation; while a transverse wave behaves differently in different planes. The phenomenon of polarization establishes the transverse nature of light waves as transverse waves. Before discussing the phenomenon of the polarization of light we will explain the meaning of the polarization of a wave. Now we shall discuss the simple experiment which helps in understanding polarization.

4.2 POLARIZATION OF LIGHT WAVES

A Mechanical Experiment to Demonstrate the Polarization of a Wave

Consider a transverse wave on a string AB passes through two parallel slits S_1 and S_2 as shown in Figure 1(*a*). The string is attached to a fixed point at B. If the slit S_1 is parallel to the direction of the transverse vibrations, the wave passes through the slit undisturbed. Now slit S_2 is rotated 90° about AB so that its axis becomes perpendicular to slit S_1. In this case, the vibrations pass through the slit S_1 undisturbed as before but are not able to transmit through S_2 as shown in Figure 1(*b*).

This means that if the vibrations are in all directions in a plane perpendicular to the directions of propagation, only those parallel to the slit S_1 are transmitted. The slit S_1 is called polarizer and the wave transmitted through it is called polarized wave. If the slit S_2 is parallel to S_1, the polarized wave is transmitted. In the intermediate positions of slit S_2, about AB, the wave is partly transmitted and partly stopped and thus the amplitude of emergent vibrations show variations. On the other hand, if the longitudinal wave is set up in the string it is freely allowed by S_2 in all positions, and rotation of any of the slits about AB as axis does not affect the passage of the wave and hence the amplitude of emergent vibrations remains unchanged.

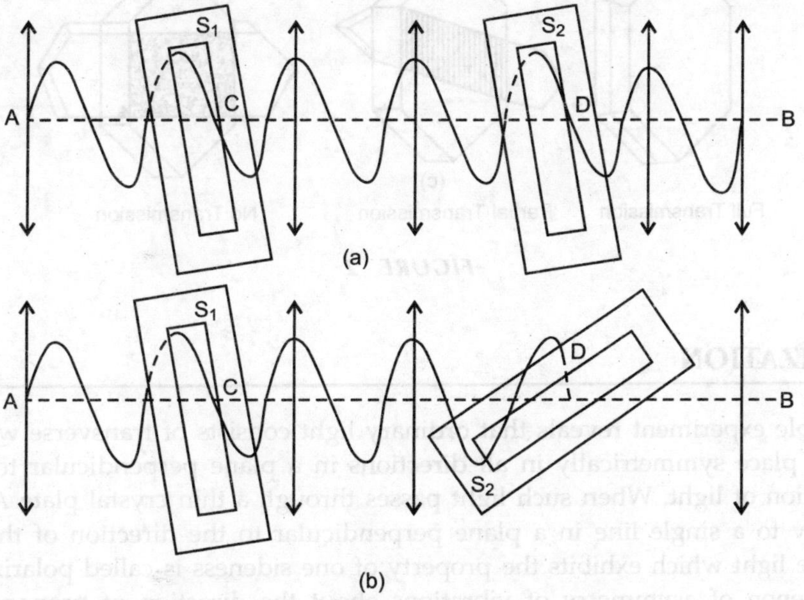

FIGURE 1

The Transverse Nature of Light Waves

Consider an optical experiment to demonstrate the transverse character of light waves. When a beam of an ordinarly light from source S is allowed to fall normally on a thin plate of tourmaline crystal, cut with faces parallel to its vertical axis, only a part of the incident light is transmitted as shown in Figure 2. If the crystal plate A is rotated about the direction of propagation of the incident light, no change is observed in the character of the transmitted light. If this beam is allowed to pass through another similar crystal plate B, placed with its axis parallel to A, the light is almost completely transmitted through B and no change is observed in the emergent beam of light (Figure 2(*a*)). If B is rotated gradually with respect to A, the transmitted light through B becomes dimmer and dimmer, and is finally cut off after a rotation of 90° (Figure 2(*b*)). For no transmission, the crystal plates are said to be in a crossed position. On further rotation of B, the brightness gradually increases and finally attains full brightness; the axes of A and B are now again mutually parallel. Thus we see that the complete darkness occurs when A and B are in a crossed position and complete brightness when they are in a mutually parallel position (Figure 2(*c*)). From the experiment it can be easily inferred that the light waves are transverse in nature.

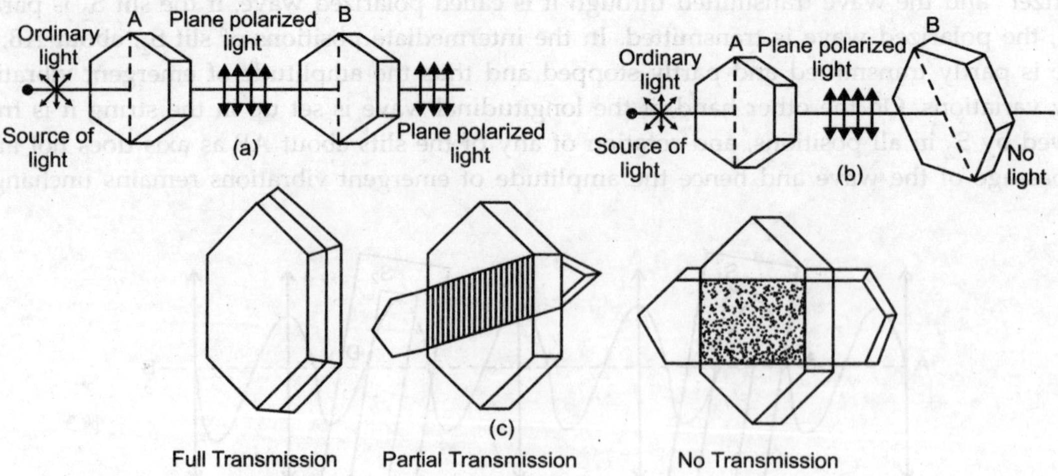

FIGURE 2

4.3 POLARIZATION

The above simple experiment reveals that ordinary light consists of transverse waves in which vibrations take place symmetrically in all directions in a plane perpendicular to the direction of the propagation of light. When such light passes through a thin crystal plate A, its vibration is confined only to a single line in a plane perpendicular to the direction of the propagation of the light. The light which exhibits the property of one sideness is called polarization of light, or the phenomenon of asymmetry of vibrations about the direction of propagation is called polarization. According to the electromagnetic theory of light, a light wave consists of electric and magnetic vectors vibrating in mutually perpendicular directions, both being perpendicular to the direction of the propagation of light. The electric vector is only responsible for the optical effects and hence called a light vector. In terms of electromagnetic theory the plane polarized light is the light in which the light vector vibrates along the vector. In terms of electromagnetic theory the plane polarized light is the light in which the light vector vibrates along a fixed straight line in a plane perpendicular to the direction of propagation of light.

Unpolarized and Polarized Light and Their Representations

Unpolarized Light

In general, an ordinary light beam having vibrations along all possible planes perpendicular to the direction of propagation of light is said to be unpolarized. When such a light is passed through a single rotating thin plane of a tourmaline crystal, there is no change in the intensity of light. The unpolarized light is represented by a star as shown in Figure 3(*a*) and also in Figure 3(*b*).

Plane Polarized Light

The light having vibrations only along a single direction perpendicular to the direction of the propagation of light is said to be plane polarized. When plane polarized light is passed through a single rotating crystal, a change in the intensity of light is observed. When the plane polarized light has vibrations in the plane of the paper they are represented by straight arrows as shown in Figure 3(*c*). When the vibrations lie in a direction perpendicular to the plane of the paper they are represented by dots as shown in Figure 3(*d*).

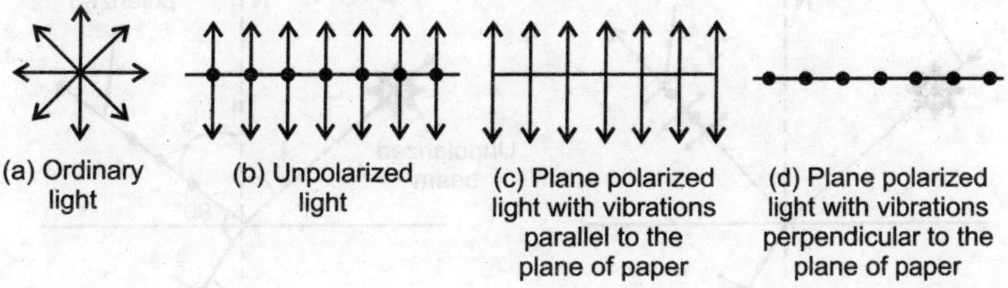

FIGURE 3

Plane of Polarization

The plane containing the direction of the propagation of light but containing no vibrations, is called the plane of polarization or in other words, the plane perpendicular to the plane of vibration is called the plane of polarization and is represented by the plane EFGH in Figure 4.

Plane of Vibration

The plane containing the direction of vibration and the direction of the propagation of light is called the plane of vibration and represented by plane ABCD in Figure 4.

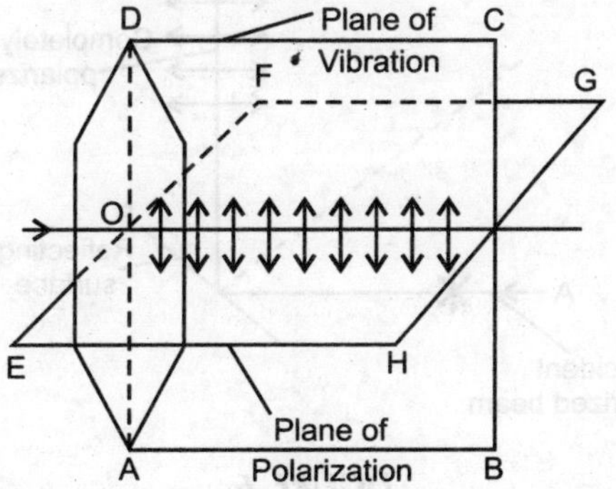

FIGURE 4

Polarization by Reflection

In 1808, Malus discovered that when a beam of unpolarized light is reflected from the surface of a transparent dielectric medium such as glass or water, the reflected light is either partially polarized (Figure 5(*a*)) or completely plane polarized, depending upon the angle of incidence. If the angle of incidence is 0° (normal incidence) or 90° (the grazing incidence), the reflected beam is unpolarized. The vibrations of the reflected beam are parallel to the reflecting surface.

The percentage of polarization varies with the angle of incidence. At a particular angle of incidence the reflected beam is completely plane polarized as shown in Figure 5(*b*). This angle of incidence at which maximum polarization occurs is called the angle of polarization.

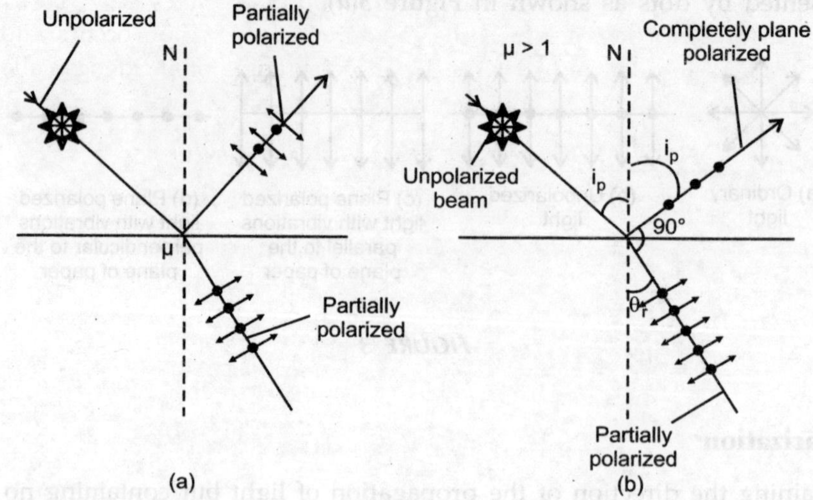

FIGURE 5

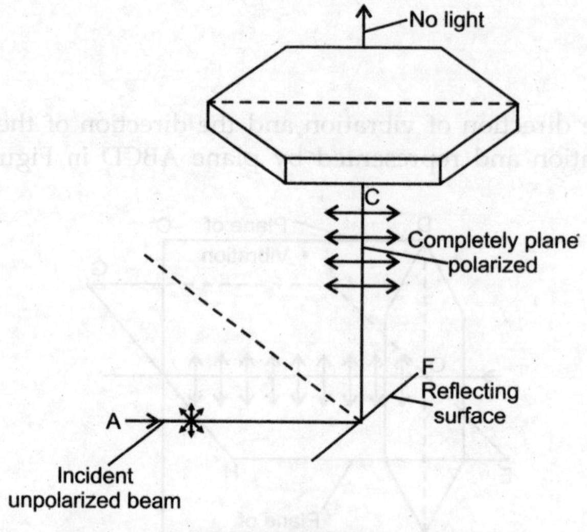

FIGURE 6

The above conclusions can be verified by passing the reflected light through a rotating tourmaline crystal. The examination of transmitted light shows the variation in intensity indicating that the reflected light is partially polarized. Keep the tourmaline crystal fixed in the position of minimum intensity and vary the angle of incidence until the intensity of the reflected beam becomes almost zero (as shown in Figure 6). The angle of incidence at which this condition occurs is called the polarizing angle. The polarizing angle is different for different reflecting surfaces: it is 57° for air-glass reflection, 33° for glass-air reflection and 53° for air-water (at 0°C) reflection.

4.4 BREWSTER'S LAW

In 1881, Sir Brewster, on the basis of his experimental observations, discovered a simple relation between the angle of incidence at which the maximum polarization occurs and the refractive index of the medium. He observed that for a particular angle of incidence called the angle of polarization, the reflected light is completely plane polarized with the plane of vibration perpendicular to the plane of incidence. It was shown by him that the tangent of the polarizing angle is equal to the refractive index of the medium, that is,

$$\mu = \tan i_p$$

This is called Brewster's law. Brewster also concluded that at the polarizing angle reflected and refracted rays are at right angles to each other. Let a beam AB of an ordinary light be incident on the glass surface at the polarizing angle i_p. A part of it is reflected along BC and a part is refracted along BD as shown in Figure 7.

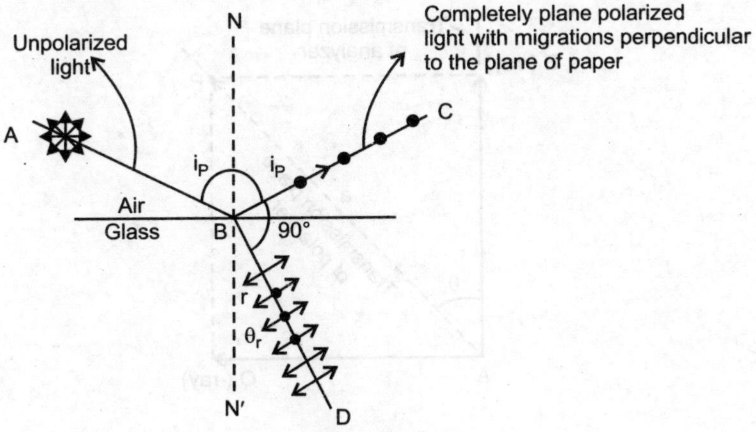

FIGURE 7

According to Brewster's law

$$\mu = \tan i_p = \frac{\sin i_p}{\cos i_p},$$

but from Snell's law $\mu = \dfrac{\sin i_p}{\sin r}$

$$\therefore \qquad \mu = \frac{\sin i_p}{\sin r} = \frac{\sin i_p}{\cos i_p}$$

or $$\frac{\sin i_p}{\sin r} = \frac{\sin i_p}{\sin\left[\frac{\pi}{2} - i_p\right]} \quad \text{or} \quad \frac{\pi}{2} - i_p = r$$

or $$i_p + r = \frac{\pi}{2}, \text{ As } \angle NBN' = \pi, \text{ we have}$$

$$i_p + \angle CBD + r = \pi \text{ or } \angle CBD = \pi - (i_p + r) = \frac{\pi}{2}$$

When the light is incident *at the polarizing angle, the reflected ray is at right angles to the refracted ray.*

Law of Malus

In 1809, Malus experimentally discovered that when a beam of completely plane polarized light is incident on an analyzer, the intensity of the emergent beam varies as the square of cosine of the angle between the planes of transmission of the analyzer and the polarizer. This law is known as the law of Malus and holds good only for a completely plane polarized incident light. To prove the law let the angle between the two planes of transmission of the analyzer and the polarizer be θ, and *a* be the amplitude of the incident plane polarized light emerging from the polarizer. The plane polarized light of amplitude *a* may be resolved in two components along and perpendicular to the plane of transmission of the analyzer as shown in Figure 8.

$$AE = a\cos\theta \text{ and } AO = a\sin\theta$$

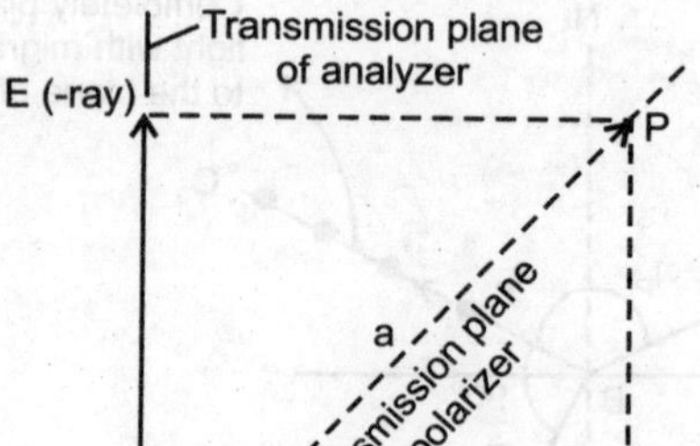

FIGURE 8

The perpendicular component ($a\sin\theta$) is eliminated in the analyzer while the parallel component ($a\cos\theta$) is freely transmitted through it. Therefore, the intensity of light emerging from the analyzer is given by $I_0 = (a\cos\theta)^2 = a^2\cos^2\theta = I\cos^2\theta$, where $I = a^2$ is the intensity of the completely plane polarized incident light. Thus the sum of the intensities of the transmitted rays is equal to the intensity of the incident polarized ray.

Example 1. *When the angle of incidence on a certain material is 60°, the reflected light is completely polarized. Find the refractive index of the material and also the angle of refraction.*

Solution. From Brewster's law, the refractive index of the second medium with respect to the first is given by

$$\tan i_p = \mu, \text{ where } i_p \text{ is the angle of polarization.}$$

In the given problem $i_p = 60°$

∴ $\mu = \tan 60° = \sqrt{3}$ or $\mu = 1.732$

the angle of refraction $r = \frac{\pi}{2} - i_p = 90° - 60° = \mathbf{30°}$.

Example 2. *A ray of light is incident on a surface of benzene of refractive index 1.50. If the reflected light is linearly polarized, calculate the angle of refraction.*

Solution. From Brewster's law we know that

$$\mu = \tan i_p$$

Here $\mu = 1.5$

$\therefore$ The angle of polarization $i_p = \tan^{-1}(\mu) = \tan^{-1}(1.5) = 56.3°$

The angle of refraction r is given by

$$r = \frac{\pi}{2} - i_p = 90° - 56.3° = \mathbf{33.7°}.$$

Example 3. *The refractive indices of glass and water are 1.54 and 1.33. Which will be greater, the polarizing angle for a beam incident from water to glass, or that for a beam incident from glass to water?*

Solution. From Brewster's law, the polarizing angle i_p is given by

$$\tan i_p = \mu$$

For a water to glass interface,

$$\mu = \frac{\mu_{glass}}{\mu_{water}} = \frac{1.54}{1.33} = 1.158$$

$\therefore$ $1.158 = \tan i_p$ or $i_p = 49.2°$

For a glass to water interface,

$\therefore$ $$\mu = \frac{\mu_{water}}{\mu_{glass}} = \frac{1.33}{1.54} = 0.864$$

$\therefore$ $0.864 = \tan i_p$ or $i_p = 40.8°$

The polarizing angle for a water to glass interface is larger.

Example 4. *The polarizing angle of a piece of glass for green light is 60°. What is the angle of minimum deviation for a 60° prism made of the same glass?*

Solution. From Brewster's law, the polarizing angle i_p is given by

$$\tan i_p = \mu$$

or $\mu = \tan 60° = 1.7321$

In terms of the angle of prism A and the angle of minimum deviation δ_m, the μ is given by

$$\mu = \frac{\sin\left(\frac{A + \delta_m}{2}\right)}{\sin\frac{A}{2}}$$

Here $\mu = 1.7321$ and $A = 60°$

$\therefore$ $$\sin\left(\frac{A+\delta_m}{2}\right) = \mu \sin\frac{A}{2}$$

or $$\sin\left(\frac{60° + \delta_m}{2}\right) = 1.7321 \times \sin 30° = 1.7321 \times 0.5$$

or $$\left(\frac{60° + \delta_m}{2}\right) = \sin^{-1}(0.86605) = 60°4'$$

or $$60° + \delta_m = 120°8' \text{ or } \delta_m = 120°8' - 60° = \mathbf{60°8'}.$$

Example 5. *Two polarizing sheets have polarizing directions parallel so that the intensity of the transmitted light is maximum. Through what angle must either sheet be turned if the intensity is to drop by one-third?*

Solution. From Malus' law, we have, $I = I_0 \cos^2 \theta$, where θ is the angle between the two polarizing sheets.

Here $$I_0 = I_m \text{ and } I = \frac{I_m}{3}$$

$$\therefore \quad \frac{I_m}{3} = I_m \cos^2 \theta$$

$$\therefore \quad \cos^2 \theta = \left(\frac{1}{3}\right)$$

or $$\theta = \cos^{-1}\left(\frac{1}{\sqrt{3}}\right)$$

$$= \mathbf{\pm 55°18', \pm 124°42'}.$$

Example 6. *A polarizer and an analyzer are oriented so that the maximum amount of light is transmitted. To what fraction of its maximum value is the intensity of transmitted light reduced when the analyzer is rotated through (a) 30°, (b) 45° and (c) 60°?*

Solution. Let I_m be the maximum intensity of the transmitted light and θ be the angle between the plane of polarizer and the plane of the analyzer, then from Malus' law,

$$I = I_m \cos^2 \theta$$

(*a*) When $\theta = 30°$, $$I = I_m \cos^2 30° = I_m\left(\frac{\sqrt{3}}{2}\right)^2 = I_m\left(\frac{3}{4}\right) \quad \therefore \frac{I}{I_m} = \mathbf{0.75}$$

(*b*) When $\theta = 45°$, $$I = I_m \cos^2 45° = I_m (1/\sqrt{2})^2 = \frac{I_m}{2}, \quad \therefore \frac{I}{I_m} = \mathbf{0.50}$$

(*c*) When $\theta = 60°$, $$I = I_m \cos^2 60° = I_m\left(\frac{1}{2}\right)^2 = \frac{I_m}{4}, \quad \therefore \frac{I}{I_m} = \mathbf{0.25}.$$

Example 7. *An analyzing Nicol examines two adjacent plane-polarized beams A and B whose planes of polarization are mutually perpendicular. In one position of the analyzer, beam B shows zero intensity. From this position a rotation of 30° shows the two beams as matched (i.e., of equal intensity). Deduce the intensity ratio I_A/I_B of the two beams.*

Solution. According to Malus' law, when a beam of a plane polarized light of intensity I_0 is incident on an analyzer, the intensity I of the emerging light is given by, $I = I_0 \cos^2 \theta$, where θ is the angle between the plane of transmission of the analyzer and the plane of vibration of the incident light.

In one position of the analyzer, beam B shows zero intensity. This means that $\theta = 90°$ for beam B. Obviously θ will be 0° for beam A. When the analyzer is rotated through 30°, then we have $\theta = 30°$ for beam A and $\theta = 60°$ for beam B. If I' be the intensity of beam A and also of beam B as transmitted by the analyzer, then we have

$$I' = I_A \cos^2 30° = I_B \cos^2 60°$$

$$\frac{I_A}{I_B} = \frac{\cos^2 60°}{\cos^2 30°} = \frac{\left(\frac{1}{2}\right)^2}{\left(\frac{\sqrt{3}}{2}\right)^2} = \mathbf{\frac{1}{3}}.$$

Example 8. *At a certain temperature the critical angle of incidence of water for total internal reflection is 48° for a certain wavelength. What is the polarizing angle and the angle of refraction for light incident on water at an angle that gives maximum polarization of the reflected light? (Given sin 48° = 0.7431).*

Solution. The relation between the critical angle and the index of refraction is

$$\mu = \frac{1}{\sin C} = \frac{1}{\sin 48°} = \frac{1}{0.7431}$$

$$\therefore \quad \mu = 1.345$$

For maximum polarization, the light must be incident at polarizing angle i_p. From Brewster's law, we have

$$\mu = \tan i_p \quad \text{or} \quad i_p = \tan^{-1}(\mu)$$

$$\therefore \quad i_p = \tan^{-1}(1.345) = 53°22'$$

If r is the angle of refraction, then we have

$$i_p + r = \frac{\pi}{2} \quad \text{or} \quad r = \frac{\pi}{2} - i_p = 90° - 53°22' = \mathbf{36°38'}.$$

4.5 DOUBLY REFRACTING CRYSTALS

There are certain crystals which exhibit the property of splitting an incident ray into two refracted rays. Such crystals are called doubly refracting crystals. They are of two types (*i*) Uniaxial crystals and (*ii*) Biaxial crystals.

In uniaxial crystals there is only a single direction known as an optic axis along which two refracted rays are transmitted with the same velocity. Out of the two refracted rays only one follows the ordinary laws of refraction. The examples of these crystals are calcite, tourmaline and quartz.

In biaxial crystals, there are two directions or two optic axes along which the velocities of refracted rays are same. None of the refracted rays obeys the laws of refraction. The examples of biaxial crystals are topaz, argonite, copper sulphate, cane sugar and mica.

Calcite crystal: The most commonly used uniaxial crystal is Iceland spar or calcite. Calcite, also known as Iceland spar, is a colorless transparent crystal. It is chemically crystallized calcium carbonate ($CaCO_3$) and occurs in nature in different forms, all of which gives a rhombohedron on cleavage as shown in Figure 9(*a*). Each of the six faces of the crystal is a parallelogram having angles of 78° and 102° approximately. At two diametrically opposite corners A and H, all the angles of three faces meeting there are obtuse (> 90°); while at the rest of the six corners two angles are acute and the remaining one is obtuse. The two corners A and H are termed as blunt corners.

Optic axis: The optic axis in a doubly refracting crystal is a direction along which all the plane waves are transmitted with a single velocity without showing the effect of double refraction. Thus the optic axis is a direction along a line passing through any one of the blunt corners and making equal angles with each of the three edges which meet at this corner. A ray of unpolarized light travelling along the optic axis or in a direction parallel to the optic axis does not split up into ordinary (O) and extraordinary

(E) rays. Ordinarily, the optic axis is not obtained by joining the two blunt corners. In a special situation when the rhombohedron is so cut that all of its edges are equal then the line joining the blunt corners coincides with the optic axis as shown in Figure 9(*b*).

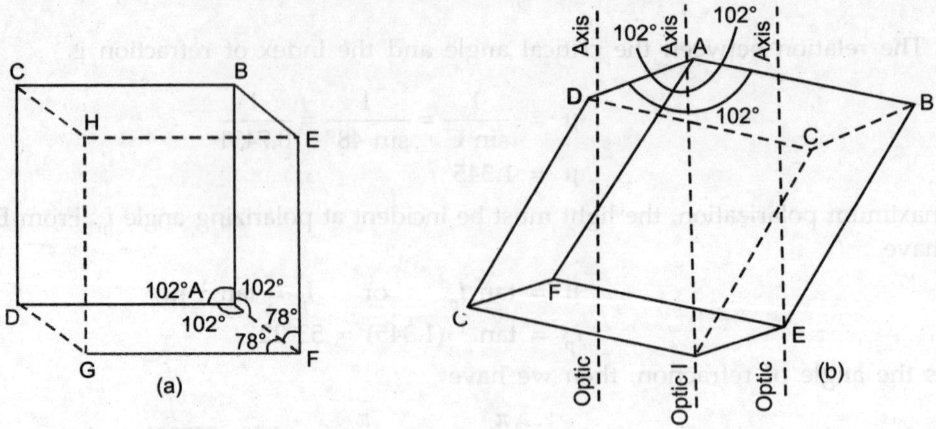

FIGURE 9

Principal section: A plane containing the optic axis of the crystal and perpendicular to the two opposite refracting faces is called the principal section of the crystal for that pair of faces Figure 10(*a*). As a crystal has six faces there are three principal sections corresponding to each pair of opposite faces. A principal section of a calcite crystal is a parallelogram having angles of 71° and 109° as illustrated in Figure 10(*b*). The principal plane or section of the ordinary ray (O-ray) is defined as a plane which contains the ordinary ray and the optic axis and the principal plane of the extra-ordinary ray (E-ray) is a plane which contains the optic axis and the extra-ordinary ray. Ordinarily, the ordinary and extra-ordinary planes do not coincide except in a case when the plane of incidence and the principal section are the same.

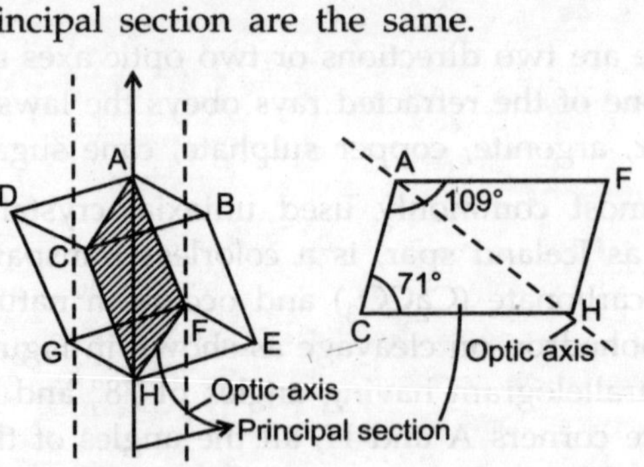

FIGURE 10

4.6 DOUBLE REFRACTION

In 1669, Erasmus Bartholinus discovered that when an unpolarized beam of light enters an anistropic crystal such as calcite or quartz, it splits into two plane polarized refracted beams travelling in different directions, one of which always obeys ordinary laws of refraction and the other, in general, does not obey them. This phenomenon is known as double refraction. The phenomenon of double refraction is easily investigated by placing a calcite crystal over a black mark on a piece of paper. The two images of the black mark are observed as shown in Figure 11(*a*) by O and E. Now if the crystal is rotated slowly about the vertical axis as shown in Figure 11(*a*) one of the two images remains fixed and other moves round the first (O). The stationary image behaves as if the crystal were an isotropic media such as water or glass and is called *ordinary image* and the ray which produces this image is called *ordinary ray (O-ray)*. The second image is called extra-ordinary image and the refracted ray corresponding to this image is called *extra-ordinary ray (E-ray)*.

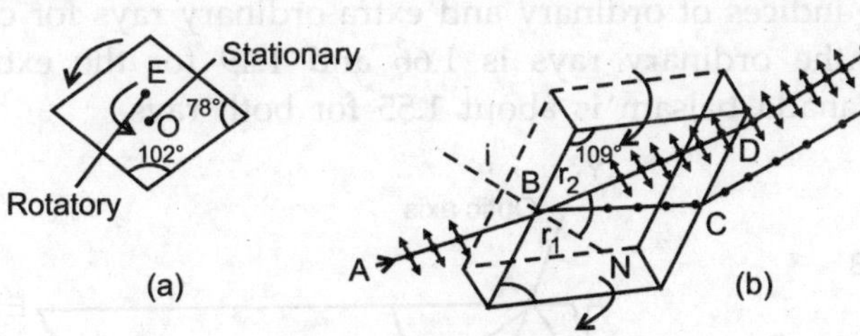

FIGURE 11

An incident ray AB is split up inside the crystal into O and E rays as shown in Figure 11(*b*). The O-ray travelling along BC makes an angle r_1 while the E-ray travelling along BD makes an angle of refraction r_2 with the normal NN′. The rays emerge from the crystal along CO and DE and are parallel to each other as well as to the incident beam. The refractive index of ordinary ray is, $\mu_o = \dfrac{\sin i}{\sin r_1}$ and that of extraordinary ray is, $\mu_e = \dfrac{\sin i}{\sin r_2}$. In calcite crystal $r_1 < r_2$, therefore $\mu_o > \mu_e$. Hence $v_o < v_e$, that is, inside the calcite crystal the E-ray travels faster as compared to the O-ray. Such crystals are called uniaxial negative crystals. There are other crystals in which $\mu_e > \mu_o$ or $v_o > v_e$, that is, inside the crystals the O-ray travels faster as compared to the E-ray. Such crystals are called uniaxial positive crystals. The example of negative crystals are tourmaline, ruby etc., and the examples of positive crystals are quartz, iron oxide, etc.

4.7 NICOL PRISM

In 1828, William Nicol designed an ingenious optical device by specially cutting calcite crystal for producing and analyzing plane polarized light. It is known as the Nicol prism. Its principle is based on the phenomenon of double refraction. Actually, the Nicol prism is not a prism but a parallelopiped.

Principle

It is well known that when an unpolarized beam enters the calcite crystal, it splits up into two plane polarized rays, as O-rays and E-rays, with vibrations in two mutually perpendicular planes. If by some optical means, we eliminate one of the beams, then we would obtain only one plane polarized beam. The Nicol prism is designed in such a way so as to eliminate the ordinary ray by total internal reflection. Hence only the extra-ordinary ray is transmitted through the prism.

Construction

The Nicol prism is constructed by taking a long rhomb of calcite so that its length AE is three times its breadth AD as illustrated in Figure 12(*a*). The end faces, ABCD and EFGH of the rhomb, are ground such that the angles C and E reduce from 71° to 68°. The crystal is then cut into two halves along the diagonal A'G' and the halves are cemented together by a thin layer of Canada balsam. The Canada balsam is a transparent substance whose refractive index lies midway between the indices of ordinary and extra-ordinary rays for calcite. The refractive index of the crystal for the ordinary rays is 1.66 and 1.49 for the extra-ordinary ray; the refractive index of the Canada balsam is about 1.55 for both rays.

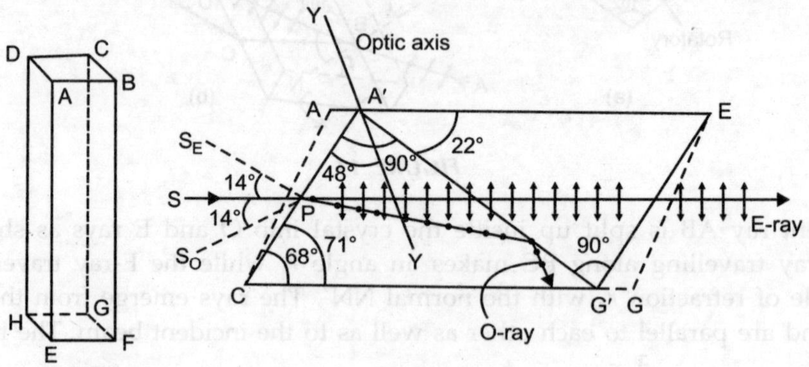

(a) (b)

FIGURE 12

Action

A ray SP of unpolarized light incident on one of the end faces A′C of the Nicol and moving nearly parallel to its length is split up into E and O rays whose vibrations are respectively perpendicular and parallel to the principal section of the Nicol as shown in Figure 12(*b*). The ordinary ray, in passing from calcite to Canada balsam, travels from a denser (μ_o = 1.66) to rarer (μ = 1.55) medium and will suffer total internal reflection only when the angle of incidence at the Canada balsam is greater than the critical angle for two media (calcite and Canada balsam), that is, the critical angle

$$C = \sin^{-1}\left(\frac{\mu}{\mu_o}\right) = \sin^{-1}\left(\frac{1.55}{1.66}\right) = \sin^{-1}(0.933) = 69°.$$

This totally reflected ray is finally absorbed by the blackend side CG′ of the prism and only the extra-ordinary ray is transmitted through the prism as it is travelling from the rarer (μ_e = 1.49) to the denser (μ = 1.55) medium. Hence in the Nicol prism, the ordinary ray is eliminated by total internal reflection and only the extra-ordinary ray with vibrations parallel to the principal section passes through, and the beam emerging from crystal is plane-polarized. Thus the Nicol prism can be used as polarizer.

Limitations

For the effective use of the Nicol prism as a polarizer, the angle of incidence on the face A′C, should be limited to a very narrow range of about 14° on either side of SP. For the angle of incidence greater than this value, the O-ray will strike the Canada balsam layer at an angle less than the critical angle and therefore will be transmitted along with the E-ray. It can be understood as follows:

When the angle of incidence on the crystal surface A′C increases, the angle of the O-ray on the calcite balsam layer decreases. When $\angle S_oPS$ becomes greater than 14°, the angle of incidence for the O-ray on the calcite balsam layer becomes less than the critical angle (69°), hence the O-ray also transmitted with the E-ray. Therefore, the light emerging from Nicol will not be plain polarized. When the angle of incidence on the crystal surface is decreased, or when $\angle S_EPS$ becomes greater than 14°, then μ_e becomes greater than μ for the balsam layer, that is, Canada balsam becomes optically less dens than calcite and the E-ray is totally internally reflected by the balsam. Therefore no light emerges from the Nicol. Thus to avoid the unwanted transmission of the O-ray and the total internal reflection of the E-ray, the incident light should be confined within an angle of 14° on either side of SP.

Nicol as an Analyzer

The Nicol prism is widely used to analyze the plane polarized light. Its analyzing action may be easily understood from Figure 13. When two Nicols are arranged coaxially, then the first Nicol prism P which produces plane polarized light is known as a polarizer whereas, the second Nicol which analyzes the polarized light coming from P is termed as an analyzer. Such a combination of two Nicols is called a polariscope. When the principal sections of the two Nicols are parallel as in Figure 13(*a*) and (*c*), then the vibrations in the E-ray, which are in the principal section of the polarizer, are also in the principal section of the analyzer. Consequently the E-ray from the polarizer is freely transmitted by the analyzer. In this position of parallel Nicols, the intensity of the emergent light is maximum.

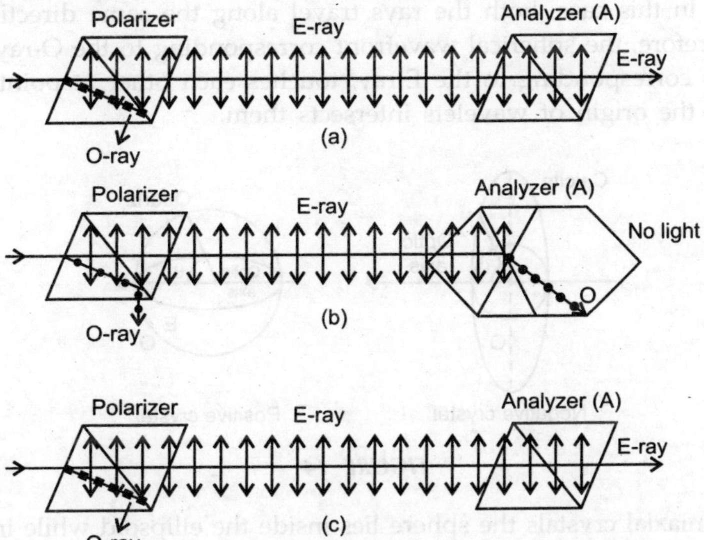

FIGURE 13

When the analyzer A is rotated about its axis from this position the intensity of the emergent E-ray decreases. When the principal section of two Nicols become mutually perpendicular, then no light emerges from the analyzer as shown in Figure 13(*b*). In this position the vibrations in the E-ray, which are in the principal section of the polarizer, become perpendicular to the principal section of the analyzer. It will thus have no component in the principal section of analyzer and will travel as an ordinary ray in it. Hence it is totally internally reflected at the balsam layer and therefore no light is transmitted though the analyzer. The above facts can be used for detecting the plane polarized light. If the given light, viewed through a rotating Nicol, shows variation in intensity with minimum zero, the given light is plane polarized.

4.8 HUYGENS' THEORY OF DOUBLE REFRACTION

Huygen explained the phenomenon of double refraction in uniaxial crystals on the basis of the wave theory of light using the principle of secondary wavelets. He made the following assumption:

- When a light wave strikes the surface of a doubly refracting crystal, every point of the crystal disturbed by the incident wavefront becomes the origin of two secondary wavelets, ordinary and extra-ordinary, which spread out into the crystal.
- For an ordinary ray (O-ray) for which the velocity is the same in all directions, the wavefront is spherical.
- For an extra-ordinary ray (E-ray) for which the velocity is different in all directions, the wavefront is an ellipsoid of revolution. (An ellipsoid of revolution is a surface obtained by rotating an ellipse about either the major or the minor axis. This axis of rotation is along the optic axis as shown in Figure 14).
- When a ray of light is incident along the optic axis, it is not split up into E and O components. In this case, both the rays travel along the same direction with the same velocity. Therefore, the spherical wavefront corresponding to the O-ray and the ellipsoid of revolution corresponding to the E-ray, touches each other at points where the optic axis through the origin of wavelets intersects them.

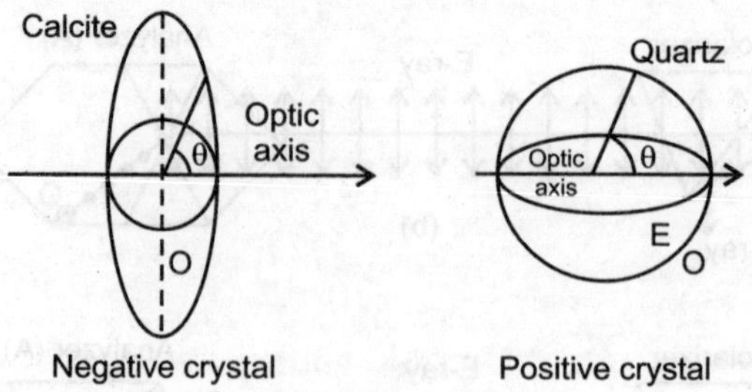

FIGURE 14

- In negative uniaxial crystals the sphere lies inside the ellipsoid while in positive uniaxial crystals the ellipsoid lies inside the sphere as illustrated in Figure 14.

The principles of Huygens' theory will now be applied to find the O and E-rays, and the refracted wavefronts in which the crystal plane is cut so that the optic axis occupies a special direction.

4.9 ELLIPTICALLY AND CIRCULARLY POLARIZED LIGHT

Plane polarized light represents the simple type of polarization. For such light the light vector vibrates the simple harmonically along a fixed straight line perpendicular to the direction of propagation. When two plane polarized beams of monochromatic light are superimposed, then under suitable conditions the resultant light vector may rotate in a plane perpendicular to the direction of propagation, and if its magnitude remains constant while the orientation varies regularly, then the tip of the vector traces a circle and the resultant light is said to be circularly polarized. If, however, the magnitude of the light vector varies periodically between its maximum and minimum values, the tip of the vector traces an ellipse and the resultant light is said to be elliptically polarized.

The Superposition of Two Plane Polarized Waves Having Perpendicular Vibrations

Let a beam of plane polarized light coming from a polarizer be incident normally on a thin plate of calcite crystal cut with faces parallel to the optic axis and oriented in such a way that the vibrations in the plane polarized light make an angle θ with the optic axis as shown in Figure 15, then the amplitude A of the vibrations in the incident plane polarized light are resolved into two components (E and O components) by the doubly refracting calcite plate as shown in Figure 16; the E-wave having vibrations parallel to the optic axis of amplitude A cos θ and the O-wave having vibrations perpendicular to the optic axis of amplitude A sin θ. As both the vibrations travel with different velocities in the same direction, a phase difference δ is introduced between the E-wave along PE and the O-wave along PO in traversing a distance of the plate. In the calcite plate $v_e > v_o$, therefore, the E-wave leads the O-wave. The displacements of E and O-waves along and perpendicular to the optic axis may be expressed by the equations.

$$x = A\cos\theta\sin(\omega t + \delta) \quad (1)$$

and $$y = A\sin\theta\sin\omega t \quad (2)$$

Substituting $A\cos\theta = a$

and $A\sin\theta = b$ in equations (1) and (2), we get

∴ $$x = a\sin(\omega t + \delta) \quad (3)$$

and $$y = b\sin\omega t \quad (4)$$

From equation (3), we have $$\frac{x}{a} = \sin\omega t\cos\delta + \cos\omega t\sin\delta \quad (5)$$

From equation (4), we have $\frac{y}{b} = \sin\omega t$ and $\sqrt{1-\left(\frac{y}{b}\right)^2} = \cos\omega t$

Substituting the sin ωt and cos ωt in equation (5), we get

$$\frac{x}{a} = \frac{y}{b}\cos\delta + \sqrt{\left(1-\frac{y^2}{b^2}\right)}.\sin\delta$$

or $$\frac{x}{a} - \frac{y}{b}\cos\delta = \sqrt{\left(1-\frac{y^2}{b^2}\right)}.\sin\delta$$

Squaring both sides, we get $\left(\frac{x}{a}-\frac{y}{b}\cos\delta\right)^2 = \left(1-\frac{y^2}{b^2}\right).\sin^2\delta$

Solving further, we get $\frac{x^2}{a^2}+\frac{y^2}{b^2}-\frac{2xy}{ab}\cos\delta = \sin^2\delta$ (6)

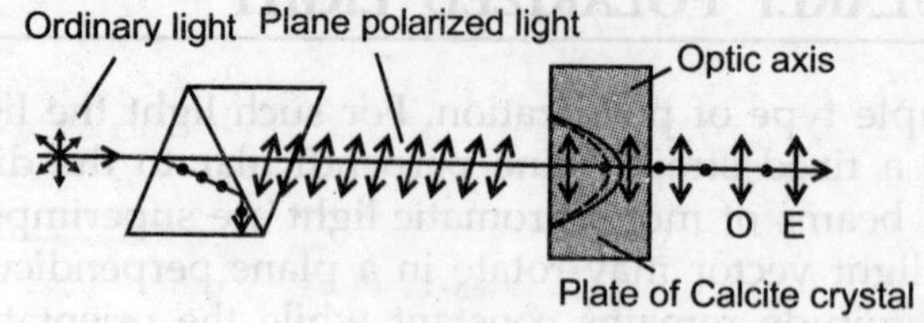

FIGURE 15

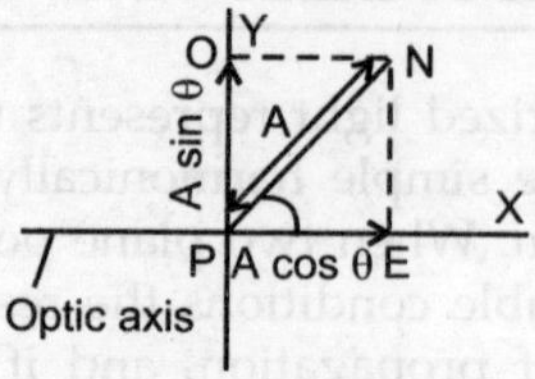

FIGURE 16

This is the equation of an oblique ellipse. Thus, in general the emergent light is elliptically polarized. The exact nature of the resultant motion and therefore of the light emerging from calcite depends upon the value of δ.

Special Cases

Case 1: When the thickness of the plate is such that

$$\delta = 0, 2\pi, 4\pi, ..., 2n\pi$$

then $\sin\delta = 0$ and $\cos\delta = 1$

then equation (6) reduces to $\frac{x^2}{a^2}+\frac{y^2}{b^2}-\frac{2xy}{ab} = 0$

or $\left(\frac{x}{a}-\frac{y}{b}\right)^2 = 0$ or $\pm y = \pm\left(\frac{b}{a}\right)x$ (7)

Thus the motion is described by a straight line, and both x and y have the same sign. Thus when two mutually perpendicular plane polarized waves are in phase then the emergent light plane is polarized with vibrations in the same plane as in the incident light as shown in Figure 17.

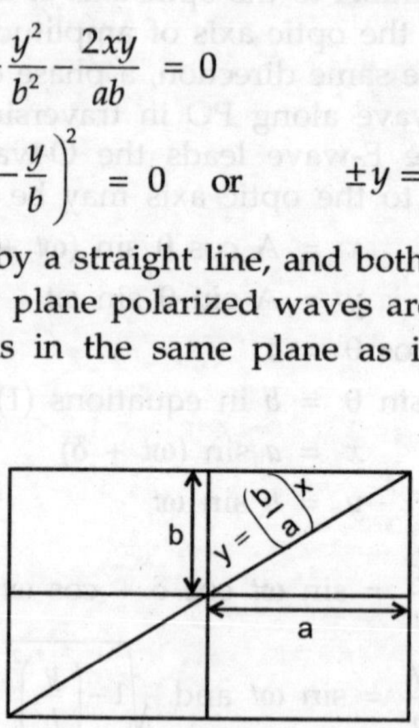

FIGURE 17

Case 2: When the thickness of the plane is such that

$\delta = \frac{\pi}{2}, \frac{3\pi}{2}, ..., \frac{(2n+1)\pi}{2}$, where n = 0, 1, 2, ... then $\sin\delta = 1$ and $\cos\delta = 0$, hence equation (6) reduces to

$$\frac{x^2}{a^2}+\frac{y^2}{b^2} = 1 \tag{8}$$

This is the equation of a simple ellipse. Thus when two mutually perpendicular plane polarized waves are in $\frac{\pi}{2}$ or $\frac{(2n+1)\pi}{2}$ phase then the emergent light is elliptically polarized (provided $a \neq b$). The plane of the ellipse being normal to the direction of propagation as shown in Figure 18.

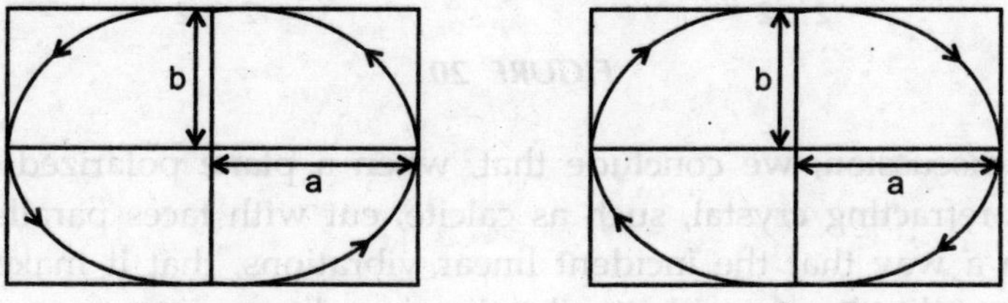

$\delta = \frac{\pi}{2}, \frac{5\pi}{2}, ...$ $\delta = \frac{3\pi}{2}, \frac{7\pi}{2}, ...$

FIGURE 18

Case 3: When the thickness of the plate is such that $\delta = \pi, 3\pi, 5\pi, ..., (2n + 1)\pi$, then $\sin \delta = 0$ and $\cos \delta = -1$, hence equation (6) reduces to

$$\frac{x^2}{a^2}+\frac{y^2}{b^2}+\frac{2xy}{ab} = 0$$

or $$\left(\frac{x}{a}+\frac{y}{b}\right)^2 = 0 \text{ or } y = -\frac{b}{a}x \tag{9}$$

This is again the equation of a straight line passing through the origin P. Thus the emergent light is again plane polarized having vibrations making an angle $2 \tan^{-1} (b/a)$ with that of incident light as shown in Figure 19.

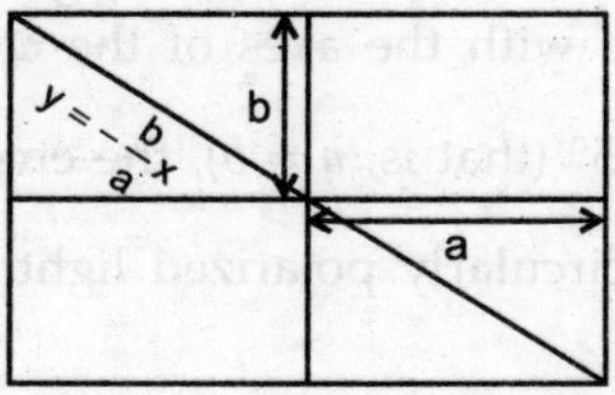

FIGURE 19

Case 4: When $a = b$ and $\delta = \frac{(2n+1)\pi}{2}$, $n = 0, 1, 2, ...$, then $\sin \delta = 1$, $\cos \delta = -1$, hence equation (6) reduces to $x^2 + y^2 = a^2$

This is the equation of a circle of radius a. Thus the emergent light is circular polarized as shown in Figure 20.

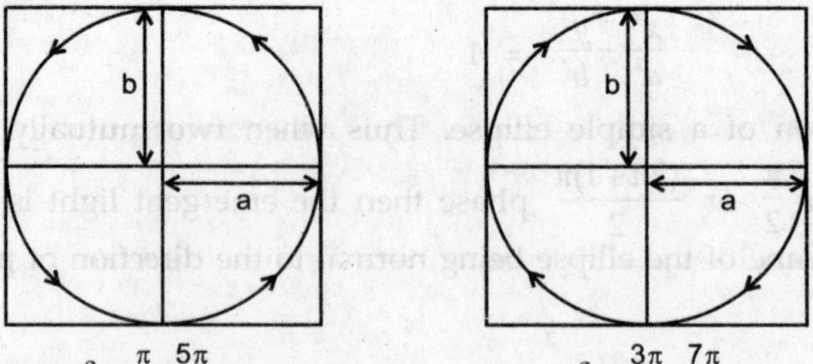

$\delta = \frac{\pi}{2}, \frac{5\pi}{2}, \ldots$ $\delta = \frac{3\pi}{2}, \frac{7\pi}{2}, \ldots$

FIGURE 20

Thus from this discussion, we conclude that, when a plane polarized light enters into a thin plate of doubly refracting crystal, such as calcite, cut with faces parallel to the optic axis but oriented in such a way that the incident linear vibrations, that it makes an angle θ with the optic axis. The amplitude of incident vibration is split up into two ordinary and extra-ordinary components which are travelling along the same direction with different velocities in the crystal and finally emerge as a single component. The shape of the resultant transmitted vibrations depend on the phase difference and hence on the thickness of the doubly refracting crystal plate.

The light emerging from the crystal plate is generally elliptically polarized, for $0 < \delta < \pi$ the emergent light is right handed elliptically polarized because in this case the rotation of the light vector is counterclockwise. Similarly for $\pi < \delta < 2\pi$, the emergent light is left handed elliptically polarized as the light vector rotates in a clockwise direction. When the thickness of the plate is such that the phase difference between the two components is an even multiple of π ($\delta = 2n\pi$), the emergent light is plane polarized with vibrations in the plane of incidence. On the other hand, if δ is an odd multiple of $\pi\,[\delta = (2n+1)\pi]$, the emergent light is plane polarized having vibrations in a direction making an angle $2\theta \left[= 2\tan^{-1}\left(\frac{b}{a}\right)\right]$ with the plane of incidence.

When $\delta = (2n + 1)\,\pi/2$, when $n = 1, 2, 3 \ldots$, the emergent light is elliptically polarized and its axes (major and minor) coincide with the axes of the component linear vibrations.

When $\delta = \frac{(2n+1)\pi}{2}$, and $\theta = 45°$ (that is, $a = b$), the emergent light is circularly polarized.

Thus the plane polarized and circularly polarized light are the special cases of elliptically polarized light.

4.10 QUARTER WAVE PLATE

A quarter wave plate, also called a phase retardation plate, is widely used for the production and the detection of elliptically and circularly polarized light. It is a plate of uniaxial doubly refracting crystal cut with the optic axis parallel to the refracting faces and whose thickness is such as to produce a path difference of $\frac{\lambda}{2}$ $\left(\text{or phase difference of } \frac{\pi}{2}\right)$ between the ordinary and extra-ordinary waves.

When a plane polarized beam is incident normally on such a plate, the vibrations in the beam break into two; one vibration (O) along the optic axis and the other vibration (E) perpendicular to it. These components travel in the same direction along the normal to the faces but with different velocities. If t is the thickness of the quarter wave plate, μ_O and μ_E are the refractive indices for the ordinary and extra-ordinary beams respectively, then the path difference introduced between the O and E-ray is given by

$$\Delta = (\mu_O \sim \mu_E)t$$

In negative uniaxial crystals like calcite, the E-ray travels faster than the O-ray so that $\mu_O > \mu_E$

$$\therefore \quad \Delta = (\mu_O - \mu_E)t$$

In positive uniaxial crystals like quartz, the O-ray travels faster than the E-ray so that $\mu_E > \mu_O$

$$\therefore \quad \Delta = (\mu_E - \mu_O)t$$

For a quarter wave plate, the path difference between the two waves on emerging must be equal to $\frac{\lambda}{4}$.

$$\therefore \quad \Delta = \frac{\lambda}{4} = (\mu_O - \mu_E)t \qquad \text{[For calcite plate]}$$

$$\therefore \quad t = \frac{\lambda}{4(\mu_O - \mu_E)} \qquad (1)$$

For positive quartz crystal

$$t = \frac{\lambda}{4(\mu_E - \mu_O)} \qquad (2)$$

Practically, the quarter wave plate has to be supported between glass plates. If the vibrations of the incident plane polarized beam make an angle of 45° with the optic axis of the quarter wave plate, then the emergent light is circularly polarized. If the vibration of the incident plane polarized light do not make an angle of 45°, 0° or 90° with the optic axis, the emergent light is elliptically polarized.

4.11 HALF WAVE PLATE

A half wave plate, also called a retardation plate, is used in the Laurntz half-shade polarimeter for the measurement of optical rotation. It is a plate of uniaxial doubly refracting crystal cut with the optic axis parallel to the refracting faces and whose thickness is such as to produce a path difference of $\lambda/2$ (or phase difference of π) between the ordinary and extra-ordinary waves. If t is the thickness of the half wave plate, and μ_O and μ_E are the refractive indices for the ordinary and extra-ordinary beams respectively, then the path difference introduced between the O and E-rays is given by $\Delta = (\mu_O - \mu_E)t$

For negative uniaxial calcite crystal, $\mu_O > \mu_E$

$$\therefore \quad \Delta = (\mu_O - \mu_E)t$$

For positive uniaxial quartz crystal, $\mu_E > \mu_O$

$$\therefore \qquad \Delta = (\mu_E - \mu_O)t$$

For a half wave plate, the path difference between the two waves on emerging must be equal to $\lambda/2$

$$\therefore \qquad \Delta = \frac{\lambda}{2} = (\mu_O - \mu_E)t$$

$$\therefore \qquad t = \frac{\lambda}{2(\mu_O - \mu_E)} \qquad \text{[For calcite crystal]}$$

and

$$t = \frac{\lambda}{2(\mu_E - \mu_O)} \qquad \text{[For quartz crystal]}$$

If the incident plane polarized beam makes an angle θ with the optic axis of the plate then the emergent light is also plane polarized with vibrations inclined at an angle of 2θ with the optic axis.

A major disadvantage with these plates is that they are true only for one particular wavelength because their thickness depends upon the wavelength of light used.

Example 9. *Find the thickness of a quarter wave plate when the wavelength of light is equal to 5890 Å and $\mu_O = 1.55$, $\mu_E = 1.54$.*

Solution. The thickness of a quarter wave plate for a crystal for which $\mu_O > \mu_E$ is given by

$$t = \frac{\lambda}{4(\mu_O - \mu_E)}$$

Here $\lambda = 5890 \text{ Å} = 5890 \times 10^{-8}$ cm, $\mu_O = 1.55$ and $\mu_E = 1.54$

$$\therefore \qquad t = \frac{5890 \times 10^{-8}}{4(1.55 - 1.54)} = \frac{5890 \times 10^{-8}}{4 \times 0.01} = \mathbf{1.47 \times 10^{-3} \text{ cm.}}$$

Example 10. *Calculate the thickness of a quarter wave plate of quartz for sodium light of wavelength 5893 Å. The ordinary and extra-ordinary refractive indices for sodium are 1.54425 and 1.55336 respectively.*

Solution. The thickness of a quarter wave plate for positive crystal like quartz is

$$t = \frac{\lambda}{4(\mu_O - \mu_E)}$$

Here $\mu_O = 1.54425$, $\mu_E = 1.55336$ and for sodium light

$$\lambda = 5893 \text{ Å} = 5.893 \times 10^{-5} \text{ cm}$$

$$\therefore \qquad t = \frac{5.893 \times 10^{-5}}{4(1.55336 - 1.54425)}$$

$$= \frac{5.893 \times 10^{-5}}{4 \times 911 \times 10^{-5}} = \mathbf{1.62 \times 10^{-3} \text{ cm.}}$$

Example 11. *A quarter wave plate is meant for $\lambda_O = 5.893 \times 10^{-5}$ cm. What phase retardation ϕ will show for $\lambda = 4.358 \times 10^{-5}$ cm? (Neglect changes of μ_O and μ_E with λ).*

Solution. If the quarter wave plate is of the smallest thickness then the thickness is given by

$$t = \frac{\lambda}{4(\mu_O - \mu_E)}$$

or $\quad$ Path difference $\Delta = (\mu_O - \mu_E)t = \dfrac{\lambda_O}{4}$

The phase retardation for $\quad \lambda = \dfrac{2\pi}{\lambda}$ (path difference)

$\therefore$ Phase retardation for $\quad \lambda = \left(\dfrac{2\pi}{\lambda}\right)\cdot\left(\dfrac{\lambda_O}{4}\right)$

In the given problem, $\quad \lambda_O = 5.893 \times 10^{-5}$ cm and $\lambda = 4.358 \times 10^{-5}$ cm

$$\therefore \quad \text{Phase retardation} = \frac{2\pi}{(4.358 \times 10^{-5})} \times \frac{5.893 \times 10^{-5}}{4}$$

$$= \mathbf{0.67\ \pi\ radian.}$$

Example 12. *A beam of linearly polarized light is changed into circularly polarized light by passing it through a slice of crystal 0.003 cm thick. Calculate the difference in the refractive index of the two rays in the crystal assuming this to be the minimum thickness that will produce the effect and that the wavelength is 6×10^{-5} cm.*

Solution. We know that the least phase difference between O and E-waves on emergence from the doubly refracting plate must be $\dfrac{\pi}{2}$ if the incident linear polarized light is changed into circularly polarized light.

Thus the slice of crystal must be a quarter wave plate of thickness

$t = \dfrac{\lambda}{4(\mu_O - \mu_E)}$ where $\mu_O \sim \mu_E$ represents the difference in refractive index, therefore

$$(\mu_O \sim \mu_E) = \frac{\lambda}{4t}$$

Here $\quad \lambda = 6 \times 10^{-5}$ cm and $t = 0.003$ cm

$$\therefore \quad \mu_O \sim \mu_E = \frac{6 \times 10^{-5}}{4 \times 0.003} = \mathbf{5 \times 10^{-3}}.$$

Example 13. *Calculate the thickness of doubly refracting crystal required to introduce a path difference of $\dfrac{\lambda}{2}$ between the O and E-rays when $\lambda = 6000$ Å, $\mu_O = 1.65$ and $\mu_E \doteq 1.48$.*

Solution. The required thickness of a doubly refracting crystal to introduce a path difference of $\dfrac{\lambda}{2}$ is given by

$$t = \frac{\lambda}{2(\mu_O - \mu_E)}$$

Here $\quad \mu_O = 1.65$, $\mu_E = 1.48$ and $\lambda = 6000$ Å $= 6000 \times 10^{-8}$ cm

$$\therefore \quad t = \frac{6000 \times 10^{-8}}{2(1.65 - 1.48)} = 0.000176 \text{ cm} = \mathbf{1.76 \times 10^{-4}\ cm.}$$

Example 14. *The faces of a quartz plate are parallel to the optic axis of the crystal (a) What is the thinnest possible plate that would serve to put the ordinary and extra-ordinary rays of $\lambda = 5890$ Å a half wave apart on their exit? (b) What multiples of this thickness would give the same result? The indices of refraction of quartz are $\mu_E = 1.553$ and $\mu_O = 1.544$.*

Solution (*a*) The thickness of the thinnest possible plate that would produce a path difference of $\frac{\lambda}{2}$ for quartz is given by

$$t = \frac{\lambda}{2(\mu_E - \mu_O)}$$

Here $\lambda = 5890 \text{ Å} = 5890 \times 10^{-8}$ cm

$\mu_E = 1.553$ and $\mu_O = 1.544$

$$\therefore \quad t = \frac{5890 \times 10^{-8}}{2(1.553 - 1.544)} = \frac{5890 \times 10^{-4}}{2 \times 0.009} = \mathbf{3.27 \times 10^{-3} \text{ cm.}}$$

(*b*) The thickness which would give the same result is *t*, 3*t*, 5*t*, ... = 3.27×10^{-3} cm, $3 \times 3.27 \times 10^{-3}$ cm, $5 \times 3.27 \times 10^{-3}$ cm, ...

Example 15. *A phase retardation plate of quartz has thickness 0.1436 mm. For what wavelength in the visible region (4500 Å – 8000 Å) will it act (i) quarter wave plate, (ii) half wave plate? Given:* $\mu_O = 1.5443$, $\mu_E = 1.5533$.

Solution. (*i*) The thickness of a quarter wave plate is given by $t = \frac{\lambda}{2(\mu_E - \mu_O)}$ for any odd multiple,

that is, $$t = \frac{(2n+1)\lambda}{4(\mu_E - \mu_O)} \quad \text{where } n = 0, 1, 2, 3, ...$$

$$\therefore \quad \lambda = \frac{4(\mu_E - \mu_O)t}{(2n+1)}$$

Here $\mu_E = 1.5533$, $\mu_O = 1.5443$ and $t = 0.1436$ mm $= 0.01436$ cm

Therefore, $$\lambda = \frac{4 \times (1.5533 - 1.5443) \times 0.01436}{(2n+1)} = \frac{51700 \text{ Å}}{(2n+1)}$$

Substituting $n = 0, 1, 2, 3, ...$, we get λ = 51700 Å, 17233 Å, 1034 Å, 7386 Å, 5744 Å, 4700 Å, ...

Hence in the visible region, λ = 7386 Å, 5744 Å and 4700 Å

(*ii*) For a half wave plate, we have

$$\lambda = \frac{2(\mu_E - \mu_O)t}{(2n+1)} = \frac{2 \times 0.009 \times 0.01436}{(2n+1)} = \frac{25850 \text{ Å}}{(2n+1)}$$

Substituting $n = 0, 1, 2, 3, ...$, we get

λ = 25850 Å, 8617 Å, 5170 Å, 3693 Å, ...

Hence in the visible region, λ = 5170 Å and 3693 Å.

4.12 PRODUCTION OF CIRCULARLY AND ELLIPTICALLY POLARIZED LIGHT

Circularly Polarized Light

Circularly polarized light is produced by recombining two mutually perpendicular vibrations of equal amplitude and time and there is a phase difference of $\pi/2$ or path difference of $\lambda/4$

between them. The experimental arrangement for producing circularly polarized light is depicted in Figure 21. The unpolarized light is allowed to pass through a Nicol P to get a plane polarized beam of light. Another Nicol A is placed at some distance from P such that the two Nicols are in crossed position. In this position of Nicols no light is transmitted from A and hence the field of view appears dark. A quarter wave plate Q mounted on a tube C is introduced between the two Nicols in the path of the plane polarized beam (without disturbing the positions of P and A) such that the polarized beam falls normally on a quarter wave plate. The amplitude of the plane polarized beam on entering the plate is split up into two mutually perpendicular components with a phase change of $\pi/2$. The field of view may now be bright. The quarter wave plate is rotated about the outer fixed tube B until the field of view is again dark. This happens when vibrations of incident light are along the optic axis of quarter wave plate. Keeping Q fixed, tube C is rotated so that the mark M on Q exactly coincides with the zero mark on C. Now by rotating the quarter wave plate the mark M is made to coincide with the 45° mark on C. Now the position of quarter wave plate Q is such that the vibrations in the incident plane polarized light make an angle 45° with the optic axis of the plate. In this position the plane polarized light entering the quarter wave plate Q is split up into O and E components of equal amplitudes. In this way the light emerging from quarter wave plate is circularly polarized. If the analyzing Nicol A is rotated, no change in intensity is appeared in the field of view.

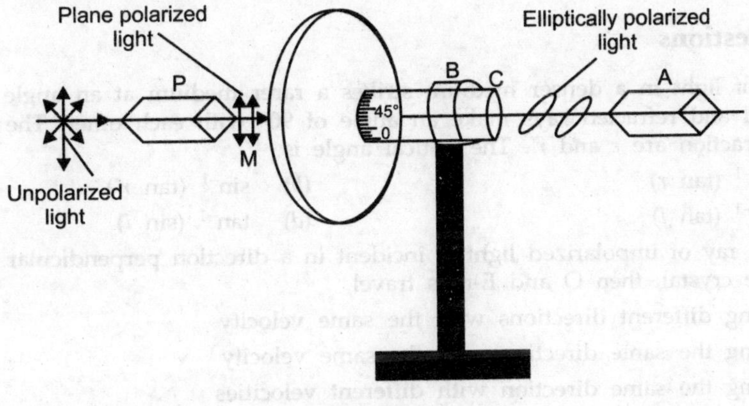

FIGURE 21

Elliptically Polarized Light

Elliptically polarized light is produced by recombining two mutually perpendicular vibrations of unequal amplitudes and there is a phase difference of $\pi/2$ of path difference of $\lambda/4$ between them. The experimental arrangement for producing elliptically polarized light is depicted in Figure 15. The unpolarized light is allowed to pass through a Nicol P to get a plane polarized beam of light. Another Nicol A is placed at some distance from P such that the two Nicols are in crossed position so that no light is transmitted from A. A quarter wave plate Q is introduced between the two Nicols as shown in Figure 21 so that the field of view appears to be illuminated. The quarter wave plate is rotated about the fixed outer tube B until the field of view is again dark. This happens when vibrations of incident light are along the optic axis. Again the quarter wave plate is rotated to such a position that vibrations in the plane polarized incident beam make an angle θ (other than 0°, 45° and 90°) with the optic axis of the plate. In this position the plane polarized light entering the quarter wave plate is split up into O and E components of unequal amplitudes. In this way the light emerging from the quarter wave plate is elliptically polarized. If the analyzing Nicol prism A is rotated, the change in intensity appears between a maximum to a minimum value in the field of view, that is, the minimum intensity is not zero.

Conversion of Elliptically Polarized Light into Circularly Polarized Light

To convert elliptically polarized light into circularly polarized light, the elliptically polarized light is first converted into plane polarized light and then into circularly polarized light.

To do this the elliptically polarized light is allowed to fall normally on a quarter wave plate. When elliptically polarized light enters a quarter wave plate, it breaks up into two unequal components vibrating mutually perpendicular and thus having a phase difference of $\pi/2$. The quarter wave plate introduces an additional phase difference of $\pi/2$ between the two components (E and O-components) and as a result, the total phase difference now becomes either π or 0.

EXERCISES

Objective Questions

1. A ray of light in a denser medium strikes a rarer medium at an angle of incidence *i*. The reflected and refracted rays make an angle of 90° with each other. The angles of reflection and refraction are *r* and *r′*. The critical angle is

(*a*) $\sin^{-1}(\tan r)$ (*b*) $\sin^{-1}(\tan r')$
(*c*) $\sin^{-1}(\tan i)$ (*d*) $\tan^{-1}(\sin i)$

2. When a ray of unpolarized light is incident in a direction perpendicular to the optic axis of a calcite crystal, then O and E-rays travel

(*a*) along different directions with the same velocity
(*b*) along the same direction with the same velocity
(*c*) along the same direction with different velocities
(*d*) along different directions with different velocities

3. When a ray of unpolarized light is incident in a direction parallel to the optic axis of a calcite crystal, the O and E-rays travel

(*a*) along different directions with the same velocity
(*b*) along the same direction with the same velocity
(*c*) along the same direction with different velocities
(*d*) along different directions with different velocities

4. When a ray of unpolarized light is passed through a uniaxial negative crystal, it is split up into two refracted rays. The relation between the velocities of E and O-rays are

(*a*) $v_o > v_e$ (*b*) $v_e > v_o$
(*c*) $v_e = v_o$ (*d*) $v_o \geq v_e$

5. When a ray of unpolarized light is passed through a uniaxial positive crystal, it is split up into two refracted rays. The relation between the refractive indices of E and O-rays is

(*a*) $\mu_o > \mu_e$ (*b*) $\mu_o < \mu_e$
(*c*) $\mu_o = \mu_e$ (*d*) $\mu_o > \mu_e$

6. In order to avoid the transmission of O-rays and the total internal reflection of E-rays, the angle within which the incident light should be confined on either side of a line parallel to the long side of the Nicol prism is
   (a) 15° (b) 30°
   (c) 14° (d) 28°
7. Unpolarized light falls on two Nicols so oriented that no light is transmitted. If a third Nicol is placed between them, not parallel to either of the two Nicols in question, then:
   (a) no light is transmitted (b) some light is transmitted
   (c) light may or may not be transmitted (d) exactly 50% light is transmitted
8. The Nicol prism is based on the phenomenon of
   (a) refraction (b) reflection
   (c) double refraction (d) scattering
9. When a plane polarized light is incident on a quarter wave plate with its vibrations making an angle of 45° with the optic axis, the emergent light is
   (a) elliptically polarized (b) plane polarized
   (c) a mixture of elliptically and circularly polarized
   (d) circularly polarized
10. When a plane polarized light is passed through a half wave plate, the emergent light is
    (a) elliptically polarized
    (b) plane polarized
    (c) a mixture of elliptically and circularly polarized
    (d) circularly polarized
11. Circularly polarized light is produced if the amplitudes of the ordinary and extra-ordinary rays are equal and there is a phase difference of
    (a) π (b) $\pi/2$
    (c) $\pi/4$ (d) zero
12. Elliptically polarized light is produced if the amplitudes of the ordinary and extra-ordinary rays are unequal and there is a phase difference of
    (a) π (b) $\pi/2$
    (c) $\pi/4$ (d) zero
13. When elliptically polarized light, after passing through a quarter wave plate, is observed through a rotating Nicol, the emergent light would have shown
    (a) the variation of intensity with minimum not zero
    (b) no variation in intensity
    (c) the variation of intensity with minimum zero
    (d) the intensity of incident an emergent light is same
14. The transverse nature of light is conclusively proved by the phenomenon of
    (a) polarization (b) interference
    (c) diffraction (d) reflection
15. A polarized beam is one in which
    (a) the electric vector vibrates in all possible directions
    (b) the magnetic vector vibrates in all possible directions
    (c) the electric vector vibrates in a direction perpendicular to the direction of the propagation of wave
    (d) none of the above

16. An unpolarized beam of a transverse wave is one whose vibrations
(*a*) are combined in a single plane
(*b*) occur in all directions perpendicular to the direction of propagation
(*c*) occur in all directions
(*d*) have not passed through a polaroid disc

17. The plane of polarization is the plane which contains
(*a*) the direction of propagation of a wave and the direction of vibration of an electric vector
(*b*) the direction of propagation of a wave but perpendicular to the direction of vibration of an electric vector
(*c*) is normal in the direction of propagation but contains the direction of vibration of an electric vector
(*d*) none of above

18. If the light is polarized by reflection, then the angle between a reflected and refracted ray is
(*a*) p (*b*) $\pi/2$
(*c*) π (*d*) $\pi/4$

19. When an unpolarized beam is incident on a parallel angle, then which incident is correct?
(*a*) the reflected light plane polarized but the transmitted light is partially polarized
(*b*) the reflected light is partially polarized but the refracted light is plane polarized

20. In the propagation of light waves, the angle between the direction of propagation and the plane of polarization is
(*a*) 0° (*b*) 45°
(*c*) 90° (*d*) 180°

21. The angle between the plane of polarization and plane of vibration is
(*a*) 180° (*b*) 90°
(*c*) 45° (*d*) 0°

22. If unpolarized light falls on a transparent surface at a polarizing angle, then the angle between reflected and refracted rays is
(*a*) 0° (*b*) 45°
(*c*) 90° (*d*) 180°

23. The intensity of the polarized light transmitted through the analyzer varies as the
(*a*) tangent of the angle between the planes of transmission of the polarizer and analyzer
(*b*) square of the tangent of the angle between the plane of transmission of the polarizer and analyzer
(*c*) square of the cosine of the angle between the planes of transmission of the polarizer and analyzer
(*d*) square of the sine of the angle between the planes of transmission of the polarizer and analyzer

24. A ray of light is incident on the surface of a glass plate at an angle of incidence equal to Brewster's angle ϕ. If μ represents the refractive index of glass with respect to air, then the angle between the reflected and refracted rays is
(*a*) $90 + \phi$ (*b*) $\sin^{-1}(\mu \cos \phi)$
(*c*) 90° (*d*) $90° - \sin^{-1}\left(\frac{\sin \phi}{\mu}\right)$

25. In a uniaxial crystal, the number of directions along which the two refracted rays travel with
   (a) the same velocity is one
   (b) the same velocities are two
   (c) the same velocities are three
   (d) none of these
26. When a light ray in air is incident on glass at 57°, the reflected ray is completely polarized. If the same ray is incident on water at an angle θ, and the reflected ray is also completely polarized, then
   (a) $\theta > 57°$
   (b) $\theta < 57°$
   (c) $\theta = 57°$
   (d) $\theta = 90°$

Short Questions for Discussion

1. Discuss what is meant by polarization of light?
2. Discuss how the transverse nature of light waves is established by the polarization of light.
3. Discuss a simple method of producing and detecting the polarization of light.
4. Explain tightness when polarized by reflection.
5. What is law of Malus of polarization of light?
6. State Brewster's law.
7. Discuss the concept of double refraction.
8. Define the blunt corners of a calcite crystal producing double refraction.
9. Define the plane of vibration and plane of polarization of a plane polarized light.
10. What is meant by the optic axis of a biaxial doubly refracting crystal?
11. What are E-rays and O-rays in the phenomena of double refraction?
12. What are quarter wave plates and half wave plates?
13. Analyze plane, circularly and elliptically polarized light by a quarter wave plate.
14. How would you produce and analyze plane polarized light by Nicol prism?
15. How would you distinguish between circularly and elliptically polarized light?

Long Questions for Discussion

1. How is the transverse nature of light waves established by the polarization of light?
2. Describe various methods of producing and detecting plane polarized light.
3. What is the polarization of light? Distinguish between plane polarized light and unpolarized light. Explain the terms plane of vibration and plane of polarization.
4. State Brewster's law. Show that when a ray is incident at the Brewster's angle the reflected ray is perpendicular to the refracted ray.
5. (a) Explain the law of Malus.
   (b) Explain how light is polarized by reflection.
6. Describe double refraction. Describe the construction, working, and use of a Nicol prism.
7. Describe the construction of a Nicol prism. Explain how it can be used as a polarizer and as an analyzer.
8. Describe the phenomenon of double refraction in uniaxial crystals. How is double refraction explained by Huygens' theory?
9. What is meant by plane polarized, circularly polarized and elliptically polarized light? Show that the plane polarized light and circularly polarized light are speed cases of circulatory polarized light.
10. What are quarter wave and half wave plates? Explain their use in the study of different types of polarized light.

Chapter 5

Nuclear Structure and Nuclear Forces

5.1 OVERVIEW

From the scattering of α-particles, Rutherford concluded that the atom of any element consists of a central core called a nucleus with electrons moving around it. The entire mass of the atom and its positive charge is concentrated inside the nucleus. The mass of the electron (9.108×10^{-81} kg) is insignificantly small. The nucleus consists of two particles, the proton and the neutron. Their mass are nearly the same as shown below:

$$\text{Mass of the proton} = 1.67161 \times 10^{-27} \text{ kg}$$

$$\text{Mass of the neutron} = 1.67492 \times 10^{-27} \text{ kg}$$

The proton is a positively charged particle while the neutron is a neutral particle, *i.e.*, it carries no charge. The charge on the proton is 1.6×10^{-19} coulomb. Both the proton and neutron together in the nucleus are called nucleons. The number of protons in the nucleus is called the atomic number and the sum of protons and neutrons is called the mass number. The stability of a nucleus depends on the relative number of protons and neutrons.

From natural radioactivity considerations it was established that different particles, like α particles, β particles as well as electromagnetic radiations in the form of γ rays, are emitted from the nucleus. In a real sense, the γ rays are not the constituents of nuclei but they are emitted when the nucleus in an excited state returns to its normal state just like the visible light. Research on artificial radioactivity has shown that many particles like α particles, protons, neutrons, positrons, β particles etc., enter into the constitution of the nuclei in one way or the other. Finally, cosmic ray studies demonstrated the existence of another fundamental particle, meson, which occurs in many forms as regards charge and mass.

Now if the nucleus consists of protons and neutrons only, how are these particles held together in a neutral atom? It appears that Coulomb's law of force is not valid in this case, first because the electric force between two protons is repulsive and second because the uncharged neutrons cannot exert any electric forces. The forces of attraction that keep the protons and neutrons together cannot be gravitational because they are too small. Now the problem is to find out which of the elementary particles enter into the actual constitution of nuclei.

5.2 PROTON-NEUTRON THEORY OF NUCLEAR COMPOSITION

The discovery of the neutron by Chadwick led Heisenberg to propose the proton-neutron hypothesis in 1932. According to the Heisenberg hypothesis, the nuclei are composed of protons and neutrons. The total number of protons and neutrons (nucleons) in the nucleus gives its mass number A. The number of protons in the nucleus is equal to the atomic number Z and the number of neutrons is equal to (A–Z). The total charge of the nucleus is due to protons only. To neutralize the positive charge of the nucleus *i.e.*, to make the atom electrically neutral, the number of extra nuclear electrons is Z. In case of carbon A = 12 and Z = 6, hence a carbon nucleus consists of 6 protons and 6 neutrons.

The stability of atomic nuclei depends upon the neutron/proton ratio. The neutron/proton for stable nuclides is always greater than one. For the heaviest stable nuclides such as $_{22}Pb^{208}$, the ratio exceeds 1:5. Thus to maintain the nuclear stability, the number of neutrons must be greater than the number of protons. As the number of protons increases, the electrostatic repulsive force also increases, but the additional short range attractive nuclear forces [(*n–n*) and (*n–p*)] compensate the growing potential repulsion.

5.3 STATIC PROPERTIES OF A NUCLEUS

Some of the important properties of an atomic nucleus are given below:

(*a*) *Nuclear Mass:* The mass of the nucleus is the sum of the masses of the neutrons and protons contained in it. This is usually expressed in terms of an atomic mass unit (a.m.u.). One a.m.u. = 1.66×10^{-27} kg. For example, $_6C^{12}$ nucleus has a mass of 12 a.m.u. so that its mass number is A = 12.

(*b*) *Nuclear Charge:* The charge on the nucleus is due to protons contained in it. The charge on each proton is $+1.66 \times 10^{-19}$ coulomb which is equal in magnitude to the charge of an electron. Taking the charge of a proton as one unit, the total charge on the nucleus is numerically equal to the number of protons. For example, a hydrogen nucleus, (*i.e.*, proton) carries a single unit charge.

(*c*) *Nuclear Radius:* Nuclear diameters can be measured in a number of ways, two of which are (1) by scattering of fast protons or neutrons and (2) by scattering of high energy electrons. Rutherford, as a result of his experiments on α particle scattering by thin metallic foils, concluded that the distance of the closest approach of the α particle to the nucleus of the scatterer can be regarded as a measure of the size of the nucleus. He found the distance to be of the order of 10^{-14} m. The radius of an atomic nucleus (assumed spherical) can be determined from the following empirical relation:

$$r = r_0 A^{1/3}$$

When r_0 is a linear constant and has an average value of 1.4×10^{-15} m, A is the mass number of the nucleus. Radius of the nuclei of few elements are given below:

Carbon (A = 12)

$$r = 1.4 \times 10^{-15} \times (12)^{1/3} = 3.21 \times 10^{-15} \text{ m}$$

Copper (A = 63)

$$r = 1.4 \times 10^{-15} \times (63)^{1/3} = 5.97 \times 10^{-15} \text{ m}$$

Uranium (A = 238)

$$r = 1.4 \times 10^{-15} \times (238)^{1/3} = 8.68 \times 10^{-15} \text{ m.}$$

(*d*) *Nuclear Density:* The density of the nucleus can be calculated as follows:

$$\text{Volume of the nucleus} = \frac{4}{3}\pi r^3$$

where r is the radius of the nucleus.

Taking $r = 1.5 \times 10^{-15} \times A^{1/3}$ meter,

we have
$$\text{Volume of the nucleus} = \frac{4}{3}\pi\ (1.5 \times 10^{-15})^3\ \text{A meter}^3$$
$$= 14.15 \times 10^{-45}\ \text{A meter}^3$$
$$\text{Mass of the nucleus} = \text{A} \times \text{mass of proton (approximately)}$$
$$= 1.673 \times 10^{-27}\ \text{A kg}$$

$$\therefore \quad \text{Density of nucleus} = \frac{\text{mass of the nucleus}}{\text{volume of the nucleus}}$$

$$= \frac{1.673 \times 10^{-27}\ \text{A}}{14.15 \times 10^{-45}\ \text{A}}\ \text{kg/meter}^3$$

$$= 1.18 \times 10^{17}\ \text{kg/meter}^3.$$

(*e*) *Nuclear Spin:* Like the electron, the proton and neutron each have a spin angular momentum = $\frac{1}{2}.(h'2\pi)$. Each contribute a spin to the nucleus and combine to give it a spin angular momentum. Due to the spin, the nucleus possesses a magnetic moment. Actually, in atomic spectra many spectral lines are found to have a fine structure which is explained on the basis of electron spin. It is further observed that the fine structure lines show a still finer structure known as hyperfine structure which has been attributed to nuclear spin.

(*f*) *Nuclear Quantum States:* From the study of artificial radioactivity and α and γ ray spectra, it would become clear that every nucleus possesses a set of quantum states in a corresponding number of discrete energy levels. Transitions betwen different nuclear states are accompained by the emission of γ rays.

5.4 THE ATOMIC MASS UNIT AND MASS ENERGY EQUIVALENCE

Atomic Mass Unit

In nuclear physics, the atomic mass unit is the common unit of mass and energy. Before 1960, the isotope of oxygen ${}_6O^{16}$ was taken as the reference for the atomic mass unit. The one atomic mass unit was taken as one sixteenth of the mass of one atom of oxygen.

$$\therefore \quad 1 \text{ a.m.u.} = \frac{{}_8O^{16} \text{mass}}{16}$$

$$= \frac{16 \text{ grams/mole}}{6.03 \times 10^{23} \text{ atoms/mole}} \times \frac{1}{16}$$

$$= \frac{1}{6.03 \times 10^{23}} \text{ grams}$$

In 1960, the General Assembly of the International Union of pure and applied physics adopted carbon ${}_6C^{12}$ as the reference of the atomic mass unit. One atomic mass unit is taken as one twelfth of the mass of a carbon atom. Accordingly 1 a.m.u. = 1.66038×10^{-27} kg. The mass of all particles and atoms are found in terms of a.m.u.

Mass Energy Equivalence

Einstein established that mass is also a form of energy. The mass m of a particle is equivalent to its energy given by

$$E = mc^2$$

where c is the velocity of light. The energy E is called mass energy. The energy may be emitted in the form of γ rays of frequency ν given by

$$E = mc^2 = h\nu$$

where h is Planck's constant.

Let us calculate energy in joule or electron volt when a mass equal to 1 a.m.u. is converted into energy.

We know that $E = mc^2$

Here $m = 1 \text{ a.m.u.} = 1.66 \times 10^{-27}$ kg

and $c = 3 \times 10^8$ m/sec

$$\therefore \quad E = (1.66 \times 10^{-27})(3 \times 10^8)^2$$

$$= 1.49 \times 10^{-10} \text{ joule}$$

$$= \frac{1.49 \times 10^{-10}}{1.6 \times 10^{-19}} \text{eV} = 931 \times 10^6 \text{ eV} = 931 \text{ MeV}.$$

5.5 THE MASS DEFECT AND PACKING FRACTION

From the study of masses of the isotopes by the spectrograph method, it is known that the mass number A of isotopes are all nearly whole numbers; the slight variation being only in the second place of decimal. This is known as the whole number rule. The whole number rule lead to the general acceptance that atomic nuclei are made up of hydrogen nuclei (protons) and neutrons having almost the same mass. Although atomic mass is close to a whole number, they differ from integers by a small amount, *i.e.*, they depart from the whole number rule systematically. The light nuclei have masses slightly higher than the nearest integer (*e.g.*, $H^1 = 1.00813$, $He^4 = 4.00408$, $Li^6 = 6.01614$); a very large number of medium nuclei have a mass somewhat lower (*e.g.*, $Ne^{20} = 19.9988$, $Cl^{35} = 34.933$, $Mo^{98} = 97.944$); while a heavy one slightly higher again *e.g.*, $Hg^{200} = 200.028$, $T^{205} = 205.036$).

The mass defect of the atom is the difference of its atomic mass and its mass number, *i.e.*,

$$\Delta m = M - A$$

where M is the measured mass and A is the mass number. Thus the mass defect gives the deviation of atomic mass from the whole number A.

The packing fraction p is the mass defect per elementary particle in the nucleus, *i.e.*,

$$p = \frac{\Delta m}{A} = \frac{M - A}{A} = \frac{\text{mass defect}}{\text{mass number}}$$

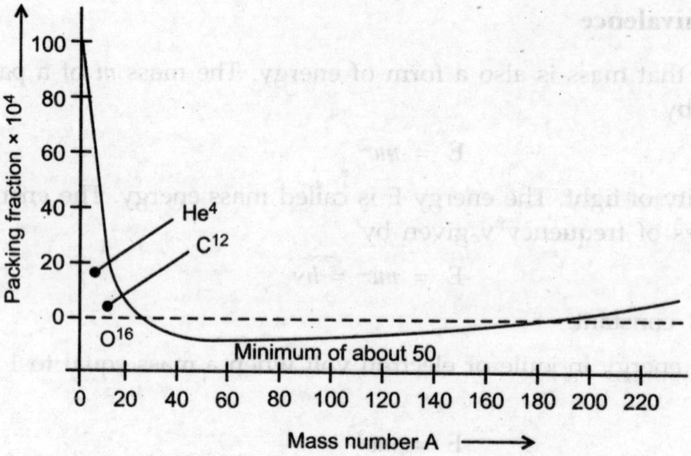

FIGURE 1 **Showing variation of packing fraction.**

Figure 1 shows the variation of the packing fraction of a large number of nuclides as a function of the mass number. The graph shows the following results:

- The packing fraction is positive for elements having a mass number below 20.
- The packing fraction is negative for elements having a mass number between 20 and 200.
- The packing fraction is positive for elements having a mass number beyond 200.

The significance of the packing fraction is that it is directly related to the availability of nuclear energy and stability. The negative packing fraction means that some mass has been converted into energy during the formation of that particular nucleus. Now in order to break this nucleus, the same amount of energy would have to be supplied. This shows that the negative packing fraction implies exceptional nuclear stability. On the other hand, the positive packing fraction implies that the nucleus is somewhat unstable.

5.6 THE MASS DIFFERENCE AND NUCLEAR BINDING ENERGY

We have seen that the mass of the nucleus is always different from the sum of the mass of its constituents. The difference in measured mass M and the mass number A is called the mass defect.

$$\therefore \qquad \text{Mass defect } \Delta m = (M - A)$$

When two nucleons are bound together into a nucleus and a certain mass disappears, it is converted into energy which is known as nuclear binding energy. In order to break the nucleus into its constituent particles, the same amount of energy is required for this purpose.

Let us calculate the binding energy of lithium. This atom has A = 7 and Z = 3, *i.e.*, the nucleus contains three protons and four neutrons.

The mass of a lithium nucleus = 7.016005

The mass of three protons = 3 × 1.007277 = 3.021831 a.m.u.

The mass of four neutrons = 4 × 1.008665 = 4.034660 a.m.u.

The total initial mass of the three protons and four neutrons

= 3.021831 + 4.034660 = 7.056491 a.m.u.

As the mass of the lithium nucleus is less than the total initial mass of the three protons and four neutrons, the decrease in mass is given by

$$\Delta m = 7.056491 - 7.016005$$

$$\Delta m = 0.040486 \text{ a.m.u.}$$

∴ Binding energy of lithium nucleus

$$\therefore \qquad = 0.040486 \times 931 \qquad (\because 1 \text{ a.m.u.} = 931 \text{ MeV})$$

$$= 37.7 \text{ MeV.}$$

Thus in order to break the lithium nucleus an energy more than 37.7 MeV must be supplied.

In the case of lithium, the binding energy per nucleon is given by

$$\frac{37.7}{7} = 5.4 \text{ MeV} \qquad (\because \text{There are 7 particles})$$

The binding energy per nucleon as a function of mass number is shown in Figure 2. It is observed from the graph that binding energy per nucleon (except for He^4, C^{12} and O^{16}) rises first sharply and reaches a maximum value 8.8 MeV in the neighborhood of A = 50. The curve falls very slowly after A = 50 and reaches at 8.4 MeV at about A = 140. For higher mass numbers, the energy decreases to about 7.6 MeV. It should also be pointed out here that the packing fraction curve (Figure 1) is also minimum at about A = 50, predicting that nuclei near A = 50 are most stable.

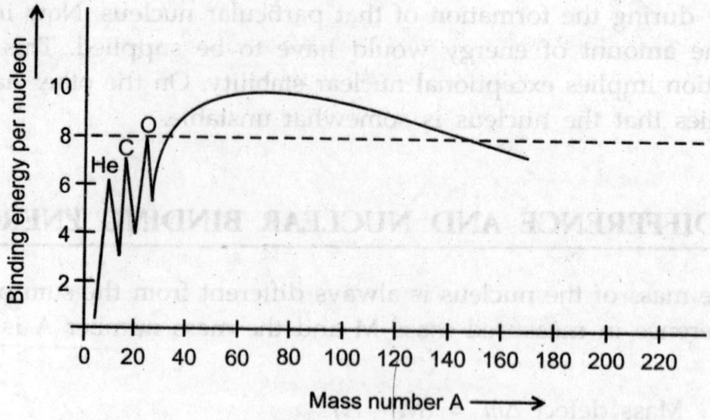

FIGURE 2 **Binding energy per nucleon as function of A.**

Example 1. *Find the binding energy of an α particle from the following data:*

Mass of the helium nucleus = *4.001265 a.m.u.*

Mass of a proton = *1.007277 a.m.u.*

Mass of a neutron = *1.008666 a.m.u.*

Take one a.m.u. = *931.4812 MeV.*

Solution. Mass of two protons = 2 × 1.007277 = 2.014554 a.m.u.

Mass of two neutrons = 2 × 1.008666 = 2.017332 a.m.u.

The total initial mass of two protons and two neutrons

= 2.014554 + 2.017332 = 4.031886 a.m.u.

Mass defect Δm = 4.031886 – 4.001265

= 0.030621 a.m.u.

∴ Binding energy of α particle = 0.030621 × 931.4812 = 28.5221 MeV

Binding energy per nucleon = 28.5221/4 = **7.10525 MeV.**

Example 2. *In a thermonuclear reaction 1.00 × 10^{-3} kg, when hydrogen is converted into 0.993 × 10^{-3} kg helium, (i) calculate the energy released in joule, (ii) if the efficiency of the generator is 5%, calculate the electrical energy in kilowatt hours.*

Solution. (*i*) Mass defect Δm = (1.000 – 0.993) × 10^{-3}

= 0.007 × 10^{-3} kg

∴ The energy released is given by energy mass relation

$$E = mc^2 = 0.007 \times 10^{-3} \times (3 \times 10^8)^2$$

$$= 63.0 \times 10^{10} \text{ joules}$$

(*ii*) Electrical energy $= \dfrac{5}{100} \times E = \dfrac{5}{100} \times 63.0 \times 10^{10}$

$$= 0.05 \times 63.0 \times 10^{10} \text{ joule}$$

$$= \frac{0.05 \times 63.0 \times 10^{10}}{36 \times 10^{5}} \text{ kilowatt hour}$$

$$= \mathbf{8.75 \times 10^{3} \text{ kilowatt hour.}}$$

Example 3. *A neutron breaks into a proton and electron. Calculate the energy produced in this reaction in MeV. Mass of an electron = 9×10^{-31} kg. Mass of proton = 1.6725×10^{-27} kg. Mass of neutron = 1.6747×10^{-27} kg. Speed of light = 3×10^{8} m/sec.*

Solution. Mass defect Δm = [Mass of neutron − (mass of proton + mass of electron)]

$$= 1.6747 \times 10^{-27} - (1.6725 \times 10^{-27} + 9 \times 10^{-31})$$

$$= 0.0013 \times 10^{-27} \text{ kg}$$

∴ Energy released $E = \Delta m.c^2$

$$= 0.0013 \times 10^{-27} \times (3 \times 10^{8})^2$$

$$= 1.17 \times 10^{-13} \text{ joule}$$

$$= \frac{1.17 \times 10^{-13}}{1.6 \times 10^{-13}} \text{ MeV} = \mathbf{0.73 \text{ MeV.}}$$

5.7 NUCLEAR FORCES

We know that the nucleus consists of protons and neutrons. Due to the positive charge on protons, there will be a repulsive electrostatic force between the two protons and the resulting repulsive force between protons will tend to push the nucleus apart. Therefore, for the nucleus to have a permanent existence, there must be some strong attractive forces between protons and neutrons inside the nucleus. These attractive forces cannot be gravitational because they are much smaller than the attractive force demanded.

Moreover, these forces cannot be electrical in nature because the strong repulsive forces between protons will lead to disruption of nucleus. Actually, these forces are short range attractive forces known as nuclear forces. The nuclear forces have the following properties:

- These forces are attractive forces between proton-neutron, proton-proton and neutron-neutron.
- These forces are strongest in nature.
- They are spin dependent.
- These forces are short range forces, (*i.e.*, they are effective within the nucleus diameter of the order of 10^{-15} meter and outside the nucleus the Coulombic repulsive force becomes predominant). A graph of the Coulombic law of repulsion and short range forces of attraction is shown in Figure 3.
- These forces are independent of charge, *i.e.*, the nuclear forces are the same for proton-proton, proton-neutron and neutron-neutron interactions.
- In the case of nuclear forces, each nucleon attracts those nuclei which are immediate neighbors, just like the molecules in a liquid or gases that interact with neighboring molecules.

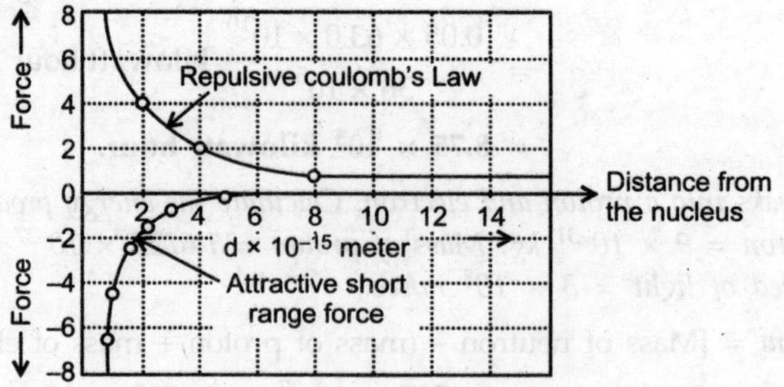

FIGURE 3 **Showing short range force.**

Yukawa in 1935 proposed a theory to explain the binding forces between neutrons and protons known as the Meson theory of nuclear forces. According to this theory, there exists a new type of field between nucleons known as the meson field when the proton and the neutron do not have their independent existence inside the nucleus. The meson may have a positive or negative charge which is equal to an electronic charge. The mesons are supposed to exchange rapidly between nucleons thereby changing their identity equally fast and are responsible for keeping them bound together. The exchange is represented as follows:

$$\text{Proton} \underset{\text{Positive meson}}{\overset{\text{Negative meson}}{\rightleftarrows}} \text{Neutron}$$

Similarly to explain the binding force between neutron and neutron the existence of a neutral meson was postulated.

5.8 NUCLEAR MODELS

Various nuclear models have been proposed to interpret the special static properties of nuclei such as stability, spin, magnetic moment etc. The more important models are gammow, liquid drop and shell models. It has been found that not a single one of these models explains satisfactorily all the nuclear properties, but the nuclear shell model has been quite successful in explaining the properties of static nature. The most stable state is the ground state in which nuclei are usually found.

1. The Liquid Drop Model

In 1936, Bohr suggested that the nucleus might be like a droplet of dense liquid, composed of sub-droplets of the neutrons and protons it contained. In many ways, close packing of the nucleons, constant density, and short range forces are similar to a drop of liquid (Figure 5). The surface tension of liquids holds a small liquid drop in its spherical form. Just as the rapid motion of liquid molecules is a measure of the liquid temperature, similarly the motion of nucleons within the nucleus is a measure of nuclear temperature, called the pseudo-temperature by Bohr.

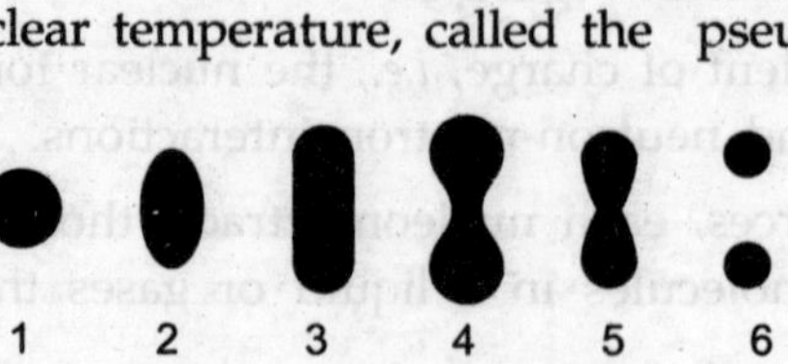

FIGURE 4 **Liquid drop model of nucleus.**

The following similarities were seen in the liquid drop and the nucleus:

- The nucleus is spherical in shape in a stable state just like a liquid drop due to symmetrical surface tension forces. The surface barrier is similar to surface tension on the liquid surface.
- The volume of nucleus and liquid drop does not depend on density.
- From a liquid drop, molecules are evaporated due to thermal agitation similarly. If a nucleus is bombarded it emits nuclear radiations immediately.
- Nuclear forces are short-range and so are intra-molecular forces in a liquid drop.
- Oscillations of a drop cause disruption by breaking into droplets similar to the fission process in a nucleus.

The forces holding the neutrons and protons together cannot be electrical. The neutrons are chargeless and also the mutual electrical repulsion of positively charged protons would pull a nucleus apart. We must conclude that there is another force acting between all nucleons, holding them together by the attraction of each nucleus for its immediate neighbors like intermolecular forces hold a liquid droplet together. Some nucleons are on the surface on the liquid drop and thus are not surrounded by attracting neighbors on all sides. A large nucleus has a small function of its nucleons on the surface since surface area to volume ratio becomes smaller. As the temperature rises, the motion of molecules of a liquid drop shows a tendency to disintegration by evaporation from a liquid drop. Similarly, the motion of nucleons within the nucleus (the average kinetic energy of its nucleons) increases in order to disintegrate. As the nucleons move about they are hit by other nucleons and give a sufficiently high velocity to come through the potential barrier on the surface of the nucleus and thus disintegrate by escaping. This explains the disintegration.

The nuclear fission is due to oscillations brought about by the impinging neutrons. This type of oscillation leads to the disruption of the drops into two or more smaller droplets as shown in Figure 5. In the liquid molecules the dissipative forces acting in the molecule are counterbalanced by the surface tension forces. The spherical nature of the liquid molecule is due to this force of surface tension.

According to this model, the nucleus is similar to a small electrically charged liquid drop, *i.e.*, the nucleus takes a spherical shape for its stability. The nucleons (protons and neutrons) move within this spherical enclosure like molecules in a liquid drop. The motion of nucleons within the nucleus is a measure of the nuclear temperature as the molecular motion of molecules in liquid is the measure of its temperature. The nucleons always remain a constant distant apart and share the total energy of the nucleus. The nucleons deep inside the nucleus are attracted from all sides by neighboring nucleons while those on the surface are attracted from one side only.

Nuclear Structure and Nuclear Forces

In this way the binding energy for the nucleons at the surface of the nucleus is smaller than the binding energy for the nucleons inside the nucleus. We know that there is a larger fraction of surface nucleons in light nuclei than in medium and heavy nuclei. This shows that the total binding energy is small for small A. So the spherical surface which encloses the nucleus may be regarded as analogus to the surface tension which holds a water drop to the evaporation of particles from a liquid surface. Similarly, the process of nuclear capture may be compared with the condensation of the vapors. The evaporation in a liquid drop takes place when Maxwellian energy distribution within the drop causes a particular molecule to have sufficient energy to overcome the intermolecular attraction and escape. Similar is the case with a compound nucleus. The nucleus remains in the excited state until the nucleon or group of nucleons happen to possess sufficiently large excitation energy to escape through the potential barrier. On the basis of Coulombian repulsive force, we can explain why heavy nuclei tend to be made up of more neutrons than protons. The protons inside the nucleus repel each other according to Coulomb's law. So the binding energy is decreased. The repulsive force increases with increasing A and thereby resulting in a smaller value of binding energy for large A. We can calculate Coulomb repulsion by assuming the nucleus to be a liquid drop in the form of a uniformly charged sphere of radius R and charge +Ze. Since Coulomb repulsion has the effect of making nuclei unstable heavy nuclei tend to made up of more neutrons than protons.

The model is based on the following assumptions:

(1) The material of the nucleus is incompressible and the density of all the nuclei is the same.

(2) The forces in the nucleus consist of (*a*) Coulomb forces between protons and (*b*) powerful attractive nuclear forces.

The following are the analogies between a liquid drop and a nucleus:

- Both are spherical in nature.
- In both the cases, the density is independent of its volume with the exception that the density of the nucleus is independent of the nucleus while the density of the liquid depends upon its type.
- The molecules in liquid drop interact over short ranges and this is also true for nucleons in a nucleus.
- As the surface tension forces act on the surface of a drop, similarly a potential barrier acts on the surface of nucleus.
- When the temperature of the molecules in a liquid drop is increased, evaporation of molecules take place. Similarly when the nucleons in the nucleus are subjected to external energy, a compound nucleus is formed which emits nucleons almost immediately. The process is known as nuclear fission.

2. The Shell Model

According to the shell model, the nucleus consists of a series of protons and neutrons placed in certain discrete levels, or shells, just like the electrons in the discrete shells of an atom. According to Pauli's exclusion principle two protons with opposite spins and two neutrons having opposite spins are accommodated in a particular shell. In this way the first shell accommodated two protons and two neutrons and is more tightly bound than other shells.

The concept of the shell model is based on the following fact. It has been observed that nuclei containing protons and neutrons number 2, 8, 20, 50, 82, 126 etc., known as magic numbers or shell numbers, are exceptionally stable as in the case of electronic shell rule of 2, 8, 18 etc. This shows that there are definite energy shells in the nuclei.

From our investigations of the nucleus, we can say that nuclear forces are strongly attractive and they extend over a very short range from the center of the nucleus. Whatever may be the form of potential, the shell model is based on the following two assumptions:

- Each nucleon moves freely in a force field described by the potential, which is a function of the radial distance from the center of the system.
- The energy levels or shells are filled according to the Pauli exclusion principle.

Consider a nucleon of mass M with angular momentum $\sqrt{l(l+1)}h$ moving in a potential V(r). Assuming that V(r) is independent of θ and ϕ, the Schrödinger wave equation can be written as

$$\frac{d^2}{dr^2}(r\text{R}) + \frac{2\text{M}}{h^2}\left[\text{E} - \text{V}(r) - \frac{l(l+l)h^2}{2\text{M}r^2}\right](r\text{R}) = 0 \qquad (1)$$

where R is the radial wave function and E is the energy eigen value. Here the same quantum numbers n, l, j, m, result in a shell model as in an atomic model.

The square well potential is represented by

$$\begin{aligned} \text{V}(r) &= -\text{V}_0 \quad \text{for } r \le \text{R} \\ &= 0 \quad \text{for } r > \text{R} \end{aligned} \qquad (2)$$

The harmonic-oscillator potential is given by

$$\text{V}(r) = -\text{V}_0 + 1/2\ kr^2 \qquad (3)$$

If we combine the square well potential and harmonic oscillator potential, the new form of the potential will be

$$\begin{aligned} \text{V}(r) &= -\text{V}_0\left(1 - \frac{r^2}{\text{R}}\right) \quad \text{for } r \le \text{R} \\ &= 0 \quad \text{for } r > \text{R} \end{aligned} \qquad (4)$$

Using the above potential in equation (1) and solving it, we get all the nuclear magic numbers (shell structure) except magic number 28. In this way the shell model of the nucleus is an attempt to account for the existence of magic numbers and certain other nuclear properties in terms of nucleon behavior in a common force field.

5.9 THE SEMI-EMPIRICAL MASS FORMULA

In 1935, Von Weizasacker expressed the atomic mass of a nuclide in terms of the series of binding energy correction terms with the main mass contribution from protons and neutrons. The modified expression of the mass is known as the semi-empirical mass formula. Later on this formula was modified by Bethe and others but the main outlines remain the same. The mass of a neutral atom can be expressed as

$$_Z\text{M}^\text{A} = \text{ZM}_p + \text{NM}_n - \text{B} \qquad (1)$$

where

$_Z\text{M}^\text{A}$ = Atomic mass of the nuclide

Z = Number of protons

M_p = Mass of the proton

N = Number of neutrons

M_n = Mass of the neutrons

B = Binding energy expressed in mass units.

The value of B is calculated empirically is made up of a number of correction terms given as

$$B = E_v + E_s + E_e + E_r + E_s \tag{2}$$

where E_v, E_s, E_e, E_r and E_s are volume energy correction, surface energy, Coulomb energy, asymmetry energy, and pairing energy respectively.

Now we shall proceed to discover empirically the values of these energies:

Volume Energy Correction E_v

It has been experimentally verified that the binding energy per nucleon is approximately constant over a wide range of mass number. To a first approximation we may assume that the total binding energy of a nucleus is proportional to the total number of nucleons A in it, *i.e.*,

$$E_v \propto A \quad \text{or} \quad E_v = a_v A \tag{3}$$

where a_v is an undetermined constant. Here subscript v is used to show that the contribution to the binding energy comes from the entire volume of the nucleus.

Surface Energy

E_s, nucleus of an atom has some nucleons on its surface. As a molecule at the surface of a liquid is attracted less strongly in one direction in comparison to a molecule in the interior (a molecule in the interior of liquid is attracted equally in all directions due to the other surrounding molecules) similarly the nucleons at the surface of a nucleus are bound less tightly.

5.10 THE PARTICLE ACCELERATOR

For producing nuclear transmutations, high energy atomic particles like protons, neutrons, deuterons, etc., are used. A particle accelerator is a device built to accelerate (*i.e.*, to increase the kinetic energy) the charged particles so as to give them the desired energy. The maximum energy that can be acquired by a particle depends upon the type of machine. In present day by accelerators, energies of hundreds of billions of electron volts (BeV) have been reached. In principle, there is no upper limit to such high energies of particles, but it is the cost of construction and the maintenance of the machine that limits the highest energy. The accelerated particles can be used as powerful projectiles to bombard various targets. Nowadays, accelerators have become basic tools in the rapidly growing branch of nuclear science known as high energy particle physics. Nuclear reactions carried out by high energy particles led to the discovery of new particles like antiprotons, antineutrons, hyperons, k-mesons, etc. High energy particles also provide valuable information about the complex problem of nuclear structure and nuclear forces. When the nuclei of light atoms such as those of lithium, beryllium, boron, carbon, and nitrogen are artificially accelerated, transuranic elements are produced. Particle accelerators are also used for the production of isotopes (this was the primary use of particle accelerators prior to the development of nuclear reactors) and X ray sources for radiography and radiotherapy.

The operation of an accelerator consists of the following steps: The particles are first produced in the source, and then injected by a suitable accelerating voltage into the machine. Then the particles are collimated into a beam and are constrained to follow a certain trajectory in a vacuum under the action of an electric field. They acquire the desired energy, and are made to hit the selected target placed inside the machine. The resultant interactions are now studied.

Accelerators may be divided into the following three classes:
(*a*) linear, (*b*) static, and (*c*) circular.

Linear Accelerators

Linear accelerators can accelerate electrons and protons up to maximum energies of 40 GeV and 50 MeV respectively.

Static Accelerators

This type of accelerator demands the production of high voltage so that an atomic particle can be given energy up to 12 electron volts. The Van de Graaff generator is an example of this type of accelerator.

Circular Accelerators

In circular accelerators, the particles are accelerated while moving in a spiral or circular path and the energy is given to the particles in a series of successive additive stages. They are further classified as:

(*i*) *Lawrence cyclotron or fixed frequency cyclotron:* It can be used for accelerating particles like protons, deuterons, and α-particles up to energies 40 MeV.

(*ii*) *Synchronicyclotron or frequency modulated cyclotron:* This is a modified form of the Lawrence cyclotron which gives comparatively higher energies to the particles.

(*iii*) *Betatron:* It is used for accelerating electrons up to energies 300 MeV.

(*iv*) *Electron cyclotron:* This is a modified form of the betatron and gives electron energies up to 6 GeV.

(*v*) *Proton cyclotron:* The proton accelerator gives proton energies up to 10 GeV.

(*vi*) *Alternating Gradient Synchrotron (AGS):* This is a proton accelerator which gives proton energies up to 70 GeV.

(*vii*) *Omnitron:* It is an accelerator which gives strong beams of accelerated particles of all masses from hydrogen to uranium.

5.11 LINEAR PARTICLE ACCELERATORS

As the name indicates, in a linear particle accelerator, the charged particles move in a straight line. The linear accelerator is shown in Figure 5. It consists of a number of hollow cylindrical tubes T_1, T_2, T_3,.... etc. arranged co-axially, called drift tubes. These tubes are of increasing lengths and are arranged in a glass vacuum chamber. The alternate tubes are connected together to form two common junctions. These junctions are connected to a high frequency power supply, *i.e.*, a radio frequency (R.F.) oscillator. Thus tubes T_1, T_3, and T_5 are negative. According to the applied frequency, the reversal of potential takes place periodically. The positive ions are produced by the source S which travel along the axis of the tube. Since a potential difference is maintained between two neighboring cylinders, the ions are accelerated in the gap between two cylinders. Here it should be remembered that the ions travel with constant velocity in field free space within cylinders themselves with increasing velocity at successive cylinders.

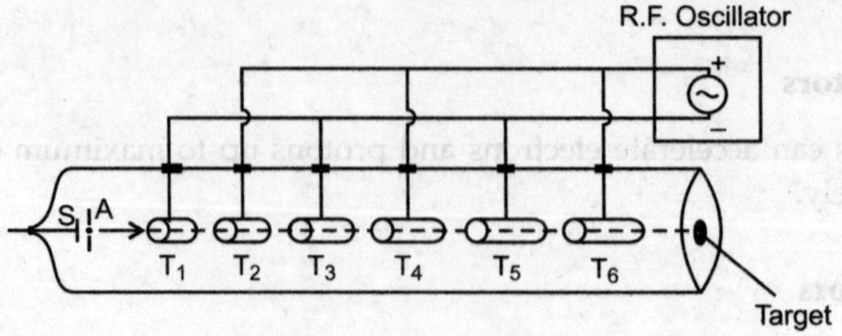

FIGURE 5 **Linear particle accelerators.**

Let a positive ion leave A when tube T_1 is negative with respect to A. The ion will be accelerated and on entering tube T_1 it travels with uniform velocity which it has acquired during acceleration between the gap A and tube T_1. The length of tube T_1 is selected in such a way with relation to the frequency of the oscillator that when the ion leaves tube T_1, tube T_2 becomes negative. (This is only possible when the time taken by the ion to travel to tube T_1 is exactly equal to half the time period of the R.F. oscillator.) The ion will again face an accelerating field, *i.e.*, it is further accelerated between the gap of tube T_1 and tube T_2. Now the ion travels through T_2 at constant but increased velocity. Thus, if the lengths of the tubes and the frequency of the R.F. oscillator are adjusted such that every time the ion faces accelerated field, it will acquire an additional energy every time it passes through the gap between two tubes. Finally, the ion is emitted from the final drift tube towards the target with extremely high velocity.

If n is the number of tubes, and V_m is the maximum voltage of the oscillator, then the energy acquired by an ion of charge q is given by

$$nqV_m = \frac{1}{2}mv_n^2 \tag{1}$$

where m is the mass of the ion and v_n its velocity while entering the nth tube.

The time interval for an ion to pass through any tube is

$$t = \text{half period} = \frac{T}{2} = \frac{1}{2f} \tag{2}$$

where f is the frequency of the oscillator.

Let l_n be the length of the nth tube and v_n the velocity of the ion when travelling through it. Then

$$t = \frac{l_n}{v_n} = \frac{1}{2f}$$

$$\therefore \quad l_n = \frac{v_n}{2f} \tag{3}$$

from equation (1) and equation (3), we have

$$l_n = \sqrt{\frac{nq V_m}{2mf^2}} \tag{4}$$

Thus the lengths of the drift tubes must be proportional to $\sqrt{1} : \sqrt{2} : \sqrt{3} : \sqrt{4}$, etc.

The chief advantage of the linear accelerator is that ions emerge as a well collimated beam. It does not require intense magnetic fields like circular accelerators. We know that charged particles (particularly electrons) moving with high speeds in circular paths (as in circular accelerators) lose energy by radiation. Such a difficulty does not arise in linear accelerators.

The linear accelerator for energizing protons is similar but a linear accelerator for electrons presents a considerable problem because the mass of the electron is very small. At 1 MeV, the electron has a velocity 0.94 c where *c* is the velocity of light. The transit times through the tubes are so short that a frequency of about 1000000 mc/s would be needed. As the velocity of the electron approaches that of light, its mass increases. For this reason, linear accelerators for electrons are of the cavity resonator or wave guide resonator type.

5.12 THE LAWRENCE CYCLOTRON

Construction

The Lawrence cyclotron is shown in Figure 6. It consists of two hollow flat semicircular metal boxes D_1 and D_2, called the dees on account of their shape like the letter D. The two dees are separated by a narrow parallel gap. A high frequency oscillator, which provides an alternating potential of the order of 10^4 to 10^5 volts at a frequency of 10^7 cycles, is connected between the two dees. The oscillator establishes an alternating electric field in the air gap, *i.e.*, the electric field is directed towards D_1 and then towards D_2. Thus D_1 and D_2 become alternately positive and negative at the same rate as the frequency of the oscillator. A source S is placed at the center of the dees which supplies the positive ions to be accelerated. These dees are mounted inside a vacuum chamber. The chamber is mounted horizontally between the pole pieces PP of a huge electromagnet capable of producing a vertical field of about 1.6 Web/meter^2.

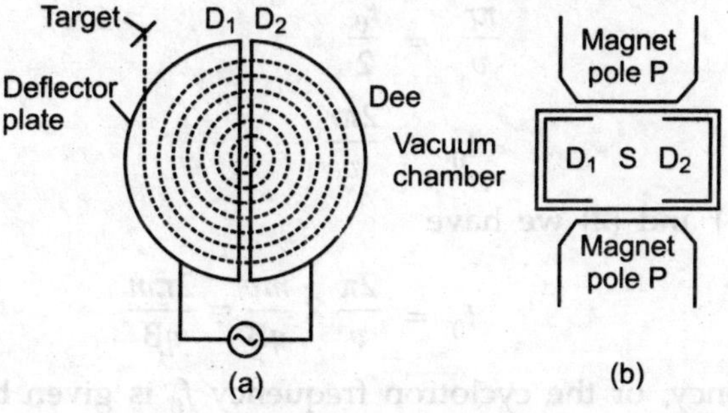

FIGURE 6 **High frequency oscillator.**

Working

The positive ions emitted from the source will be accelerated in the gap towards the dee which is negative at that time. Let it be D_2. Since there is no electric field inside the dees, the positive ions move with constant velocity along circles of constant radius under the influence of the magnetic field which is perpendicular to the dees. If by the time the ions emerge from D_2, the polarity of the applied potential is reversed, (*i.e.*, the dee D_1 now becomes negative), the positive ions will again face the negative dee and thus will be again accelerated by the field in the gap. Since their velocity is increased, they will now move through D_1 along the circular arc of the greater radius as shown in the Figure 6(*a*). Here the time of passage to complete the semi-circle in the dee remains the same as in D_2. If the time of travel in D_1 is equal to half the time period of the oscillator voltage, the positive ions after coming from D_1 will find the reversed field and hence they are accelerated again in the gap $D_1 D_2$. In this way, the positive ions move faster and faster moving in every expanding circles until they reach the outer edge of the dees where they are deflected by the deflector plate and strike the target. Here it should be remembered that the time required for the positive ions to make a complete turn within the dees is the same for all speeds and is equal to the time period of the oscillator.

Theory

When a particle of mass m and charge q moves with a velocity v in the magnetic field then force Bqv acts on the particle perpendicular to its direction of motion due to which it moves in a circular path of radius r. The equation of motion of the particle is written as

$$\text{Magnetic force} = \text{Centrifugal force}$$

or

$$Bqv = \frac{mv^2}{r}$$

$$r = \frac{mv}{qB} \tag{1}$$

For resonance between the applied a.c. voltage and the moving particle, the time taken by the particle to travel the circular arc within any of the dees must be equal to half the time period t of the applied oscillator voltage. Time taken in moving the semicircular path is $\pi r/v$

$$\therefore \qquad \frac{\pi r}{v} = \frac{t_0}{2}$$

or

$$t_0 = \frac{2\pi r}{v} \tag{2}$$

From equations (1) and (2) we have

$$t_0 = \frac{2\pi}{v} \times \frac{mv}{qB} = \frac{2\pi m}{qB} \tag{3}$$

now the orbital frequency, or the cyclotron frequency f_0 is given by

$$f_0 = \frac{1}{t_0} = \frac{qB}{2\pi m} \tag{4}$$

The value of q/m is fixed for an ion hence the value of f_0 is adjusted corresponding to B or vice-versa.

Now the ion will have maximum energy when it will travel at the boundary of dee. If the outside radius of dee is R then according to equation (1), the maximum velocity v_m of the ion may be written as

$$v_m = \frac{RqB}{m} \quad (5)$$

and so the maximum kinetic energy of the ion will be given by

$$E_m = \frac{1}{2}mv_m^2 = \frac{R^2q^2B^2}{2m} \quad (6)$$

We know that 1 MeV = 1.6×10^{-13} joule

$$E_m = \frac{R^2q^2B^2}{2m \times 1.6 \times 10^{-13}}$$

$$= \frac{3.12 \times 10^{12} R^2q^2B^2}{m} \text{ MeV}$$

The maximum available particle energy is limited due to the following factors:

- the limited power and frequency of the oscillator,
- the maximum strength of the magnetic field which can be produced, and
- the energy of charged particles emerging from the cyclotron is limited due to the variation of mass with velocity, *i.e.*,

$$m = \frac{m_0}{\sqrt{1 - \frac{v^2}{c^2}}}$$

where m_0 is the rest mass, m the mass in motion, when velocity is v and c is the velocity of light.

The frequency of rotation of the charged particle becomes

$$f_0 = \frac{qB}{2\pi m} = \frac{qB\sqrt{1 - \frac{v^2}{c^2}}}{2\pi m_0}$$

The frequency of rotation of the charged particle decreases with an increase of velocity. Consequently, the charged particle takes a longer time to complete the semi-circular path. The particle continues to lag behind the alternating potential differences until a stage is reached when it can no longer be accelerated further.

The frequency of the charged particle can be kept constant by making $B\sqrt{1 - v^2/c^2}$ constant. To achieve this, the magnetic field B is increased as the velocity of the particle increases. Alternatively, the frequency of the applied alternating potential difference may be varied so that it is always equal to the frequency of the rotation of the charged particle.

Example 4. *Calculate the magnetic field in which the cyclotron dees should be placed to accelerate protons, the frequency applied being 8 × 10⁶ c/sec, e = 4.8 × 10⁻¹⁰ e.s.u., mass of proton = 1.67 × 10⁻²⁴ gm.*

Solution. We know that

$$f_0 = \frac{qB}{2\pi m} \quad \text{or} \quad B = \frac{2\pi m f_0}{q}$$

Here $m = 1.6 \times 10^{-24}$ gm, $f_0 = 8 \times 10^6$ c/s, $q = 4.8 \times 10^{-10}$ e.s.u.

$$q = \frac{4.8 \times 10^{-10}}{3 \times 10^{10}} = 1.6 \times 10^{-20} \text{ e.m.u.}$$

Putting these values in the preceding equation

$$B = \frac{2 \times 3.14 \times (1.67 \times 10^{-24})(8 \times 10^6)}{1.6 \times 10^{-20}}$$

$$= 0.524 \times 10^4 \text{ gauss} = \mathbf{0.524\ weber/m^2}.$$

Example 5. *In a certain cyclotron, the maximum radius that the path of a deuteron may have before it is deflected out of the magnetic field is 20 cm.*

(i) Calculate the velocity of the deuteron at this radius.

(ii) What is the energy of the deuteron in MeV? Given a magnetic field = 1500 Gauss, the mass of the deuteron = 3.34 × 10⁻²⁷ kg.

Solution. (*i*) The velocity of the deuteron is given by

$$v = \frac{qBr}{m}$$

Here

$$q = 1.6 \times 10^{-19} \text{ coulomb}, \quad m = 3.34 \times 10^{-27} \text{ kg}$$

$$r = 20 \text{ cm} = 0.2 \text{ meter}$$

$$B = 15000 \text{ gauss} = 15000 \times 10^{-4} \text{ weber/m}^2$$

$$= 1.5 \text{ weber/m}^2$$

$$v = \frac{1.6 \times 10^{-19} \times 1.5 \times 0.2}{3.34 \times 10^{-27}}$$

$$= 1.437 \times 10^7 \text{ m/sec}$$

(*ii*) The energy of the deuteron $= \frac{1}{2} mv^2$

$$= \frac{1}{2} \times 3.34 \times 10^{-27} \times (1.437 \times 10^7)^2$$

$$= 3.449 \times 10^{-13} \text{ joules}$$

$$= \frac{3.449 \times 10^{-13}}{1.6 \times 10^{-19}} \text{ eV}$$

$$= 2.156 \times 10^6 \text{ eV} = \mathbf{2.156\ MeV}.$$

5.13 THE SYNCHROCYCLOTRON OR FREQUENCY MODULATED CYCLOTRON

The resonance frequency of a cyclotron is given by

$$f_0 = \frac{qB}{2m}$$

We know that the variation of mass with velocity is given by

$$m = \frac{m_0}{\sqrt{1 - \frac{v^2}{c^2}}}$$

if $v = 0.8$ c, $v/c = 0.8$ and $m = 1.66$ m.

Thus at high velocities due to a relativistic increase in mass, the orbital frequency of the particle is decreased. The particle gradually goes out of phase with the applied a.c. voltage whose frequency is constant. It is, therefore, necessary to provide corrective measures to compensate the decrease in frequency and to keep the particle in phase with the applied a.c. voltage. Thus compensation for the decrease in frequency can be made in the following two ways:

- Keeping the value of B constant, the oscillator frequency is decreased in step with the decrease in orbital frequency of the particle. A device which adopts this principle is called synchrocyclotron or frequency modulated cyclotron.
- The magnetic flux density B may be increased so that $B\sqrt{1 - \frac{v^2}{c^2}}$ remains constant. The frequency of the high frequency oscillator is kept fixed. A device which adopts this principle is called a synchrotron.

In a synchrocyclotron, the change in frequency should be done continuously and slowly compared with the rapid alterations of high frequency dee voltage. If this is not done, the phase stability will be disturbed. For a continuous change in frequency, a frequency modulated supply should be used. This modulation is generally performed at 50 to 60 cycles per second and arranged by rotating the movable vanes of a variable capacitor forming part of a tuned circuit of a valve oscillator which is used to supply the dee voltage.

We have seen that the output of a cyclotron was continuous but in case of a synchrocyclotron, as the particles starting from the center are subjected to frequency modulation, when they approach the periphery, and hence they come out in bursts of few hundred per second, each burst lasting for about 100 microseconds.

Advantages of a Synchrocyclotron Over a Cyclotron

The following are the advantages of a synchrocyclotron over a cyclotron:

- A single dee may be used. Instead of a second dee, an earthed sheath is used opposite the opening of the dee and a.c. potential is applied between the single dee and the earth.
- Comparatively small voltages may be needed across the dee gaps. Thus greater efficiency of a high frequency oscillator may be obtained.

- The gap between the magnetic pole pieces can be reduced thus permitting high flux densities to be obtained.
- The pole pieces can be suitably shaped to produce a flux density decreasing outwards from the center, thus achieving good focussing of accelerated particles.

Disadvantages of a Synchrocyclotron Over a Cyclotron

The following are the disadvantages of a synchrocyclotron over a cyclotron:

- Only a small number of particles can be brought to the orbit of maximum radius and energy.
- The output of the particles is in pulses.

The Berkeley synchrocyclotron is shown in Figure 7. It has a large magnet with a pole diameter of 4.7 m and weighing 4000 tons.

The modulation frequency of the Berkeley synchrocyclotron is 120 Hz. This is produced by a rotating condenser that forms the part of capacitance in the RF oscillatory circuit. To reduce the scattering of ions by residual gas, the molecule of pressure in the chamber is reduced to about 10^{-6} cm of mercury. Each ion makes over 10^4 revolutions before attaining maximum energy. Here a single dee is employed. This feature makes the ion beam much more accessible for experiment. An ejector plate with a high positive voltage pulse deflects the groups on to the target to be bombarded. The synchrocyclotron can accelerate deuterons to 200 MeV, α-particles to 400 MeV, and protons to 350 MeV.

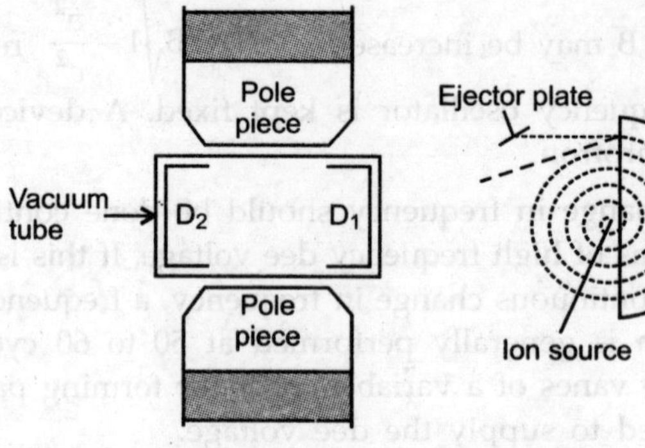

FIGURE 7 **Synchrocyclotron.**

5.14 THE BETATRON DEVICE

The betatron is a device for speeding up electrons to extremely high energies with the help of an expanding magnetic field. The cyclotron principle was incapable for accelerating the electrons to high energies. The first betatron was constructed by Kerst at the University of Illinois (U.S.A.) in 1940, and was capable of accelerating electrons to energies of 2 MeV. A model constructed in 1945 was capable of accelerating electrons to energies of 100 MeV. The betatron differs from the cyclotron in the following two fundamental respects.

- In the betatron, the electrons are accelerated by the expanding magnetic field.
- The circular orbit of the betatron has a constant radius.

Construction

The construction of a betatron is shown in Figure 8. It consists of a highly evacuated annular tube D known as the doughnut chamber. This chamber is placed between the poles of an electromagnet excited by alternating currents usually of a frequency of 60 Hz or 180 Hz. The poles of the electromagnet are constructed in such a way as to provide a stronger field in the central space. Electrons are produced by an electron gun and are injected into the doughnut at the beginning of each cycle of alternate currents in the electromagnet, thus increasing the magnetic flux giving rise to a voltage gradient (electric field) round the doughnut which accelerates the orbiting electrons. This increases the energy of the electrons.

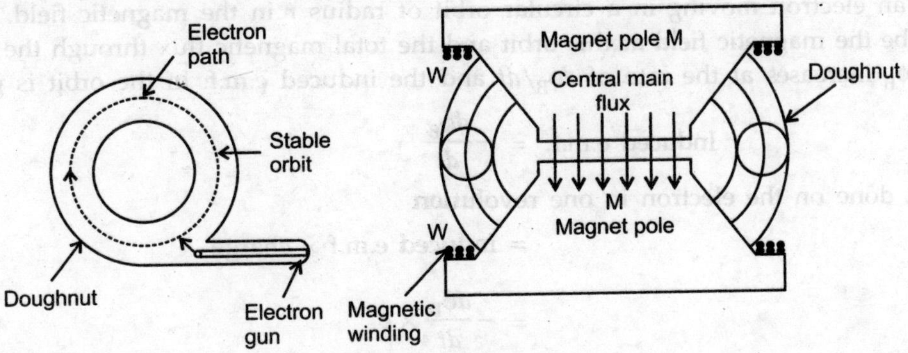

FIGURE 8

The Principle of the Betatron

The principle of the betatron is the same as that of a transformer in which an alternating current applied to the primary coil induces an alternating current in the secondary coil. In the betatron the secondary coil is replaced by a doughnut vacuum chamber. When the electron is injected into the doughnut, the alternating magnetic field has two effects, (*i*) an electromotive force is produced in the electron orbit by changing magnetic flux that gives an additional energy to the electrons and (*ii*) a radial force is produced by the action of the magnetic field whose direction is perpendicular to the electron velocity which keeps the electron moving in a circular path.

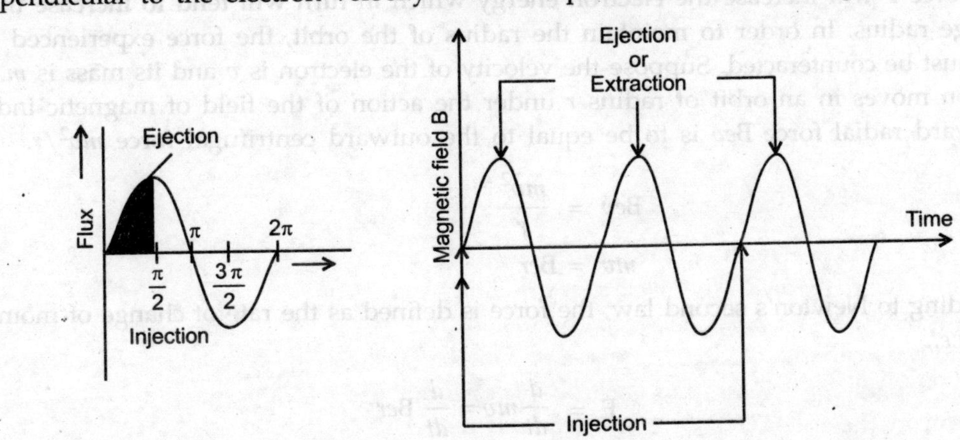

***FIGURE* 9 Variation of the magnetic field with time showing injection and extraction of electrons.**

The Operation

Electrons from the electron gun are injected into the doughnut shaped vacuum chamber when the magnetic field is just rising from its zero value in the first quarter cycle as shown in Figure 9. The electrons now make several thousand revolutions and gain energy. When the magnetic field has reached its maximum value, the electrons are pulled out from their orbit. Either they strike a target and produce X rays or emerge from the apparatus through a window.

Betatron Condition

Consider an electron moving in a circular orbit of radius r in the magnetic field. Let at any instant B be the magnetic field at this orbit and the total magnetic flux through the orbit is ϕ_B. The flux ϕ_B increases at the rate of $d\phi_B/dt$ and the induced e.m.f. in the orbit is given by

$$\text{Induced e.m.f.} = -\frac{d\phi_B}{dt} \tag{1}$$

Work done on the electron in one revolution

$$= \text{induced e.m.f} \times \text{charge}$$

$$= -\frac{d\phi_B}{dt} \times e$$

This work must be equal to the tangential force F acting on the electron multiplied by the length of the orbit path, *i.e.*,

$$\text{Work done} = \text{force} \times \text{distance}$$

$$= F \times 2\pi r$$

$$\therefore \quad F \times 2\pi r = -\frac{d\phi_B}{dt} \times e$$

or

$$F = -\frac{e}{2\pi r} \times \frac{d\phi_B}{dt} \tag{2}$$

This force F will increase the electron energy which in turn will tend to increase the orbit of the large radius. In order to maintain the radius of the orbit, the force experienced by the electron must be counteracted. Suppose the velocity of the electron is v and its mass is m. When the electron moves in an orbit of radius r under the action of the field of magnetic induction B, the inward radial force Bev is to be equal to the outward centrifugal force mv^2/r.

$$\therefore \quad Bev = \frac{mv^2}{r}$$

or

$$mv = Ber \tag{3}$$

According to Newton's second law, the force is defined as the rate of change of momentum ($p = mv$), *i.e.*,

$$F = \frac{d}{dt} mv = \frac{d}{dt} Ber$$

or

$$F = er\frac{dB}{dt} \tag{4}$$

To maintain the radius constant, the value of F given in equation (3) and equation (4) should numerically be equal, hence

$$\frac{e}{2\pi r}\frac{d\phi_B}{dt} = er\frac{dB}{dt}$$

or $$\frac{d\phi_B}{dt} = 2\pi r^2 \frac{dB}{dt}$$

Integrating, we get $$\phi_B = 2\pi r^2 B$$

This is known as the betatron condition.

Example 6. *The magnetic flux within a stable orbit in a betatron changes at a constant rate of 15 Wb/s. What would be the energy of an electron which undergoes 10^6 revolutions?*

Solution. We know that increase in energy per revolution

$$= e\frac{d\phi_B}{dt} \text{ joules}$$

$$= \frac{d\phi_B}{dt} \text{ electron-volt}$$

$$= 15 \text{ eV} = 15 \times 10^{-6} \text{ MeV}$$

Final energy of an electron after 10^6 revolutions

$$= 15 \times 10^{-6} \times 10^6 = 15 \text{ MeV}.$$

Example 7. *In a certain betatron, the maximum magnetic field at the electron orbit is 0.5 Wb/m². The diameter of the stable orbit is 1.5 m. If the frequency of the alternating current through electromagnet coils is 59 Hz, calculate for the electrons:*

(i) Final energy

(ii) Average energy gained per revolution

Solution. (*i*) In a betatron, the acceleration time is equal to a quarter time period, *i.e.*, $1/4f$ or $\pi/2\omega$ second. The total distance travelled by the electron is $c \times \pi/2\omega$ because the electron velocity is nearly equal to the velocity of light *c*. If R is the radius of the stable orbit, then the number of revolutions N made is given by

$$N = \frac{c\pi/2\omega}{2\pi R} = \frac{c}{4\omega R} = \frac{3\times10^8}{4\times(2\pi\times50)\times1.5}$$

The momentum of electron $= E/c$ (treated relativistically)

$$\therefore \quad mv = E/c$$

Moreover, $$\frac{mv^2}{R} = Bev$$

$$\therefore \quad mv = BeR \quad \text{or} \quad E/c = BeR$$

or $$E = BeRc$$

or $$E = \frac{0.5 \times 1.6 \times 10^{-10} \times 0.75 \times 3 \times 10^2}{1.6 \times 10^{-12}} \text{ MeV}$$

$$= 112.5 \text{ MeV}.$$

(*ii*) The average energy per revolution is

$$= 112.5 \times 10^6/3.1 \times 10^5$$

$$= \mathbf{363\ eV.}$$

5.15 THE PROTON SYNCHROTRON

The proton synchrotron is a machine which accelerates protons to energy in the GeV range (1 GeV = 1 billion electron volt). The device uses a ring shaped or doughnut vacuum tube. This tube is circular in the installation at Birmingham, England, and has a racetrack design in the installation in America. Here we shall describe the American design.

The Construction of the American Synchrotron

The american synchrotron consists of a ring shaped magnet which is in four quadrants as shown in Figure 10. The four quadrants are spaced by straight sections. A time varying magnetic field is produced by applying a.c. voltage to the copper winding of the magnet. The protons are injected by a linear accelerator or Van de Graaff generator into the doughnut at energy between 1 MeV and 10 MeV. This injection is tangential to the orbit or ring shaped doughnut. A high frequency resonator cavity accelerator (two electrodes connected to a high frequency generator) is used in one of the straight parts.

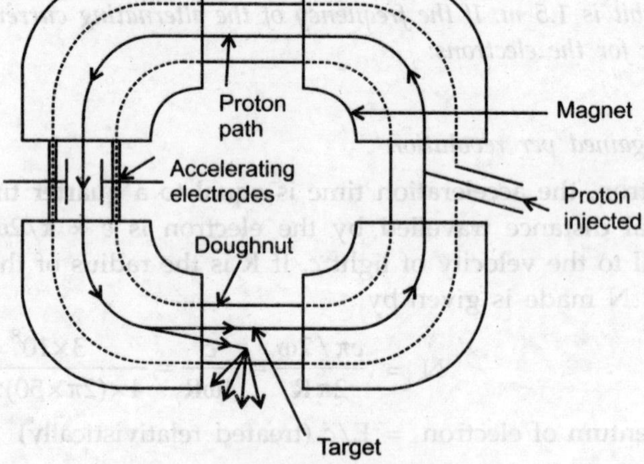

FIGURE 10

Operation

The pre-accelerated protons (01–10 MeV) are injected into the doughnut with the help of a linear accelerator. Under the influence of the magnetic field, the protons are made to circulate in a ring shaped vacuum tube. In each revolution the protons are accelerated once when they pass through electrodes connected to the high frequency generator. The field strength and the frequency of the oscillator are simultaneously increased in such a way that the protons travel in a circular path, or constant radius, and arrive at the electrodes when applied voltage is in the right phase for acceleration. When protons reach their maximum energy, the oscillator frequency is distorted. Now the proton's orbit contracts or expands such that they strike the target.

The proton synchrotron can accelerate protons to energies of the order of 10 GeV. The Berkeley machine can accelerate protons to 6 BeV and is called a betatron. The Brookhaven machine is called a cosmotron and can energize protons to 2.2 BeV. The machine is known as a cosmotron because particles of such energies can only be found in energetic cosmic rays.

Example 8. *In a betatron B_{max} applied 0.5 Wb/m² with orbital diameter 1.5 m. If the electromagnet is energized at 50 Hz, calculate the average energy gained by electrons in a revolution.*

Solution. The number of revolutions

$$= \frac{c}{4\omega R}$$

$$= \frac{3 \times 10^8}{4 \times (2\pi \times 50) \times 1.5}$$

$$= 3.1 \times 10^5$$

$$mv = \frac{E}{c} \text{ in electric field}$$

$$E = BeRc$$

$$= 0.5 \times 1.6 \times 10^{-19} \times 0.75 \times 3.0 \times 10^8 \text{J}$$

$$E = \frac{0.5 \times 1.6 \times 10^{-19} \times 0.75 \times 3.0 \times 10^8}{1.6 \times 10^{-13}} \text{ MeV}$$

$$= 112.5 \text{ MeV}$$

The energy/revolution

$$= \frac{112.5 \times 10^5}{3.1 \times 10^5}$$

$$= \mathbf{36.3 \ eV.}$$

5.16 NUCLEAR REACTION

In radioactivity, the nucleus of one element spontaneously changes into the nucleus of another element and the process is known as a nuclear decay reaction. It is also possible to induce a nuclear reaction by bombarding a nucleus with high energy particles from the particle accelerator. Such reactions which change the identity of a target nucleus are called artificial nuclear reactions.

When energetic particles such as protons, neutrons, etc. are shot at a target:

- The incident projectile particle may be simply deviated so that when the incoming and outgoing particles are the same, the process is scattering.
- The incident particle may be completely absorbed without emission of any particle. The process is radioactive capture.

■ The incident and the emergent particles may be different. The process is nuclear reaction.

It is impossible to know which processes have taken place and thus we have to depend on the probability of occurrence. The probability that a bombarding particle interacts with the target is expressed in terms of a nuclear cross-section.

5.17 NUCLEAR REACTION CROSS-SECTION

The probability that a bombarding particle will interact with the target material and cause a nuclear reaction is characterized by a parameter known as nuclear reaction cross-section. It is denoted by σ (sigma). It is basically an effective circular area which a target nucleus presents to the incident particle undergoing a particular nuclear process. It is not just a geometrical area (πr^2) of the nucleus. But a nuclear reaction cross-section is defined as

$$\sigma = \frac{N}{N_0 N_t} \text{ m}^2/\text{nucleus} \qquad (1)$$

When N_t is the number of target nuclei per unit area.

N_0 is the number of incident particles striking per second and N is the number of nuclei undergoing interaction per second. When $N_t = 1$ (unity), then the nuclear reaction cross-section becomes:

$$\sigma = \frac{N}{N_0} \qquad (2)$$

Hence, a nuclear reaction cross-section may be defined as the ratio of the number of nuclei undergoing interaction to the number of incident particles striking per unit area, when the target nuclei per unit area is one.

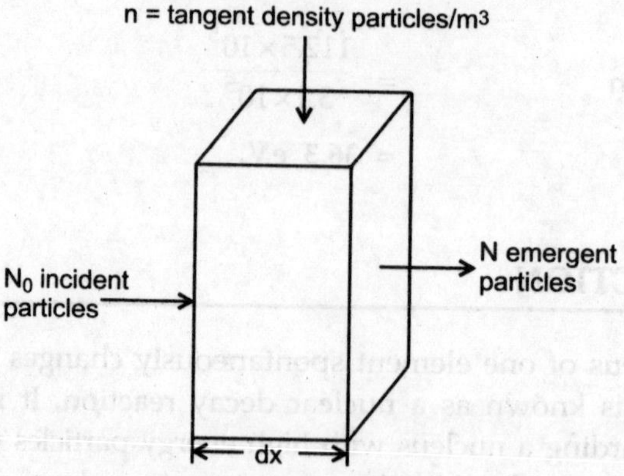

FIGURE 11(a) **Explanation of σ.**

Determination of σ

Let us consider a sheet of thickness x and an area of cross-section A containing n nuclei per unit volume. Let us consider the nuclei to be rigid spheres with an effective cross-sectional area σ. A nuclear reaction can take place only when the incident particle strikes the area σ [See Figures 11(a) and (b)].

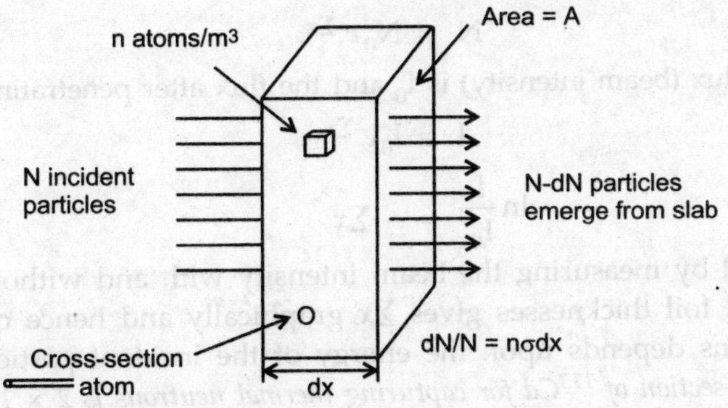

***FIGURE 11 (b)* The relationship between cross-section and beam intensity.**

The probability that an incident particle interacts with one of the target nuclei is therefore proportional to σ. The probability of a collision interaction is given by:

$$P = \sigma n x \tag{3}$$

Let N_0 be the number of particles in the incident beam, and N be the number of particles which emerge out of the sheet. Further, let dN be the number of particles reduced from the beam in passing through a thickness dx of the sheet.

$$dN = -N\sigma n dx$$

$$-\frac{dN}{N} = \sigma n dx \tag{4}$$

Integrating the above equation, we get

$$-\int_{N_0}^{N} \frac{dN}{N} = \sigma n \int_0^x dx$$

$$-[\ln N]_{N_0}^{N} = \sigma n x$$

$$-\ln \frac{N}{N_0} = \sigma n x$$

or

$$\ln \frac{N}{N_0} = -\sigma n x$$

$$N = N_0 e^{-n\sigma x} \tag{5}$$

Thus, the effective cross-sections that the target nucleus offers for an interaction is given by x

$$\sigma = \frac{1}{nx} \ln\left[\frac{N_0}{N}\right] \tag{6}$$

Nuclear cross-sections, which have the dimensions of area, are very small.

Thus units of nuclear cross-sections are barns

$$1 \text{ barn} = 10^{-28} \text{ m}^2$$

The product $n\sigma$ is called the microscopic cross-section and is denoted by Σ (it has the dimension of L^{-1}).

$$N = N_0 e^{-\Sigma x} \tag{7}$$

If the intensity flux (beam intensity) is I_0 and the flux after penetrating a distance x is I, then

$$I = I_0 e^{-\Sigma x}$$

or

$$\ln \frac{I}{I_0} = -\Sigma x \tag{8}$$

Σx can be found by measuring the beam intensity with and without targets. A plot of ln against x for several foil thicknesses gives Σx graphically and hence σ. The cross-section for most nuclear reactions depends upon the energy of the incident particle.

Example 9. *The cross-section of ^{113}Cd for capturing thermal neutrons is 2×10^4 barn, the mean atomic mass or natural cadmium is 112 u, and its density is $8.64\ gm/cm^3 = 8.64 \times 10^3\ kg/m^3$. (a) what fraction of an incident beam of thermal neutrons is absorbed by a cadmium sheet 0.1 mm thick? (b) What thickness of cadmium is needed to absorb 99 percent of an incident beam of thermal neutrons?*

Solution. (*a*) Since ^{113}Cd constitutes 12 percent of natural cadmium, the number of ^{113}Cd atoms per cubic meters is

$$n = (0.12)\left[\frac{8.64\times10^3\ \text{kg/m}^3}{(112\,\text{u/atom})(1.66\times10^{-27}\ \text{kg/u})}\right]$$

$$= 5.58 \times 10^{27}\ \text{atoms/m}^3$$

The capture cross-section is

$$\sigma = 2 \times 10^4\ \text{barn} = 2 \times 10^{-24}\ \text{m}^2$$

So

$$n\sigma = (5.58 \times 10^{27}\ \text{m}^{-3})\ (2 \times 10^{-24}\ \text{m}^2)$$

$$= 1.12 \times 10^4\ \text{m}^{-1}$$

As

$$N = N_0 e^{-n\sigma x}$$

So the fraction of incident neutrons that is absorbed is

$$\frac{N_0 - N}{N_0} = \frac{N_0 - N_0 e^{-n\sigma x}}{N_0} = 1 - e^{-n\sigma x}$$

since $x = 0.1$ mm $= 10^{-4}$m

Here

$$\frac{N_0 - N}{N_0} = 1 - e^{(-1.12 \times 10^4\ \text{m}^{-1})\ (10^{-4})} = 0.67$$

Two-thirds of the incident neutrons are absorbed.

(*b*) Since it is given that 1 percent of the incident neutrons pass through the cadmium sheet, $N = 0.01\ N_0$, and

$$\frac{N}{N_0} = 0.01 = e^{-n\sigma x}$$

$$\ln 0.01 = -n\sigma x$$

$$x = \frac{-\ln 0.01}{n\sigma} = \frac{-\ln 0.01}{1.12\times10^4\ \text{m}^{-1}}$$

$$= 4.1 \times 10^{-4}\ \text{m} = 0.41\ \text{mm}$$

Cadmium is evidently a very efficient absorber of thermal neutrons.

5.18 NUCLEAR FISSION

In 1938, two German scientists Hahn and Strassman, showed that when uranium is bombarded by neutrons, one of the radioactive elements produced is an isotope of barium, Z = 56 ($_{56}Ba^{141}$). They found that barium was accompanied by a radioactive isotope of the gas Krypton (Z = 36). Meitner and Frisch called the phenomenon nuclear fission, since it resembled the division of cells in biology.

Definition

The process in which a heavy nucleus, after capturing a neutron, splits up into two lighter nuclei of comparable mass with the release a large amount of energy, is known as nuclear fission.

When uranium is bombarded with neutrons, a uranium nucleus captures a slow neutron, forming an unstable compound nucleus. The compound nucleus splits into two nearly equal parts. Some neutons are also released in this process. The fission process is:

$$_{92}U^{235} + {_0n^1} \rightarrow {_{92}U^{236*}} \rightarrow X + Y + \text{neutrons} \quad (1)$$

$_{92}U^{236*}$ is a highly unstable isotope, and X and Y are the fission fragments. The fragments are not uniquely determined, because there are various combinations of fragments possible and *n* number of neutrons are given off. Typical fission reactions are:

$$_{92}U^{235} + {_0n^1} \rightarrow {_{92}U^{236*}} \rightarrow {_{56}Ba^{141}} + {_{36}Kr^{92}} + 3\,{_0n^1} + Q \quad (2)$$

$$_{92}U^{235} + {_0n^1} \rightarrow {_{92}U^{236*}} \rightarrow {_{54}Xe^{140}} + {_{38}Sr^{94}} + 2\,{_0n^1} + Q \quad (3)$$

where Q is the energy released in the reaction.

According to equation (2) when $_{92}U^{235}$ is bombarded by a slow moving neutron, the nucleus becomes unstable ($_{92}U^{236*}$) and splits into $_{56}Ba^{141}$ and $_{36}Kr^{92}$ releasing 3 neutrons and energy Q.

Energy Released in Fission

In the nuclear fission process a large amount of energy is released. This energy is produced because the original mass of the nucleus is greater than the sum of the mass of the products produced after fission. The difference between the mass before and after fission is converted into energy according to Einstein's equation

$$E = mc^2$$

Consider the Nuclear Fission Reaction

$$_{92}U^{235} + {_0n^1} \rightarrow {_{92}U^{236*}} \rightarrow {_{56}Ba^{141}} + {_{36}Kr^{92}} + 3\,{_0n^1} + Q$$

Let us estimate the actual mass before and after the fission reaction.

$$\begin{aligned} \text{Mass of } {_{92}U^{235}} &= 235.045733 \text{ u} \\ + \text{ Mass of } {_0n^1} &= 1.008665 \text{ u} \\ \text{Total initial mass} &= 236.054398 \text{ u} \\ \text{Mass of } {_{56}Ba^{141}} &= 140.9177 \text{ u} \\ \text{Mass of } {_{36}Kr^{92}} &= 91.8854 \text{ u} \\ \text{Mass of 3 neutrons} &= 3.025995 \text{ u} \\ \text{Total final mass} &= 235.829095 \text{ u} \end{aligned}$$

Mass decrease = 236.054398–235.829095 = 0.2253 u

This decrease in mass is converted into energy

Energy released = 0.2253 × 931 = 209.8 MeV (1u = 931 MeV)

Thus in the process of fission of one nucleus of uranium, about 200 MeV energy is released.

Energy released by 1 kg of uranium

$$\text{Number of atoms in 1 kg of uranium} = \frac{(6.023 \times 10^{23})}{235}$$

Energy released in one fission = 200 MeV

Energy produce by 1 kg of uranium during fission

$$= \frac{(6.023 \times 10^{26})}{235} \times 200$$

$$= 5.128 \times 10^{26} \text{ MeV}$$

$$= (5.128 \times 10^{26}) \times (1.6 \times 10^{-13}) \text{ J}$$

$$(\because 1 \text{ MeV} = 1.6 \times 10^{-13} \text{ J})$$

$$= \frac{(5.128 \times 10^{26}) \times (1.6 \times 10^{-13})}{3.6 \times 10^{6}} \text{ kWh}$$

$$(\because 1 \text{ KWh} = 3.6 \times 10^{6} \text{ J})$$

$$= 2.26 \times 10^{7} \text{ kWh}$$

Thus the energy released by fission of 1 kg of U^{235} is 2.26×10^{7} kWh. This is the reason why nuclear energy is being used for the generation of electricity.

Bohr and Wheeler's Theory of Nuclear Fission

We know that U^{238} can be fissioned by fast neutrons only while U^{235} is fissionable by fast as well as by slow neutrons. An explanation of this observation was given by Bohr and Wheeler in terms of the liquid drop model of the nucleus.

The uranium nucleus can be considered to be similar to a spherical liquid drop (See A in Figure 12). If some energy is supplied to the drop to excite it, it starts oscillating and changes its shape to assume the form of an ellipsoid (See B in Figure 12). When the surface area increases, then, according to the liquid drop model, the surface energy also increases. However, with the deformation of the nucleus, the average separation between the protons in it increases, thus decreasing the Coulomb energy. Therefore, the competing influences at work are the nuclear surface tension, which tends to keep the nucleus spherical, and the Coulomb repulsion, which tends to deform it. If the excitation is sufficient, the Coulomb force succeeds in making the nucleus acquire a dumb-bell shape (See C in Figure 12).

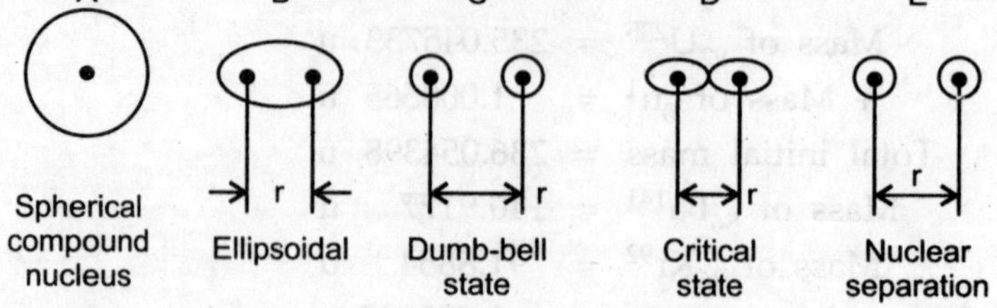

FIGURE 12 **Possible changes in shape during the fission process.**

With such a large distortion, the surface tension fails to restore the nucleus to sphericity, and the Coulomb force increases the separation between the two ends of the nucleus (See D in Figure 12), until the ends split into two fragments (See E in Figure 12).

The nuclei that result from fission are called fission fragments. Usually fission fragments are of unequal size. A heavy nucleus undergoes fission when it acquires enough excitation energy to oscillate violently. The ${}_{92}U^{235}$ isotope excited by the mere absorption of an individual neutron while ${}_{92}U^{238}$ requires more excitation energy for fission than the binding energy released when another neutron is absorbed. ${}_{92}U^{238}$ undergoes fission only by reaction with fast neutrons whose kinetic energies exceed about 1 MeV.

5.19 CHAIN REACTION

When a neutron fissions a uranium nucleus then besides the fission fragments, a few fast neutrons are also emitted. If one or more of the emitted neutrons are used to fission other nuclei, further neutrons are produced and the process is repeated. The reaction thus becomes self-propagating and is known as a chain reaction (see Figure 13).

The chain reactions may be of two types :

(i) *Uncontrolled Chain Reaction:* In this type of chain reaction, the number of neutrons is allowed to multiply indefinitely and the entire energy is released all at once as a violent explosion. Such a chain reaction takes place in a nuclear bomb.

(ii) *Controlled Chain Reaction:* In this type of chain reaction, the reaction is first accelerated so that the neutrons are built up to a certain level and thereafter the number of fission producing neutrons is kept constant by some means. Such a controlled chain reaction is used in nuclear reactors.

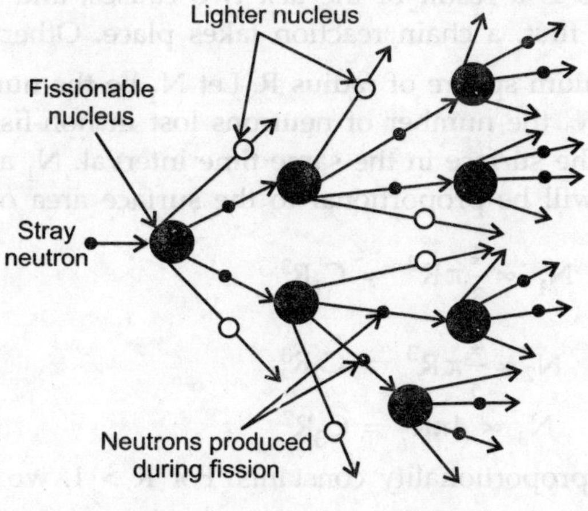

FIGURE 13 Sketch of a chain reaction. The reaction is self-sustaining if at least one neutron from each fission event on the average induces another fission event. If more than one neutron per fission on the average induces another fission, the reaction is explosive.

The Neutron Multiplication Factor

The ratio of secondary neutrons produced to the original neutrons is called the neutron multiplication factor. It is denoted by K.

$$K = \frac{\text{Number of neutrons present in one generation}}{\text{Number of neutrons present in the previous generation}}$$

If $K < 1$, the chain reaction will slow down and stop. This condition is termed as "subcritical".

If $K = 1$, the chain reaction will proceed at a steady rate. This condition is termed as "critical". Example: nuclear reactor.

If $K > 1$, the chain reaction will grow to an explosive rate. This condition is termed as "super-critical". Example: nuclear bomb.

The Critical Size for Maintenance of a Chain Reaction

In the fission of uranium nuclei, 2.5 neutrons are, on the average, emitted per fission. Some of them escape through the surface of the uranium, while many are lost in non-fission processes. The maintenance of the chain reaction depends upon a favorable balance of neutrons among the three processes given below:

- The fission of uranium nuclei which produce more neutrons than the number of neutrons used for inducing fission.
- Non-fission processes, including the radioactive capture of neutrons by the uranium, and the parasitic capture, by the different substances in the system and by impurities.
- Escape or leakage of neutrons through the surface of the system.

If the loss of neutrons is a result of the last two causes, and is less than the surplus of neutrons produced in the first, a chain reaction takes place. Otherwise it cannot take place.

Let us consider a uranium sphere of radius R. Let N_1 be the number of neutrons produced in a given time interval, N_2 the number of neutrons lost in non-fission processes, and N_3 the number escaped through the surface in the same time interval. N_1 and N_2 will be proportional to the volume, while N_3 will be proportional to the surface area of the sphere.

Thus
$$N_1 \propto \frac{4}{3}\pi R^3 = C_1R^3$$

$$N_2 \propto \frac{4}{3}\pi R^3 = C_2R^3$$

$$N_3 \propto 4\pi R^2 = C_3R^2$$

where C_1, C_2 and C_3 are proportionality constants. For $K > 1$, we must have

$$N_1 > N_2 + N_3$$

or
$$C_1R^3 > C_2R^3 + C_3R^2$$

or
$$C_1R > C_2R + C_3$$

$$(C_1 - C_2)\, R > C_3$$

or
$$R > \frac{C_3}{C_1 - C_2} = C \text{ (say)}$$

C is known as the critical size of the sample. Thus, in order to achieve a self-sustained chain reaction, the size (or mass) of the sample must be greater than a critical value C. Below the critical size (or critical mass) the chain reaction would not occur.

Critical size of a system containing fissionable material is defined as the minimum size for which the number of neutrons produced in the fission process just balance those lost by leakage and non-fission capture. The mass of the fissionable material at this size is called the critical mass. If the size is less than the critical size, a chain reaction is not possible.

5.20 NUCLEAR REACTORS

A nuclear reactor is a device in which a self-sustaining controlled chain reaction is produced in a fissionable material. It is thus a source of controlled energy. The first nuclear reactor was built by Enrico Fermi and his collaborators in 1942. The main parts of a typical nuclear reactor are as follows:

(*i*) *Fuel:* An important feature of a nuclear reactor is that the fuel here is not highly enriched. The uranium fuel (U^{235} isotope) is generally used in the form of rods about a few cm in diameter.

(*ii*) *Moderator:* An interesting feature about the fission of U^{235} is that it is much more probable when it is bombarded with slow neutrons than when it is bombarded with fast neutrons. This aspect becomes a very important feature in the design of a nuclear reactor. The slowing down of neutrons in a reactor is achieved by moderators. Thus fast neutrons are slowed down by mixing suitable moderators with uranium fuel. Commonly used materials for moderators are heavy water, graphite and beryllium. In general a moderator should have a light weight and its chances of absorbing neutrons should be as low as possible.

(*iii*) *Control Rods:* The control of reactions in a reactor is achieved by control rods. These rods are generally of those materials whose nuclei possess the property of readily capturing slow neutrons. Commonly used material for control rods are cadmium and boron. When the rods are pushed into the reactor, the fission rate decreases, and when they are pulled out, the fission grows.

(*iv*) *Shield:* Due to neutron bombardment, some dangerous biological effects might arise. To avoid this the reactor is surrounded by a concrete wall about 2 meters thick and containing a high proportion of elements like ion.

(*v*) *Coolant:* The energy is released inside the reactor in the form of heat which is removed by means of a cooling agent, known as coolant. The carbon dioxide (CO_2) gas is the main coolant in a reactor. It is circulated through the interior of the reactor by a pumping system.

(*vi*) *Safety Device:* Special sets of control rods (known as shut-off rods) are designed for an emergency that might arise whenever the reactor begins to accelerate. These rods are designed to automatically drop in and stop the reaction altogether, by immediately absorbing the neutrons.

Construction

One type of nuclear reactor is shown in Figure 14. It consists of a large number of uranium rods placed in a calculated geometrical lattice between layers of pure graphite (moderator) blocks. To prevent oxidation of uranium and also to preserve the gaseous fission products, the rods are covered by close fitting aluminium cylinders. The control rods are so inserted in the lattice that they can be raised or lowered between the uranium rods whenever necessary. The whole reactor is surrounded by a concrete shield.

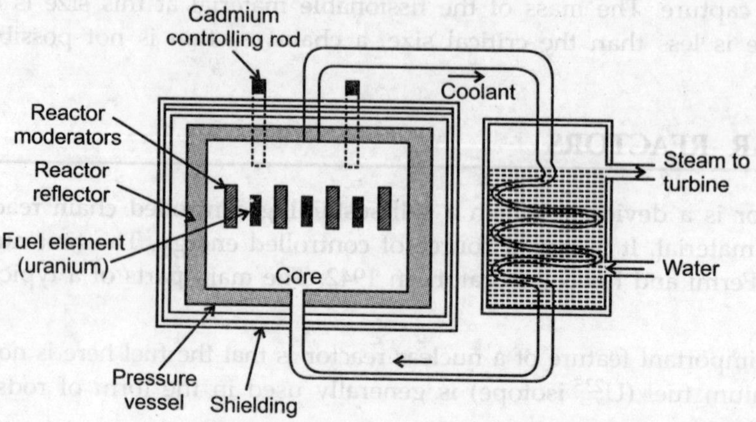

FIGURE 14 **Nuclear reactor.**

Working

As a neutron strikes a U^{235} nucleus and fission occurs, two or three fast neutrons are emitted. These neutrons are slowed down form energies of several MeV to energies of less than 1 eV by collision with moderator nuclei, after which they induce further fission in U^{235}. The reaction, once started, is controlled by moving the control rods in and out.

Coolant (CO_2) is pumped rapidly through the reactor to carry away the heat generated by the fission of the uranium nuclei. The hot carbon oxide gas passes through a heat exchanger where the heat is absorbed in water and is converted into steam. This steam drives the turbines and generates electric power.

Breeder Reactor

If a thermal reactor core with U^{235} fuel is surrounded by a blanket of a fertile material like U^{238}, U^{238} can be converted into fissionable fuel. Reactors of these types are called fuel producing reactors. These reactions are as follows:

$$_{92}U^{238} + {_0}n^1 \longrightarrow {_{92}}U^{239} + \gamma$$

This is followed by $$_{92}U^{239} \longrightarrow {_{93}}Np^{239} + \beta^-$$

$_{93}Np^{239}$ is also radioactive. It emits a β–particle to form plutonium.

$$_{93}N^{239} \longrightarrow {_{94}}Pu^{239} + \beta^- + \gamma$$

This process of producing one type of fissionable material (Pu^{239}) from a non-fissionable material (U^{238}) is called breeding and the reactor is called a breeder reactor.

The breeding reactions for the fertile material ${}_{90}Th^{232}$ are

$$ {}_{90}Th^{232} + n \longrightarrow {}_{90}Th^{232*} \xrightarrow[22\,min]{\beta^-} {}_{91}Pa^{233} \xrightarrow[27\,days]{\beta^-} {}_{92}U^{233} $$

Uses of Nuclear Reactors

Nuclear reactors are used mainly for the following purposes:

(i) *Production of Pu^{239}*: An ordinary uranium reactor is used to produce plutonium (Pu^{239}) which is a better fissionable material than U^{235}. Pu^{239} is used in nuclear reactors and in nuclear bombs in place of U^{235}. The critical size of Pu^{239} is smaller compared to that of U^{235} for the fission chain reaction. Therefore, the use of Pu^{239} is economical compared to U^{235}.

(ii) *Production of a Neutron Beam:* During the fission process, fast moving neutrons are emitted. By converging these neutrons into a fine beam, artificial disintegration of other elements are studied.

(iii) *Production of Radio Isotopes:* Artificial radioactive isotopes of many elements are produced in the reactor. For this, the element is placed in the reactor and bombarded with fast moving neutrons. These radio isotopes are utilized in medicines, biology, agriculture, industry and scientific discoveries. Sodium-24 is manufactured in this way:

$$ {}_{11}Na^{23} + {}_{0}n^{1} \longrightarrow {}_{11}Na^{24*} $$

(iv) *Production of Energy:* The energy produced in reactors is used to run electric generators. Thus nuclear energy is converted into electrical energy. This energy is used at power stations and can also be used as an alternate source for driving engines, for the propulsion of ships, submarines, and aircraft.

Example 10. *In a nuclear reactor, fission is produced in 1 gm of U^{235} (235.0439 amu) in 24 hours by a slow neutron (1.0087 amu). Assuming that ${}_{36}Kr^{92}$ (91.8973 amu) and ${}_{56}B^{141}$ (140.9139 amu) are produced in all reactions and no energy is lost, write the complete reaction and calculate the total energy produced in MeV and in kilowatt-hours. Given 1 amu = 931.5 MeV.*

Solution. The nuclear fission reaction is

$$ {}_{92}U^{235} + {}_{0}n^{1} \longrightarrow {}_{56}Ba^{141} + {}_{36}Kr^{92} + 3\,{}_{0}n^{1} $$

The sum of the mass before reaction is

$$ 235.0439 + 1.0087 = 236.0526 \text{ a.m.u.} $$

The sum of the mass after the reaction is

$$ 140.9139 + 91.8973 + 3(1.0087) = 235.8373 \text{ a.m.u.} $$

The mass loss in the fission is

$$ \Delta m = 236.0526 - 235.8373 = 0.2153 \text{ a.m.u.} $$

The energy equivalent of 1 amu is 931.5 MeV. Therefore, the energy released in the fission of a U^{235} nucleus

$$ = 0.2153 \times 931.5 = 200 \text{ MeV} $$

The number of atoms in 235 gm of U^{235} is 6.02×10^{23} (Avogadro number). Therefore, the number of atoms in 1 gm.

$$= \frac{6.02 \times 10^{23}}{235} = 2.56 \times 10^{21}$$

Hence the energy released in the fission of 1 gm of U^{235} *i.e.*, in 2.56×10^{21} fission is

$$E = 200 \times 2.56 \times 10^{21} = 5.12 \times 10^{23} \text{ MeV}$$

Now, $1 \text{ MeV} = 1.6 \times 10^{-13}$ joule

$\therefore$ $E = 5.12 \times 10^{23} \times (1.6 \times 10^{-13}) = 8.2 \times 10^{10}$ joule

Now, $1 \text{ kW-hour} = 1000 \text{ watt} \times 3600 \text{ sec} = 3.6 \times 10^{6}$ joules

$$E = \frac{8.2 \times 10^{10}}{3.6 \times 10^{6}} = 2.28 \times 10^{4} \text{ kW-hour}$$

Example 11. *A reactor is developing nuclear energy at a rate of 32,000 kilowatts. How many atoms of U^{235} undergo fission per second? How many kg of U^{235} would be used up in 1000 hours of operation? Assume an average energy of 200 MeV released per fission. Take Avogadro's number as 6×10^{23} and 1 MeV = 1.6×10^{-13} joule.*

Solution. The power developed by the reactor is 32000 kilowatts *i.e.*, 3.2×10^{7} watts. Therefore, the energy released by the reactor per second.

$$= 3.2 \times 10^{7} \text{ joules} \qquad [\because 1 \text{ watt} = 1 \text{ joule/sec}]$$

$$= \frac{3.2 \times 10^{7}}{1.6 \times 10^{-13}} \text{ MeV}$$

$$[\because 1.6 \times 10^{-13} \text{ joule} = 1 \text{ MeV}]$$

$$= \mathbf{2.0 \times 10^{20} \text{ MeV.}}$$

The energy released per fission is 200 MeV. Therefore, the number of fissions occuring in the reactor per second

$$= \frac{2.0 \times 10^{20}}{200} = 1.0 \times 10^{18}$$

The number of atoms (or nuclei) of U^{235} consumed in 1000 hours

$$= 1.0 \times 10^{18} \times (1000 \times 3600)$$

$$= 36 \times 10^{23}$$

Now, 1 gm-atom (*i.e.*, 235 gm) of U^{235} has 6×10^{23} atoms. Therefore, the mass of U^{235} consumed in 1000 hours

$$= \frac{36 \times 10^{23}}{6 \times 10^{23}} \times 235$$

$$= 1410 \text{ gm} = \mathbf{1.41 \text{ kg.}}$$

Example 12. *If in a certain fission process, the mass loss is 0.1%, then calculate the energy liberated by the fission of 1 kg of the substance. How much kilowatt-hour electric energy can be generated from it?*

Solution. The mass loss in 1 kg of the substance is

$$\Delta m = \frac{0.1}{100} \times 1 \text{ kg} = 0.001 \text{ kg}$$

According to Einstein's mass-energy relation the energy liberated is:

$$\Delta E = (\Delta m)\, c^2$$
$$= 0.001 \text{ kg} \times (3.0 \times 10^8 \text{ m/sec})^2$$
$$= 9.0 \times 10^{13} \text{ joule}$$

Now, 1 kilowatt hour = 3.6×10^6 joules

$$\therefore \quad \Delta E = \frac{9.0 \times 0^{13}}{3.6 \times 10^6}$$
$$= \mathbf{2.5 \times 10^7 \text{ kilowatt-hour}}$$

Example 13. *Calculate the fission rate of U^{235} required to produce 2 watts and the amount of energy that is released in the complete fission of 0.5 kg of U^{235}. The energy released per fission of U^{235} is 200 MeV.*

Solution. The power to be produced is

$$P = 2 \text{ watts} = 2 \text{ joule/sec}$$
$$= \frac{2}{1.6 \times 10^{-13}} \text{ MeV/sec}$$
$$[\therefore 1 \text{ MeV} = 1.6 \times 10^{-13} \text{ joule}]$$
$$= 1.25 \times 10^{13} \text{ MeV/sec}$$

The energy released per fission is 200 MeV.

Required number of fissions per second

$$= \frac{1.25 \times 10^{13}}{200} = 6.25 \times 10^{10}$$

The fission rate is 6.25×10^{10} per second.

The number of atoms in 235 gm of U^{235} is 6.02×10^{23}. Therefore, the number of atoms in 0.5 kg (500 gm) of U^{235}

$$= \frac{6.02 \times 10^{23}}{235} \times 500$$
$$= 1.28 \times 10^{24}$$

The energy released per fission is 200 MeV. Therefore, the total energy released in the complete fission of 0.5 kg of U^{235}

$$= 200 \text{ MeV} \times (1.28 \times 10^{24})$$
$$= 2.56 \times 10^{26} \text{ MeV.}$$

5.21 NUCLEAR FUSION

When two or more light nuclei moving at very high speed are fused to form a single heavy nucleus, then this process is known as *nuclear fusion*. For example, when four hydrogen nuclei are fused together, a helium nucleus is formed. The mass of the single nucleus formed is always less than the sum of the masses of the individual light nuclei. The difference in mass is converted into energy according to Einstein's equation.

$$E = mc^2$$

Consider a single helium nucleus formed by the fusion of two deuterium nuclei. Mass of ${}_1H^2$ = 2.014102 u; mass of ${}_2He^4$ = 4.002604 u

$${}_1H^2 + {}_1H^2 \longrightarrow {}_2He^4$$

$$\begin{aligned} \text{The initial mass of 2 deuterium atoms} &= 2 \times 2.014102 \\ &= 4.028204 \text{ u} \\ \text{Mass of helium atom} &= 4.002604 \text{ u} \\ \text{Decrease in mass} &= 4.028204 - 4.002604 \\ &= 0.025600 \text{ u} \\ \therefore \quad \text{Energy released} &= 0.025600 \times 931.3 \text{ MeV} \\ &= 23.84 \text{ MeV} \end{aligned}$$

Thus the energy released in fusion is 23.84 MeV.

The energy output in a fusion reaction is much less than the energy released in the fission of a U^{235} nucleus which is about 200 MeV. But this does not mean that fusion is a weaker energy source than fission. The number of deuterium in 1 gram of heavy hydrogen is much larger than the number of U^{235} nuclei in 1 gram of uranium. Therefore, the energy output per unit mass of the material consumed is much greater in case of the fusion of light nuclei than in case of the fusion of heavy nuclei.

It is very difficult to carry out the fusion process because the nuclei to be fused are positively charged and these nuclei repel one another strongly. They must be brought very close together by high pressure and high kinetic energies of about 0.1 MeV. For this a temperature of the order of 10^8 Kelvin is required. Such high temperatures are available in the sun and stars. As very high temperatures are required for the occurrence of the fusion process, the process is called a thermonuclear reaction and the energy released is called thermonuclear energy.

Example 14. *Certain stars obtain part of their energy by the fusion of three α particles to form a ${}_6C^{12}$ nucleus. How much energy does each such reaction involve? The mass of a helium atom is 4.00260 amu while the mass of an electron is 0.00055 amu. The mass of a ${}_6C^{12}$ atom is 12.000 amu by definition. (1 amu = 931.5 MeV)*

Solution. The mass of a ${}_2He^4$ atom is 4.00260 amu and it has 2 electrons. Therefore, the mass of its nucleus (α-particle)

$$\begin{aligned} &= 4.00260 - \text{mass of 2 electrons} \\ &= 4.00260 - (2 \times 0.00055) \\ &= 4.00260 - 0.00110 \\ &= 4.00150 \text{ amu} \end{aligned}$$

The mass of a ${}_6C^{12}$ atom is 12.000 amu and it has 6 electrons. Therefore, the mass of ${}_6C^{12}$ nucleus

$$\begin{aligned} &= 12.000 - (6 \times 0.00055) \\ &= 12.000 - 0.00330 \\ &= 11.99670 \text{ amu} \end{aligned}$$

When 3 α-particles fuse in a ${}_6C^{12}$ nucleus, the mass loss is

$$(3 \times 4.00150) - 11.99670 = 0.00780 \text{ amu}$$

The equivalent energy is 0.00780 × 931.5 MeV = 7.26 MeV.

5.22 THE GEIGER-MULLER COUNTER

The Geiger-Muller counter is the most common and versatile charged particle detector. It is a modified form of a cathode ray tube.

Construction

It consists of a rugged metal case enclosed in a thin glass tube. The hollow metal case acts as a cathode. A fine tungsten wire is streched along the axis of the tube and is insulated from it by ebonite plugs. It acts as an anode. The tube is evacuated and then partially filled with a mixture of 90% argon at 10 cm pressure and 10% ethyl alcohol vapor at 1 cm pressure. The fine tungsten wire is connected to the positive terminal of a high tension battery (about 1000 to 3000 volts) through a high resistance R (about 100 megaohms) and the negative terminal is connected to the metal tube. The D.C. voltage is kept slightly less than that which will cause a discharge between the electrodes. At one end of the tube a thin window of mica is arranged to allow the entry of radiation into the tube. The schematic of a Geiger-Muller counter is shown in Figure 15.

Principle

The basic principle of the Geiger-Muller counter can be understood as follows. We know that when an ionizing particle passes through the gas in an ionizing chamber, it produces a few ions. If the applied potential difference is strong enough, these ions will produce a secondary ion avalanche whose total effect, to start with, will be proportional to the energy associated with the primary ionizing event.

However, if the applied electrode potential difference is very high, the secondary ionization phenomenon becomes so dominant that the primary ionizing event loses its importance. In other words, the size of the final pulse produced depends only on the triggering off of ionization by an ionizing particle but independent of the energy of this particle.

A high energy particle entering through the mica window will cause one or more of the argon atoms to ionize. The electrons and ions of argon thus produced, cause other argon atoms to ionize in a cascade effect. The result of this one event is a sudden, massive electrical discharge that causes a current pulse. The current through R produces a voltage pulse of the order of 10 μV. An electronic pulse amplifier accepts the small pulse voltages and amplifiers them to about 5 to 50 V. The amplified output is then applied to a counter. As each incoming particle produces a pulse, the number of incoming particles can be counted.

FIGURE 15 **A schematic of a Geiger-Muller counter.**

Figure 16 shows the variation of the count rate with potential difference across a Geiger-Muller counter. It is seen that there is a threshold below which the tube does not work. This can be several hundred volts. After this, the number of pulses are proportional to the voltage. This region is known as the proportional region. If the applied voltage is increased further, then a point will be reached after which the count rate remains constant over a certain region. This region is known as the plateau or Geiger region. This region is used for the Geiger-Muller counter operation. Beyond the plateau region the applied electric field is so high that a continuous discharge takes place in the tube as shown in Figure 16 and the count rate increases very rapidly. It does not require any ionizing even for this to happen so that the tube must not be used in this region.

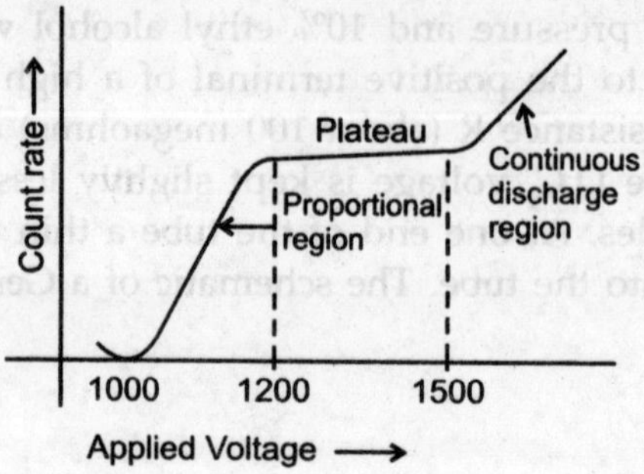

FIGURE 16 **Plateau graph of Geiger-Muller counter tube.**

The Geiger-Muller counter can count about 500 particles per second. It is to be noted that a Geiger-Muller counter will not register those particles that pass through it in the 'dead time'. Dead time refers to the time taken by the tube to recover between counts. It requires about 200 μs for the tube to recover. If a lot of particles enter the Geiger-Muller tube at a rapid rate, the tube will not have time to recover and some particles may not be counted.

The efficiency of the counter is defined as the ratio of the observed counts/sec to the number of ionizing particles entering the counter per second. Counting efficiency is defined as the ability of its counting, if at least one ion pair is produced in it:

$$\text{Counting efficiency } \eta = 1\text{-exp } (spl)$$

where s = specific ionization at one atmosphere

p = pressure in atmosphere

l = path length of the ionization particle in the counter.

Example 15. *A Geiger-Muller counter wire collects 10^8 electrons per discharge when the counting rate is 500 counts/min. What will be the average current in the circuit?*

Solution. Counting rate = 500 counts/min.

Total number of electrons collected per min.

$$= 500 \times 10^8 \times 10^{10}$$

$$\text{Charge/min.} = 5 \times 10^{10} \times 1.6 \times 10^{-19} \text{ coulomb/min}$$

$$\text{Charge/second} = \frac{5 \times 1.6 \times 10^{-9}}{60} = 1.33 \times 10^{-10} \text{ coulomb/sec}$$

$$\text{Average current} = \mathbf{1.33 \times 10^{-10} \text{ A.}}$$

5.23 MASS SPECTROGRAPHS

A mass spectrograph is an instrument that measures the nuclear (atomic) mass. Mass spectrographs were designed by F.W. Aston, A. J. Dempster, and K. T. Bainbridge using different properties. The initial mass spectrographs measured atomic masses with a precision of 1 part in 10^3 whereas modern instruments can yield a precision of 1 part in 10^7. The spectrograph designed by K. T. Bainbridge has already been discussed in chapter 1 in detail. In the Bainbridge mass spectrograph a velocity selector was used to produce a monovelocity ion beam and a transverse magnetic field was employed to discriminate between ions of different mass.

Aston's Mass Spectrograph

In 1919 F. W. Aston designed a mass spectrograph. This spectrograph enables the measurement of the mass of single atomic ions and is useful for the investigation of isotopes.

Principle

The deflection of the beam of positive rays in the electrostatic field depends on the velocity of ions and how they are dispersed. The dispersed beam is then subjected to a magnetic field which is perpendicular to electrostatic field.

Working

Figure 17 shows the apparatus used by Aston. The stream of positive ions is rendered into a fine beam by passing through slits S_1 and S_2. Due to the action of the electric field, all positive ions having the same value of charge per unit mass are not only deviated by an angle θ but are also dispersed by an angle $d\theta$ due to their different velocities. The magnetic field deviates the ions by an angle ϕ and reconverges them by $d\phi$. The field is so adjusted that it produces a deviation of the beam in the opposite direction and brings all ions having the same value of charge per unit mass, even though differing in velocities, to a focus on the plate. Ions having different values of e/m are brought to focus at different points on the photographic plate.

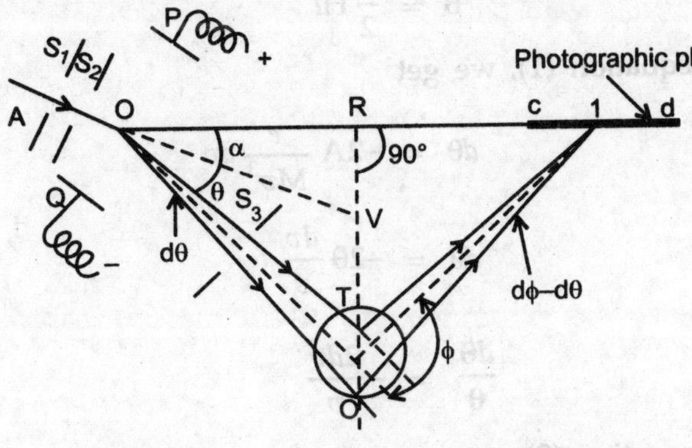

FIGURE 17 **Aston's mass spectrograph.**

Theory

Let us consider the following parameters:

E = electric field

e = charge on the ion

m = mass of the ion

v = velocity of the ion

θ = mean angle of deviation in the electric field

ϕ = mean angle of deviation in the magnetic field

$d\theta$ = dispersion angle in the electric field

$d\phi$ = convergence angle in the magnetic field

S = linear displacement of the ions in the electric field

l = length of the path in the electric field

Now,
$$\theta = \frac{S}{l}$$

and
$$S = \frac{1}{2}\frac{Ee}{M}\cdot\frac{l^2}{v^2}$$

$\therefore$
$$\theta = \frac{1}{2}\cdot\frac{Ee}{M}\cdot\frac{l}{v^2} = A.\frac{e}{Mv^2} \qquad (1)$$

where
$$A = \frac{1}{2}El$$

Similarly, in the magnetic field

$$\phi = \frac{L'}{l}$$

and
$$S' = \frac{1}{2}\cdot\frac{Hv}{M}\cdot\frac{l'^2}{v^2}$$

$\therefore$
$$\phi = \frac{1}{2}\frac{Hev}{M}\cdot\frac{l'}{v^2} = B.\frac{e}{Mv} \qquad (2)$$

where
$$B = \frac{1}{2}Hl'$$

By differentiating equation (1), we get

$$d\theta = -2A\frac{e}{Mv^3}dv$$

or
$$d\theta = -2\theta\frac{dv}{v} \qquad \left(\because \theta = A.\frac{e}{Mv^2}\right)$$

or
$$\frac{d\theta}{\theta} = -\frac{2dv}{v} \qquad (3)$$

By differentiating equation (2), we get

$$d\phi = -B.\frac{e}{Mv^2}dv$$

or
$$d\phi = -\phi.\frac{dv}{v} \qquad \left(\therefore \phi = B.\frac{e}{Mv}\right)$$

or $$\frac{d\phi}{\phi} = -\frac{dv}{v} \quad (4)$$

From equations (3) and (4), we get

$$\frac{d\theta}{\theta} = 2\frac{d\phi}{\phi}$$

or $$\frac{d\theta}{d\phi} = \frac{2\theta}{\phi} \quad (5)$$

Let a be the distance between the centers of electric and magnetic fields and b be the distance between O′ and the photographic plate.

In the absence of the magnetic field the width of the beam = $(a + b)\, d\theta$.

And in the absence of the electric field the width of the beam = $b.d\phi$.

The condition of focussing is given by

$$(a + b)\, d\theta = b.d\phi$$

or $$\frac{a+b}{b} = \frac{d\phi}{d\theta}$$

or $$\frac{a}{b} + 1 = \frac{\phi}{2\theta} \quad \text{[From equation (5)]}$$

or $$\frac{a}{b} = \frac{\phi}{2\theta} - 1$$

or $$\frac{a}{b} = \frac{\phi - 2\theta}{2\theta}$$

or $$2a.\,\theta = b.\,(\phi - 2\theta) \quad (6)$$

This is the required condition for focussing.

If $$b = \infty$$

then $$0 = \frac{\phi - 2\theta}{2\theta}$$

or $$\phi = 4\theta \quad (7)$$

This gives the position of the photographic plate. The modified apparatus gives an accuracy of 1 part in 10^4 and resolvability of 600.

Advantages

The following are the advantages of Aston's Mass Spectrograph:

- All the particles having the same value of $\frac{e}{m}$ are focussed at a single point. Hence, the intensities of the lines of the photographic plate in Aston's mass spectrograph are large.
- The intensity of a line in the mass spectrum is proportional to the total number of particles of that mass. Hence a rough idea of the relative abundance of various isotopes of an element can be made by this method.

Limitations

The following are the limitations of Aston's Mass Spectrograph:

- The mass scale is not linear.
- Owing to the polarization of the electrodes of the electric field, the traces are not quite straight but slightly curved.

Example 16. *The distance between traces corresponding to masses 12 and 16 in an Aston's mass spectrograph is 4.8 cm. Calculate the mass of the particle whose trace is at a distance of 8.4 cm from the trace of mass 16.*

Solution. Let x be the distance of the mark from the trace of mass 16.

$$\therefore \quad x \propto 16$$

Also
$$x - 4.8 \propto 12$$

Hence
$$\frac{x}{x-4.8} = \frac{16}{12}$$

or
$$x = 19.2 \text{ cm.}$$

If M is the mass of the particle whose trace is at a distance of 8.4 cm from the trace of mass 16, we have

$$19.2 \pm 8.4 \propto M$$

and
$$19.2 \propto 16$$

Hence
$$\frac{M}{16} = \frac{19.2 \pm 8.4}{19.2} = 1 \pm \frac{8.4}{19.2} = 1 \pm \frac{7}{16}$$

Hence
$$M = 16 \pm 7$$
$$= \mathbf{23 \text{ or } 9.}$$

EXERCISES

Objective Questions

1. What percentage of original radioactive atoms is left after five half-lives?
   (*a*) 20% (*b*) 10%
   (*c*) 5% (*d*) 3%
2. Out of the following which is not emitted by radioactive substances?
   (*a*) electrons (*b*) positions
   (*c*) neutrons (*d*) neutrinos
3. Consider the following nuclear reaction $^{235}_{92}U + ^{1}_{0}n \rightarrow ^{140}_{54}Xe + ^{A}_{Z}Sr + 2^{1}_{0}n + \text{energy}$. The value of Z and A are:
   (*a*) Z = 39, A = 92 (*b*) Z = 37, A = 93
   (*c*) Z = 38, A = 95 (*d*) Z = 38, A = 94

4. Regarding the sun, the production of energy is due to
(*a*) nuclear fission (*b*) nuclear fusion
(*c*) chemical reaction (*d*) thermal reaction

5. Control rods used in a nuclear reactor are made of
(*a*) cadmium (*b*) steel
(*c*) beryllium (*d*) copper

6. If, in a nuclear fission reaction involving neutron capture, the multiplication factor is greater than 1, then
(*a*) the process will ultimately come to a halt
(*b*) an explosion may result
(*c*) the reaction will be self sustaining with no explosion possible
(*d*) it is impossible to predict whether the reaction will halt or continue

7. In a nuclear reaction, boron rods may be inserted to:
(*a*) slow down neutron (*b*) speed up the reaction
(*c*) absorb gamma radiation (*d*) absorb excess neutrons

8. The mass per nucleon in an ordinary hydrogen atom is
(*a*) the same as the mass per nucleon in an oxygen atom
(*b*) 1/16 the mass per nucleon in an oxygen atom
(*c*) slightly greater than the mass per nucleon in an oxygen atom
(*d*) slightly smaller than the mass per nucleon in an oxygen atom

9. The mass of a nucleus is less than the sum of the mass of the individual particles that comprise it. This mass defect is accounted for by the
(*a*) binding energy (*b*) energy levels
(*c*) atomic mass (*d*) mass number

10. Neutrons are more effective as bombarding particles than protons because neutrons
(*a*) have more kinetic energy (*b*) emit beta particles
(*c*) are unaffected by electric fields (*d*) exert a stronger nuclear force

11. Which of the following is a good nuclear fuel?
(*a*) uranium-236 (*b*) plutonium-239
(*c*) neptunium-239 (*d*) thorium-236

12. In order to sustain a chain reaction, a mass of uranium-235 must be
(*a*) very hot (*b*) bombarded with neutrons
(*c*) at least of a certain size (*d*) exposed to gamma rays

13. In a fission-fusion-fission bomb, uranium-238 nuclei are split by
(*a*) high temperatures (*b*) fast neutrons
(*c*) alpha particles (*d*) uranium-235 nuclei

14. Binding energy per nucleon on $_1H^2$ and $_2He^4$ are 1.1 eV and 7.0 MeV respectively. Energy released in the process $_1H^2 + _1H^2 = _2He^4$ is
(*a*) 20.0 MeV (*b*) 16.6 MeV
(*c*) 25.2 MeV (*d*) 23.6 MeV

15. A particle of mass $3m$ at rest decays into two particles of mass m and $2m$ having non-zero velocities. The ratio of the de Broglie wavelengths of the particles (λ_1/λ_2) is

(*a*) 1/2 (*b*) 1/4

(*c*) 2 (*d*) none of these

Short Questions for Discussion

1. Describe the static properties of a nucleus.
2. Explain why the Gammow model is also known as α-particle model.
3. Why do we say that the nucleus behaves like a liquid drop?
4. Give the silent features of the nuclear shell model.
5. State clearly your understanding of binding energy and mass defect with respect to a nucleus.
6. What is an atomic mass unit? Derive the relation between an atomic mass unit and a million electron volt (MeV).
7. Calculate the energy equivalent of 1 a.m.u.
8. How is the concept of binding energy related to the stability of an atomic nucleus?
9. What is a Q-value in a nuclear reaction?
10. Distinguish between nuclear fission and nuclear fusion.
11. Explain the nuclear fission reaction with the help of a liquid drop model.
12. What is meant by a chain reaction? What are the factors which adversely affect the chain reaction? How can we sustain a chain reaction?
13. What is the energy balance in an atomic nucleus?
14. What are particle accelerators? Name and classify them.
15. Can an electron be accelerated by a cyclotron? Explain.
16. What are the limitations of a cyclotron? How are these limitations overcome by the synchrocyclotron?
17. What is a nuclear reaction cross-section?
18. What is a mass spectrograph? Name different types of mass spectrographs.

Long Questions for Discussion

1. State the proton-neutron theory of a nucleus and then explain the static properties of a nucleus.
2. Define the following terms: nuclear density, nuclear spin, and nuclear quantum state.
3. What do you mean by packing fraction and Nuclear binding energy? How does the binding energy varies as the mass number of the element.
4. Why do we say that a nucleus behaves like a drop of liquid? What are the essential features which are common in a drop of liquid and a nucleus?
5. Explain the postulates of the liquid drop modes.
6. Give salient features of the nuclear shell model and point out its success and failures.
7. Give the construction and working of a linear particle accelerator.
8. Describe the construction and action of a cyclotron. Discuss its limitations.

9. What is the difference between a cyclotron and a synchrocyclotron? Explain the theory and principle of a synchrocyclotron.
10. Describe the working of a betatron. Show that in order to keep the electron in a constant radial motion, the magnetic induction in the areas enclosed by the path must be maintained at any instant, equal to double the induction over the circular orbits.
11. Give an account of the theory, construction, and working of a modern synchrotron. State its uses.
12. Show that the radius of curvature R of the path of a particle inside the dees of a cyclotron is proportional to the $N^{1/2}$ where N is the number of times the particle has been accelerated across the space between the dees.
13. Define a nuclear reaction cross-section. Find an expression for the effective cross- section that the target nucleus offers for an interaction.
14. What do you understand by the processes of nuclear fission and nuclear fusion?
15. What are the differences and similarities between nuclear fission and nuclear fusion of U^{238}?
16. Why is it possible to produce the fission of U^{235} with slow neutrons whereas it is necessary to use fast neutrons to produce the fission of U^{238}?
17. Write a short note on the principles and workings of a nuclear reactor.
18. Explain the principles and workings of an Aston's mass spectrograph. Derive the condition for its focussing.
19. Discuss the construction and working of a Bainbridge mass spectrograph.
20. What is a mass spectrograph? Describe the construction of Aston's mass spectrograph and show how it can be used in the detection of isotopes. State the limitations of Aston's mass spectrograph.

Chapter 6

SEMICONDUCTORS AND NANO-PHYSICS

6.1 FREE ELECTRON THEORY

Electrical and Magnetic Properties/Electrical Properties of Metals/ Free Electron Model of Metals

We know that metals are good conductors of heat and electricity (*i.e.*, the metals have high thermal and electrical conductivities). The metallic conductors obey Ohm's law according to which in metallic conductors, the current density in the steady state is directly proportional to the intensity of electric field (*i.e.*, $J = \sigma E$, where j is the current density, E is the intensity of electric field and σ is the electrical conductivity of conductor. The resistivity of metals at ordinary temperatures is directly proportional to the absolute temperature $(p\mu T)$, but at extremely low temperatures, it is proportional to the fifth power of absolute temperature $(p\mu T^5)$. The ratio of thermal conductivity to the electrical conductivity is same at a given temperature for all the metals (*i.e.*, K/σ = constant for all metals at a given temperature or $K/\sigma T$ = constant for all metals at all temperatures). To explain these properties of metals, Lorentz and Drude, in the beginning of twentieth century, propounded the free electron model. They postulated that:

1. The outermost electrons of the constituent atoms of a metal are most weakly bound with the atoms. Hence these electrons get separated from their atoms and move freely inside the substance. These electrons are called the free electrons or the conduction electrons.
2. There are large numbers of free electrons inside the metals and they behave like molecules of a gas enclosed in a vessel. Hence it is also called the free electron gas. The thermal and electrical conductivities of a metal depend on the number of free electrons present

inside the metal. These free electrons are responsible for the thermal and electrical conduction inside the metal.

3. The free electrons in thermal equilibrium obey the Maxwell-Boltzmann statistics. Each electron has 3 degrees of freedom.
4. According to Maxwell-Boltzmann statistics, mean energy per electron at an absolute temperature T is ½ kT, where k is Boltzmann's constant.
5. Inside the metal, the free electron move randomly with a high speed (≈ 105 m/s) and their speed depends on the temperature of metal. During motion, when these free electrons collide with the positive ions (*i.e.*, the atoms from which electrons are detached) of the metal, their speed and direction of motion changes such that the number of electrons passing through an imaginary surface of metal, at any instant in one direction is equal to the number of electrons passing through that surface at that instant in the opposite direction. As a result the rate of direction of flow of electrons in a particular direction is zero.
6. When the metal is kept in an external electric field, the free electrons get accelerated in direction opposite to the external electric field, but there is no increase in their velocity due to collisions with the positive ions of the metal again and again but they begin to move with a constant average velocity in direction opposite to external electric field. This average velocity is called the drift velocity. This velocity is of the order of 10^{-1} m/s, which is very small as compared to their random velocity.

On the basis of above model the electrical conductivity and thermal conductivity of metals can be successfully explained. From this model we find that electrical conductivity of metals decreases with increase in their temperature and the ratio of thermal conductivity to electrical conductivity of each metal at any temperature is constant. But this model could not explain the variation of resistivity of metal with temperature particularly at low temperatures. Moreover, this model could not explain the specific heat of metals and magnetic susceptibility of paramagnetic metals.

Cause of Failure: According to the free electron to the free electron model, if N is the number of free electrons per gm mole in a metal, the total energy of these electrons will be $\frac{3}{2}NkT$ and the specific heat will be $\frac{3}{2}Nk = \frac{3}{2}R$ per *gm* mole. But according to the classical theory, the lattice energy of 1 gm mole due to vibrations of lattice of N molecules is 3NkT *i.e.* ,the contribution to specific heat due to lattice vibrations is 3Nk = 3R per gm mole. Thus total molar specific heat of metal must be equal to $\frac{3}{2}R + 3R = \frac{9}{2}R$ (which is independent of temperature). But experimentally it is found that the specific heat of metal depends on its temperature and at room temperature and at high temperatures, the molar specific heat of metal is nearly 3R (≈ 6 cal/gm mol K) which can be considered to be due to the lattice vibrations only *i.e.,* the contribution to specific heat of metal due to free electrons is negligible.

Similarly, according to free electron model, each free electron due to its spin motion has some magnetic moment due to which the magnetic susceptibility of a paramagnetic substance is inversely proportional to the absolute temperature of the substance. But experimentally it is found that the magnetic susceptibility of paramagnetic metals is nearly independent of temperature.

To explain the above facts Summerfield modified this model on the basis of quantum statistics. The modified model is called the Summerfield's free electron model.

Summerfield's Free Electron Model

According to Summerfield, each free electron inside metal experiences an electrostatic attractive force due to all the positive ions and an electrostatic repulsive force due to the other electrons. Thus each free electron inside metal is in the attractive electric field due to positive ions and a repulsive electric field due to the remaining all electrons. The force of repulsion due to mutual interactions of electrons can be assumed to be negligible and the attractive field due to positive ions can be considered to be uniform everywhere inside the crystal. Thus each free electron inside the metal is in an attractive potential field. Since the crystal structure of solid is periodic *i.e.,* in a solid crystal, each positive ion is at a definite distance from each other; therefore this potential field inside the metal must also be periodic. But for the sake of convenience, Summerfield assumed that this potential inside the metal is constant. Since no electron is emitted from the metal at an ordinary temperature, therefore it can be assumed that the electron inside the metal is more stable than outside the metal *i.e.,* the potential energy of a stationary electron inside the metal is less than the potential energy of a stationary electron outside the metal. Thus inside the metal, the free electrons can be assumed to be inside a potential box or potential well of depth E_s as shown in Figure 1. Here E_s is the difference in potential energies of electron when inside and outside the metal. The electron present inside this well has so much energy that it can move inside the well, but it can not come out of the metal surface after crossing the potential barrier. Inside the well, all the energy states from zero to some energy E_F are failed by the electrons (*i.e.,* the maximum kinetic energy of electrons inside the metal is E_F). The energy E_F is called *Fermi energy.* Thus Fermi energy is the maximum kinetic energy of electrons inside the metal. If an energy $\Phi = E_s - E_F$ is imparted to the electron of the maximum kinetic energy, the electron will cross the potential well and will get emitted out of the metal surface. This energy Φ is called the work function or threshold energy of the metal; the electron found in the highest energy level inside the metal gets emitted from the metal.

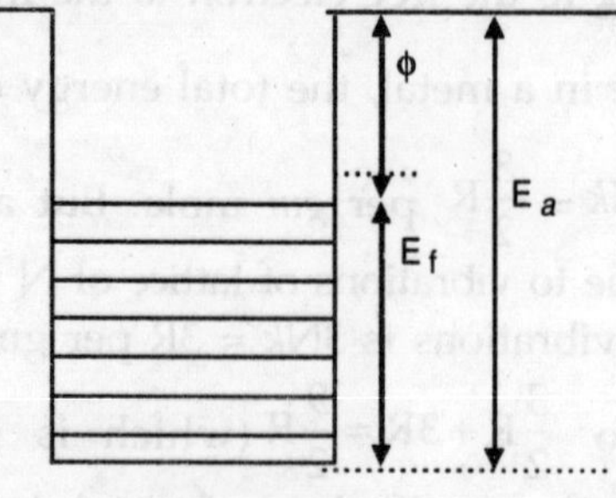

FIGURE 1

For the sake of convenience if it is assumed that the energy of free electron inside the metal is zero, then taking it as the reference level, the number of possible energy levels in the energy range E and E + dE is given by the following function according to the quantum statistics:

$$g(dE) = \frac{4\pi V(2m)^{3/2} E^{1/2}\, dE}{h^3} \tag{1}$$

But the spin of electron is ½, hence the free electron inside the metal obeys Pauli's exclusion principle, according to which there cannot be more than one electron in one state *i.e.*, inside the metal, free electrons obey the Fermi-Dirac statistics. Thus the free electrons present inside the metal can remain in several possible energy states and these electrons occupy energy levels up to a definite energy limit (say up to E_F). According to Fermi-Dirac statistics, the free electrons obey the following energy distribution function:

$$f\,(E) = \frac{1}{e^{\alpha+(E/kT)}+1} \qquad \text{where } \alpha \text{ is a constant} \qquad (2)$$

Thus the number of free electrons in energy ranges E and E + dE at any temperature T is given as

$$N_{(E)}\,dE = g(E)\,f\,(E)\,dE = \frac{4\pi V}{h^3}(2m)^{3/2}\,E^{1/2}\,\frac{dE}{e^{\alpha+(E/kT)}}+1 \qquad (3)$$

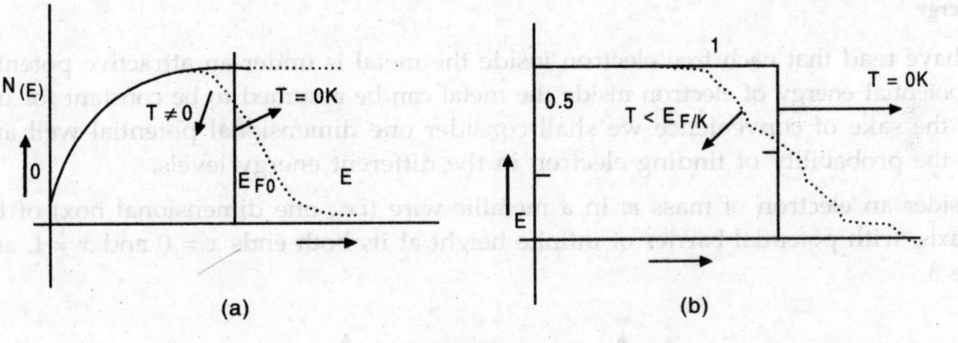

FIGURE 2

Figure 2 (a) shows the variation of $N_{(E)}$ with energy E and Figure 2 (b) shows the variation of Fermi function $F_{(E)}$ with energy E. If there is one free electron per atom in the metal, then at ordinary temperatures the value of e^{α} is very small as compared to 1. (*i.e.*, $e^{\alpha} << 1$). If e^{α} is the Fermi energy, then Fermi function can be expressed as

$$f\,(E) = \frac{1}{e^{(E-E_F)/kT}+1} \qquad (4)$$

At absolute zero (*i.e.*, T = 0K), if E < E_{FO}, then f(E) = 1 and if E > E_{FO}, then f(E) = 0, where E_{FO} = 0, where E_{FO} is the Fermi energy at absolute zero.

Thus at absolute zero Fermi energy E_{FO} is that energy below which all the energy levels are completely occupied by the electrons (*i.e.*, number of electrons in each energy level is equal to the number of electrons allowed by the Pauli's exclusion principle) and the energy levels above it are completely empty. Hence/Fermi energy E_{FO} is the maximum possible energy of electrons at absolute zero. At absolute zero, the value of Fermi energy E_{FO} is of the order of 1 eV. It may be mentioned here that classically the energy of electron at absolute zero is zero.

At temperatures below the melting point of metals, the value of kT is very small in comparison to the Fermi energy E_F (*i.e.*, kT << E_F), then from equation (4), at E = E_F, 'f(E) = ½. Hence at any temperature, Fermi energy is the energy corresponding to which the probability for electrons is ½. In other words, at any temperature T, Fermi energy is the energy below which the number of free electrons in the energy levels is equal to half the total number of free electrons.

Form the above discussion we conclude that even at extremely low temperatures most of the electrons inside metal have some kinetic energy due to which all the energy states from zero to the Fermi energy in metal are occupied by the electrons and the energy states above the Fermi energy are all empty (Figure 1). This is the reason why electrons are not emitted by themselves at ordinary temperatures. The emission of electrons becomes possible only when the free electrons are excited to the empty energy levels above the Fermi energy by imparting energy $\Phi = E_s - E_F$ in form of heat or any other form. This thermal excitation is possible only for the electrons close to the Fermi energy level. To excite the electrons from the energy states lower than this, very large excitation energy is required.

Now on the basis of the model, we shall solve the Schrodinger equation for the free electron gas in a one-dimensional potential box of constant potential and then calculate the value of Fermi energy and density of energy states.

Solution of one Dimensional Schrodinger Equation in a Constant Potential; Density of States and Fermi Energy

We have read that each free electron inside the metal is under an attractive potential field and the potential energy of electron inside the metal can be assumed to be constant (or uniform). Here for the sake of convenience we shall consider one dimensional potential well and shall calculate the probability of finding electron in the different energy levels.

Consider an electron of mass m in a metallic wire (*i.e.*, one dimensional box) of length L along X-axis, with potential barrier of infinite height at its both ends $x = 0$ and $x = L$ as shown in Figure 3.

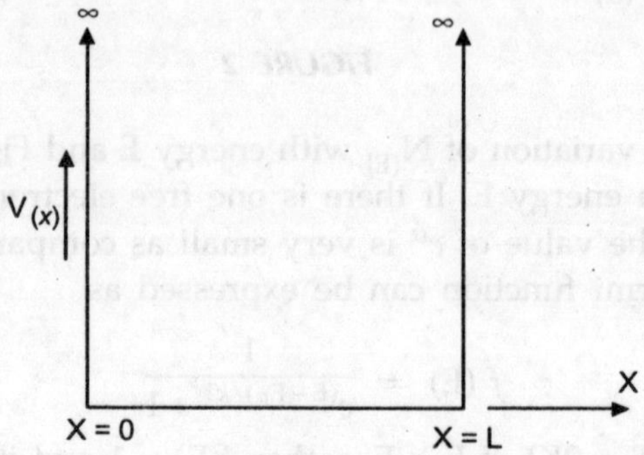

FIGURE 3

Let potential energy of electron everywhere inside the box is constant (or zero). *i.e.*,

$$V_{(x)} = 0, \text{ if } 0 < x < L$$

and

$$V_{(x)} = \infty, \text{ if } x \leq 0 \text{ and } x \geq L \quad (5)$$

Inside the box (*i.e.*, in the region $0 < x < L$), where potential V is zero, the kinetic energy of electrons will be from zero to the Fermi energy E_F *i.e.*, different electrons will be in different energy states. Hence inside the box (when V = 0), the Schrodinger wave equation for the wave function Ψ_n associated with n^{th} energy state E_n of the electron will be

$$\frac{d^2 \Psi_n}{dx^2} + \frac{2m}{\hbar^2} E_n \Psi_n = 0 \quad (6)$$

$$k^2 = \frac{2mE_n}{\hbar^2}, \text{ then } \frac{d^2\Psi_n}{dx^2} + k^2\Psi_n = 0$$

The general solution of above differential equation is $\Psi_n = Ae^{ikx} + Be^{-ikx}$ (7)

Here A and B are constants which can be determined by the boundary conditions. According to the boundary conditions, since electron is found only between $x = 0$ to $x = L$, hence at $x = 0$, $\Psi_n = 0$ and at $x = L$, $\Psi_n = 0$. *i.e.,*

from equation (3.7), at $x = 0$, $0 = A + B$

and at x = L, $0 = Ae^{ikL} + Be^{-ikL}$ (8)

on solving, $A = -B$

and $Ae^{ikL} - Ae^{-ikL} = 0$ (9)

or $A(e^{ikL} - e^{-ikL})$,

i.e., $A = 0$ or $e^{ikL} - e^{-ikL} = 0$

Since $A \neq 0$, hence $e^{ikL} - e^{ikL} = 0$

i.e., $\sin kL = 0$ or $kL = n\pi$

i.e., $k = \frac{n\pi}{L}$ where $n = 1, 2, 3, ...$

Above expression gives the wave function of nth energy state. Obviously this is the equation of a stationary wave. Substituting the different values of n, we can get the different wave modes in between $x = 0$ to $x = L$. The values of n corresponding to different states are called quantum number of those states.

The value of the constant A can be determined by the condition of normalization according to which the total probability of finding the electron from $x = 0$ to $x = L$ is unity. *i.e.,*

$$\int_0^L \Psi_n * \Psi_n \, dx = 1 \text{ or } \int_0^L |\Psi_n|^2 \, dx = 1 \quad (10)$$

or

$$\int_0^L 4A^2 \sin^2\left(\frac{n\pi x}{L}\right) dx = 1 \text{ or } \int_0^L \frac{4A^2}{2}\ 1 - \cos\frac{2n\pi x}{L}\ dx = 1$$

or

$$2A^2\left\{x - \frac{\sin(2n\pi x/L)}{2n\pi/L}\right\}_0^L = 1$$

or

$$2A^2 L = 1 \quad (11)$$

i.e.,

$$2A = \sqrt{\frac{2}{L}}$$

∴ The normalised wave function of the Schrodinger equation (3.6) is

$$\Psi_n = \sqrt{\frac{2}{L}} \sin\left(\frac{n\pi x}{L}\right)$$

Figure 4 shows the wave functions for $n = 1, 2, 3, 4$ and 5. it is clear from the diagram that for $n = 1$, $L = \lambda/2$ or $\lambda = 2L$; for $n = 2$, $L = 2\,\lambda/2$ or $\lambda = 2L/2 = L$; for $n = 3$, $L = \lambda = 2L/3$; for $n = 4$, $L = 4\,\lambda/2$ or $\lambda = 2L/4 = L/2$; and for $n = 5$, $L = 5\,\lambda/2$ or $\lambda = 2L/5$.

In general for any value of n, the allowed wavelength is $\lambda = 2L/n$.

Eigen Values of Energy: from equations (9) and (7),

$$k^2 = \left(\frac{n\pi}{L}\right)^2 = \frac{2mE_n}{\hbar^2}$$

$$E_n = \frac{\hbar^2}{2m}\left(\frac{n\pi}{L}\right)^2 = \frac{h^2}{2m}\left(\frac{n}{2L}\right)^2 = \frac{n^2h^2}{8mL^2} \quad (12)$$

From the above equation, it is clear that $E_n \propto n^2$. Hence a graph plotted between the energy E_n and the quantum number n is a parabola (Figure 5)

For $\quad n = 1, E_1 = \dfrac{h^2}{8mL^2}$

For $\quad n = 2, E_2 = \dfrac{(2)^2 h^2}{8mL^2}\, 4E_1$

For $\quad n = 3, E_3 = \dfrac{(3)^2 h^2}{8mL^2}\, 9E_1$

For $\quad n = 4, E_4 = \dfrac{(4)^2 h^2}{8mL^2}\, 16E_1$

For $\quad n = 5, E_5 = \dfrac{(5)^2 h^2}{8mL^2}\, 25E_1$

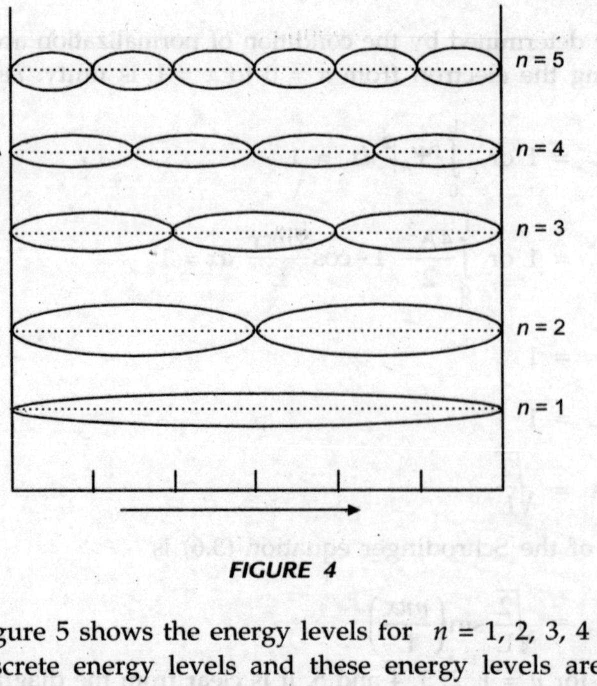

FIGURE 4

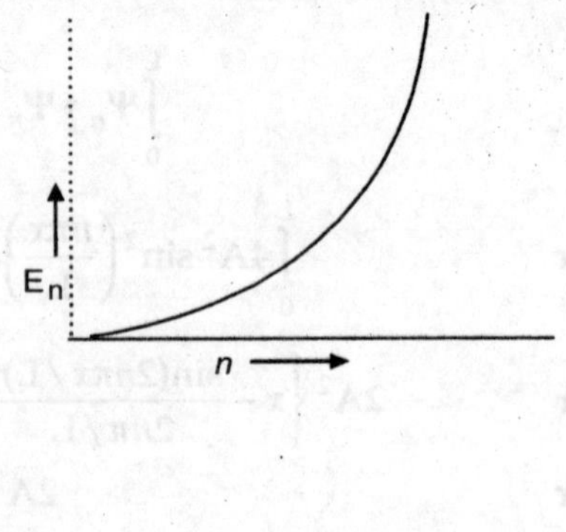

FIGURE 5

Figure 5 shows the energy levels for $n = 1, 2, 3, 4$ and 5. Obviously the electron has only the discrete energy levels and these energy levels are not equispaced. The energy interval between the two consecutive levels increases as the value of n increases. If L is very large, the energy interval is very less (*i.e.*, the energy levels are very close). For example, for L = 1cm, the energy gap between the nth and $(n - 1)^{th}$ energy levels is

$$E_n - E_{n-1} = \frac{n^2h^2}{8mL^2} - \frac{(n-1)^2h^2}{8mL^2} = \frac{h^2}{8mL^2}\left[n^2 - (n-1)^2\right]$$

$$= \frac{h^2}{8mL^2}(2n-1) = \frac{(6.6\times10^{-34})^2(2n-1)}{8\times9.1\times10^{-31}\times(0.01)^2} \text{ Joule}$$

$$= 5.98 \times 10^{-34}\ (2n - 1) \text{ Joule}$$

$$3.74 \times 10^{-15}\ (2n - 1)\ eV$$

Which is very small. Hence the energy gap between the two consecutive energy levels is visible only when the value of L is of the atomic order.

Density of States: Equation (12) gives the energy of quantum states corresponding to different quantum numbers, out of which some of the initial states (from $n = 1$ to $n = 5$) are shown in Figure 4 with their corresponding waves. Actually the Figure 4 is an extended form of Figure 1. The horizontal lines shown in Figure 1 represent the different energy states in the potential well.

The number of energy levels (or quantum states) in a unit energy range is called the density of states. *i.e.*, density of states = $\frac{dn}{dE}$

From equation (12), energy of nth state $E_n = \frac{n^2h^2}{8mL^2}$

On partial differentiation, $dE_n = \frac{h^2}{8mL^2}(2n\,dn) = \frac{h^2 n\,dn}{4mL^2}$

Number of energy states in unit energy range $\frac{dn}{dE_n} = \frac{4mL^2}{nh^2}$

Since there are two quantum states of electron present in each energy level corresponding to the spin $s = +\frac{1}{2}$ and $s = -\frac{1}{2}$, therefore density of energy states of electron in the metal

$$g\left(E_n = 2\times\frac{dn}{dE_n} = 2\times\frac{4mL^2}{nh^2}\right)$$

$$= \frac{8mL^2}{nh^2} = \frac{4L}{h}\sqrt{\frac{m}{2E_n}} \quad (13)$$

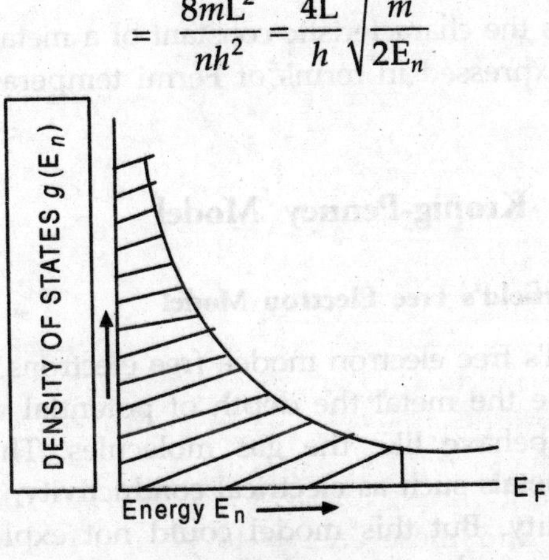

FIGURE 6

Figure 6 shows the graph between the density of states $g(E_n)$ and energy E_n.

Fermi Energy

Fermi energy E_F is the maximum possible energy of electrons at any temperature *i.e.,* at any temperature T, Fermi energy. E_F is the energy corresponding to that highest state upto which all the states are occupied by the electrons and the states above which are completely empty.

To calculate the Fermi energy, let there be N free electrons in the one dimensional box under consideration. The energy state of each free electron depends on its kinetic energy and the state of spin *i.e.,* the energy state of free electron depends on the quantum number n (since from equation 12, $E_n \propto n^2$) and spin quantum number s. For a given value of n, two orientations of electron spin ($s = \pm \frac{1}{2}$) are possible and according to Pauli's exclusion principle, not more than one electron can be in the same state, hence there can be two electrons in an energy level corresponding to each value of n (because for each value of n, corresponding to the spin $s = +\frac{1}{2}$ and $s = -\frac{1}{2}$ there are two states possible). For example, if N = 8, then energy states corresponding to n = 1, 2, 3 and 4 will be occupied by the electrons and all the states corresponding to $n > 4$ will be empty.

Let n_F be the quantum number corresponding to the highest filled energy state (*i.e.,* the energy states from $n = 1$ to $n = n_F$ are occupied by N electrons). If N is even, then $N = 2n_F$ (because there can be two electrons in each energy state).

∴ Energy corresponding to the highest filled state (*i.e.,* Fermi energy)

$$E_F = \frac{n_F{}^2 h^2}{8mL^2} \quad \text{(from equation 12)}$$

or

$$E_F = \frac{(N/2^2\, h^2)}{8mL^2} = \frac{N^2 h^2}{32\, mL^2} \qquad (14)$$

For example, if N/L = 0.4 electrons/angstrom = $0.4 \times 10^{10}\ m^{-1}$ (or $4 \times 10^9\ m^{-1}$), then

$$E_F = \frac{(6.6\times10^{-34})^2\times(0.4\times10^{10})^2}{32\times(9.1\times10^{-31})} \approx 2.39\times10^{19}\, J = 1.5 eV$$

Thus to fill 4×10^9 electron in a one dimensional box of length $1m$, the kinetic energy of the electron of the highest state should be of the order of 1.5 eV.

Obviously Fermi energy is the characteristic constant of a metal. For the sake of convenience the Fermi energy E_F can be expressed in terms of Fermi temperature T_F, as $E_F = kT_F$ where k is Boltzmann constant.

Energy Bands in a Solid; Kronig-Penney Model

Merits and Demerits of Summerfield's Free Electron Model

According to Summerfield's free electron model, free electrons move inside the metal under a constant potential (*i.e.,* inside the metal the depth of potential well can be considered to be same everywhere) and they behave like the gas molecules. This model could successfully explain several properties of metals such as electrical conductivity, thermal conductivity, specific heat and magnetic susceptibility. But this model could not explain the fact that some solid crystals are good conductors of electricity while some are insulators and some are semi-conductors (*i.e.,* the electrical conductivity of solids changes from 10^8 *mho*/cm to 10^{-22} *mho*-cm.).

Moreover the electrical conductivity of semi-conductors increases with increase in temperature. In other words, the free electron model could not explain the distinction between conductors, insulators and semi-conductors. Hence this model was modified and a new model called as band model was propounded.

Energy Band Model of Solids

According to band model, the Summerfield's assumption that the free electrons inside the metal move in a constant potential well, was considered to be wrong. Actually a free electron inside the metal moves in the electric field of positive ions and of other free electrons. Since the crystal structure itself is periodic, therefore the potential energy of electron also changes periodically with the position coordinate. *i.e.*, the motion of electron inside the metal is in a periodic potential well.

We know that inside a metal (atomic number Z), the potential energy of a free electron at a distance x in the potential field of an atom i is given as $U_{(x)} = U_{(x)} = \frac{-Ze^2}{4\pi\varepsilon_0 x}$. Hence a graph plotted for $U_{(x)}$ versus x is a rectangular hyperbola as shown in Figure 7. From the graph it is clear that for different values of x, the value of $U_{(x)}$ is negative and at $x = \infty$, the value of $U_{(x)}$ is zero. Since inside the metal, atoms are arranged in a definite order, therefore the potential energy of electron in the combined field of atoms i and j can be represented by the complete curve as shown in Figure 8, the dotted curves represent the potential energy of electron in the field of atom i and atom j separately while the complete curve represents the resultant potential energy. But in a metal, there are number of atoms arranged in a definite order in any direction, hence in the length L of the metal, the resultant potential energy of electron in the fields of atoms i, j, k, l etc, can be represented as shown in Figure 9. From Figure 9, it is clear that at the boundaries of metal (*i, e.,* at its free surfaces) the potential energy suddenly rises and becomes zero, while inside the metal, the potential well is not of uniform depth everywhere, but is periodic.

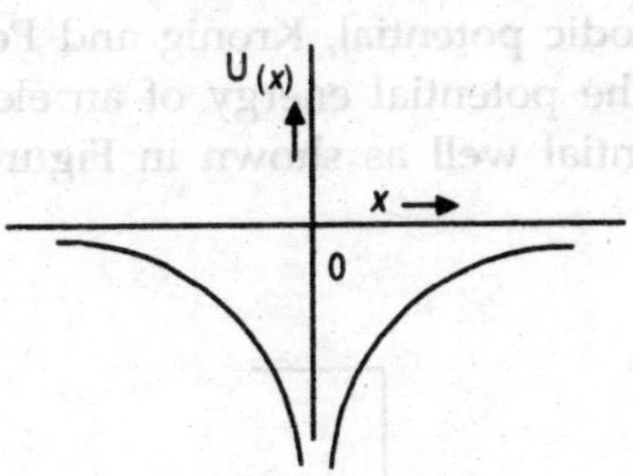

FIGURE 7

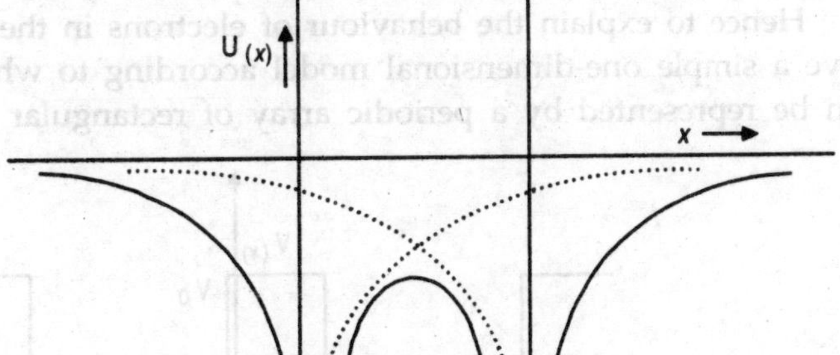

FIGURE 8

Figure 9 also shows the different energy levels inside the metal. All the electrons from the lowest energy to E_b are bound with their atoms and they can vibrate only with a very small amplitude, they can not leave their atoms. The electrons of energy higher than this, with energy in between E_b and E_F can move anywhere within the metal but on reaching at the boundary (or metal surface), they have to face a surface barrier and hence they can not emerge out of the metal surface. Here E_F is the Fermi energy level (*i.e.*, the level of maximum kinetic energy of electrons). The depth of potential well is E_s and

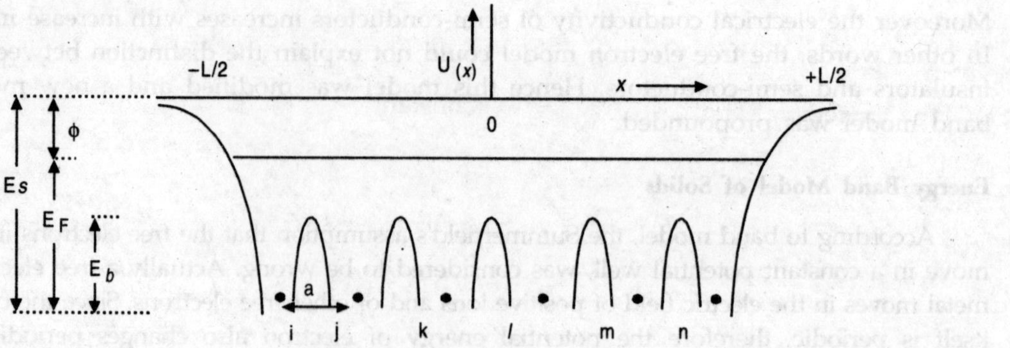

FIGURE 9

$E_S - E_F = \Phi$ is the work function of the metal. The separation between the two consecutive atoms (*i.e.*, lattice constant) is a.

Thus according to band model:

1. Each electron in metal is in the electric field produced due to charge distribution of positive ions and remaining electrons.
2. Electron is associated with the entire crystal, and not only with an atom.
3. Electron moves in a periodic potential produced by the ion cores and other electrons inside the crystal (Figure 9). This periodicity vanishes at the free surfaces of the crystal. The motion of electron inside the crystal is like the elastic waves in a continuous medium.

To obtain the different energy states of electron in the periodic potential well, again we will have to find the solution of wave equation corresponding to the wave function Ψ associated with the electron; as we have done in the last section for the constant potential well. But it is not so easy to define the periodic potential as the constant potential well was defined.

Hence to explain the behaviour of electrons in the periodic potential, Kronig and Penney gave a simple one-dimensional model according to which the potential energy of an electron can be represented by a periodic array of rectangular potential well as shown in Figure 10.

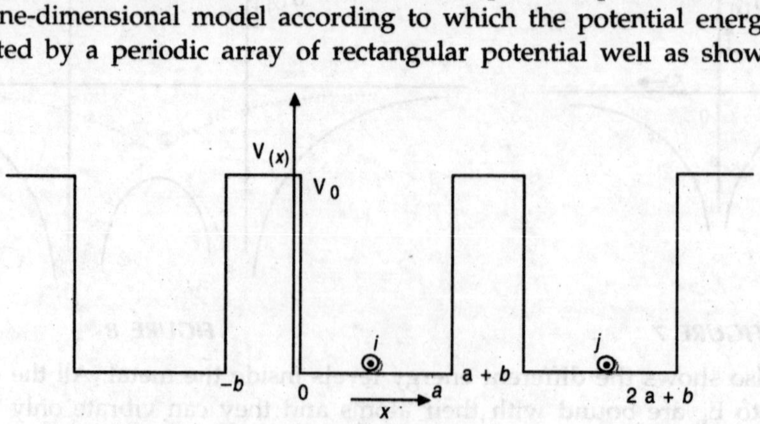

FIGURE 10

Here the potential peaks obtained from the hyperbolic curves have been assumed to be in form of rectangular peaks. Each potential well represents the potential near an atom. If the time period of potential is $(a + b)$, then potential energy is zero in $0<x<a$ and potential energy is constant ($= V_0$) in $-b < x < 0$. *i.e.,*

In region $0 < x < a$, $V_{(x)} = 0$

And in region $-b < x < 0$, $V_{(x)} = V_0$ (constant)

In both these regions, the Schrodinger wave equations for the wave function Ψ_n associated with n^{th} energy state E_n of electron are

In region $0 < x < a$, $\frac{d^2\Psi_n}{dx^2} + \frac{2m}{\hbar^2} E_n \Psi_n = 0$ Since $V = 0$

and in region $-b < x < 0$, $\frac{d^2\Psi_n}{dx^2} + \frac{2m}{\hbar^2}(E_n - V_0)\Psi_n = 0$ Since $V = V_0$ (15)

Here the energy of electron E_n is very small in comparison to the potential V_0. Assuming that as b tends to zero, V_0 becomes infinite, Kronig and Penney obtained the following condition for the allowed wave function on solving the above equation.

$$(mV_0 b/\hbar^2 \alpha) \sin \alpha + \cos \alpha = \cos k$$

Where $\alpha = \sqrt{2mE_n/\hbar^2}$ and k is the wave vector (16)

Now if we assume $P = mV_0 ba/\hbar^2$, then it is clear that the quantity P is the measure of area $V_0 b$ of the potential barrier. In other words, increase in P means increase in the area of potential barrier *i.e.,* increase in the depth of potential well. This implies that with increase in P, the binding of electron with its potential well increases.

Substituting the value of P in equation (16), the condition for the allowed wave function is

$$\frac{p \sin \alpha a}{\alpha a} + \cos \alpha a = \cos ka \quad (17)$$

Figure 11 shows a graph between the quantity on the left side of above equation and αa for $P = 3\pi/2$. Since the maximum and minimum values of the term $\cos \alpha$ are respectively + 1 and – 1, hence if two horizontal lines are drawn on the Y-axis at $y = +1$ and $y = -1$, we find that these lines cut the positive Y-axis at points *a, b, c, d, a', b', c', d'*,.... and the negative Y-axis at points *e, f, g, h, e', f', g', h'*,.... It means that the solution of equation (17) is possible only in some specific ranges of αa. In Figure 11, these ranges are shown by thick lines from p to π, from q to 2π, from r to 3π from s to 4π,....in the positive directions and from p' to $-\pi$, from q' to -2π, from r' to -3π, from s' to -4π,.... in the negative direction. It is clear that only in these specific ranges of αa, the allowed energy levels are continuously obtained. *i.e.,* energy bands are obtained in these specific ranges and not the descrete energy states. Obviously there are no energy states possible for electrons in the ranges π to q, 2π to r, 3π to s,.... *i.e.,* these ranges represent the *forbidden energy gap.*

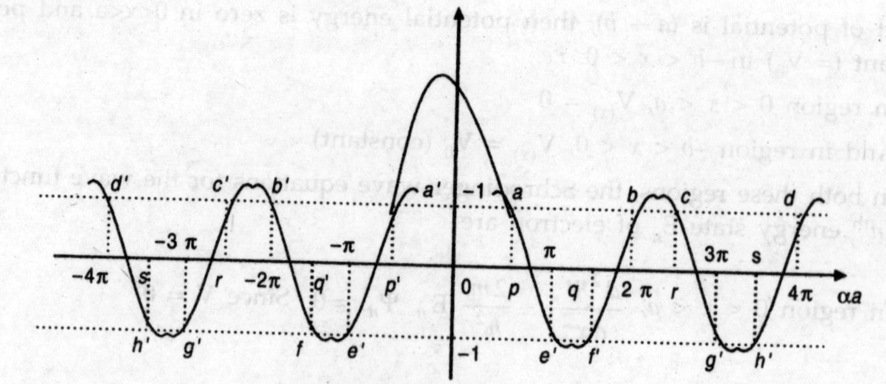

FIGURE 11

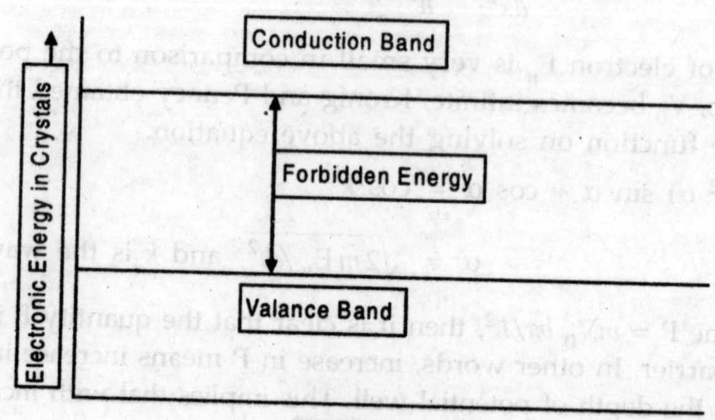

FIGURE 12

Figure 13 shows the energy spectrum with P. at the left corner P = 0 (or P/4 π = 0, in the middle P = 4 π (or P/4 π = 1) and at the right corner P = ∞ (or 4 π/P = 0) is shown. It is clear from the diagram that at P = 0, we get continuous energy modes like the potential well of free electron. When P = ∞, we get descrete energy levels E_1, E_2,.... . Between these two limits for any value of P, the allowed energy band is shown by the shaded part and the forbidden energy gap is shown by the empty space. For example at P = 1.6 π (or P/4 π = 0.4), the widths of the allowed energy bands are respectively from *q* to *r*, from *s* to *t*,...., while the forbidden energy bands are from *p* to *q*, form *r* to *s*, from *t* to *u*,....

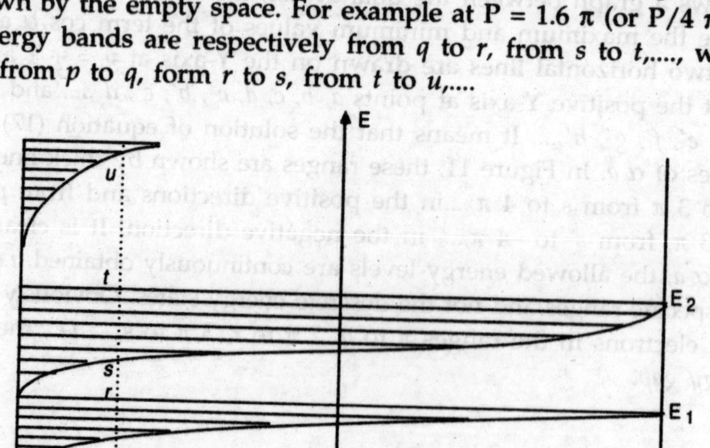

FIGURE 13

Thus from the Kronig-Penney model, we get the following conclusions:

1. In the energy spectrum of all the electrons present in the metal, there are several energy bands separated by the forbidden energy region. The energy band completely filled with electrons is called valence band and the energy band which is either completely empty or is partially filled, is called the conduction band. (Figure 12)
2. As the value of αa increases or energy increases (because $\alpha = \sqrt{2mE_n}/\hbar$ the width of the allowed energy band increases.
3. With increase in the binding energy V_0 of electrons or with increase in the value of P (since $P \propto V_0 b$), the width of a particular allowed energy band decreases and ultimately when the binding energy becomes infinite, the allowed region becomes very narrow. *i.e.*, the energy spectrum become the line spectra. In other words, when $P = \infty$, the energy levels becomes descrete just as that of a particle enclosed in a constant potential box of atomic size. When the binding energy becomes zero, the energy spectrum becomes continuous.
4. At the wave vector $k = \pm n\pi/a$ (where n = 1, 2, 3, ...) the energy is discontinuous. These values of k give the boundaries of Brillouin zones. Figure 14 shows the E – k curve. For $n = \pm 1$, we get first Brillouin zone from $k = -\pi/a$ to $k = +\pi/a$.
5. The energy in an energy band is a periodic functions of k.
6. The number of total possible wave functions in an energy band is equal to the number of unit cells.
7. The velocity of free electron is zero at the top and bottom of an energy band and the velocity of free electron is maximum at the point of inflexion of energy band (*i.e.*, at the point P in Figure 14). After this point, the velocity of electron decreases with increase in energy.
8. At T = 0K, the effective number of electrons in a completely filled band is zero, while the effective number of electrons is maximum in a band filled upto the point of inflexion, where the velocity of electrons is maximum. At absolute zero, the energy level completely filled by the electrons upto $k = \pm k_1$ is called the Fermi energy level E_F.

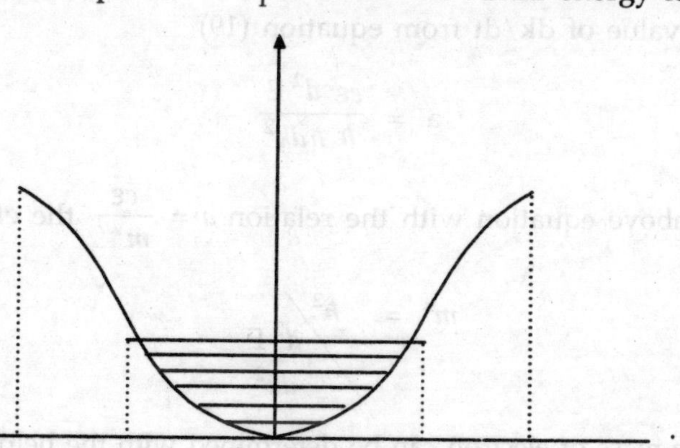

FIGURE 14

Thus conduction is not possible in the energy band completely filled with electrons. The reason is that for conduction, the electron will leave its allowed state only when there is any other allowed empty state available for it. Since in a completely filled energy band there will be no allowed state empty, therefore conduction will not be possible. Similarly id an energy band is completely empty, then also conduction will not be possible because no electron is available in it for conduction. Hence conduction will be possible only in those substances in which there are partially filled energy bands.

Effective Mass of an Electron

According to Kronig-Penny model, the motion of electron inside a solid is in a periodic rectangular potential well and there are some allowed energy bands of the energy of electrons which are separated from each other by a forbidden energy gap. According to wave mechanics, the velocity of electron corresponding to the wave vector k is equal to the group velocity of waves representing it. *i.e.*, $v = d\omega/dk$, where ω is the angular frequency of the de-Broglie waves.

If E is the energy of electron, then $E = \hbar\omega$ or $d\omega = dE/\hbar$

$$v = \frac{1}{\hbar}\frac{dE}{dk} \tag{18}$$

Let there be only one electron initially in the *k* state in the first Brillouin zone. Now if an external electric field ε is applied on the electron for a very short duration *dt*, the gain in energy of electron.

$$dE = e\varepsilon v dt = \frac{e\varepsilon}{\hbar}\frac{dE}{dk}dt$$

$$\left(\frac{dE}{dk}\right)dk = \frac{e\varepsilon}{\hbar}\frac{dE}{dk}dt \tag{19}$$

or Rate of change in wave vector $\frac{dk}{dt} = \frac{e\varepsilon}{\hbar}$

From equation (18), acceleration of electron $a = \frac{dv}{dt} = \frac{1}{\hbar}\frac{d^2E}{dk^2}\cdot\frac{dk}{dt}$

Substituting the value of dk/dt from equation (19)

$$a = \frac{e\varepsilon}{\hbar}\frac{d^2E}{\hbar dk^2} \tag{20}$$

Comparing the above equation with the relation $a = \frac{e\varepsilon}{m^*}$, the effective mass of electron

$$m^* = \hbar^2 \Big/ \frac{d^2E}{dk^2} \tag{21}$$

Thus the effective mass of electron can be determined with the help of graph plotted between energy E and wave vector *k*. we find that upto $E < E_F$ *i.e.*, in the lower half part of energy band, the value of m^* is positive and in the upper half part of the band (in $E > E_F$) the value of m^* is negative. At the points of inflexion of the curve (*i.e.*, at $E = E_F$), the value of m^* becomes infinite. It means that when $E > E_F$ the electron behaves like a positively charged particle.

In the next section we shall see that this model could successfully explain the distinction between conductors (metals), insulators and semi-conductors.

Metals, Insulators and Semiconductors

The difference in metals, insulators and semiconductors can be explained on the basis of band theory in solids. According to the theory of energy bands, the motion of electrons inside the metal is under a periodic potential well. The allowed energy levels of electrons are in energy bands separated by the forbidden energy gap. The electrical conduction is not possible in the energy bands completely filled by the electrons (*i.e.*, the energy bands in which there are two electrons in each energy level) because for conduction, an electron can leave its allowed energy state only if any other allowed empty energy state is available for it. Similarly, there is no conduction possible in a completely empty energy band.

Figure 15 shows the distribution of electrons in the various energy bands for metals, insulators and semiconductors at absolute zero (T = 0 K). The shaded part represents the region filled with electrons. The electrons first occupy the lowest energy band and then they occupy successively the upper bands. The upper most energy band which is completely filled by electrons is called the *valence band* and the partially filled or completely empty band is called the *conduction band*. The energy gap between the upper most completely filled energy band and its next completely empty or partially filled energy band is called the *forbidden gap*. It is that minimum energy by imparting which electron moves from the valence band to the conduction band *i.e.*, electrical conduction becomes possible through the substance. This energy is imparted by the thermal agitations or light radiations.

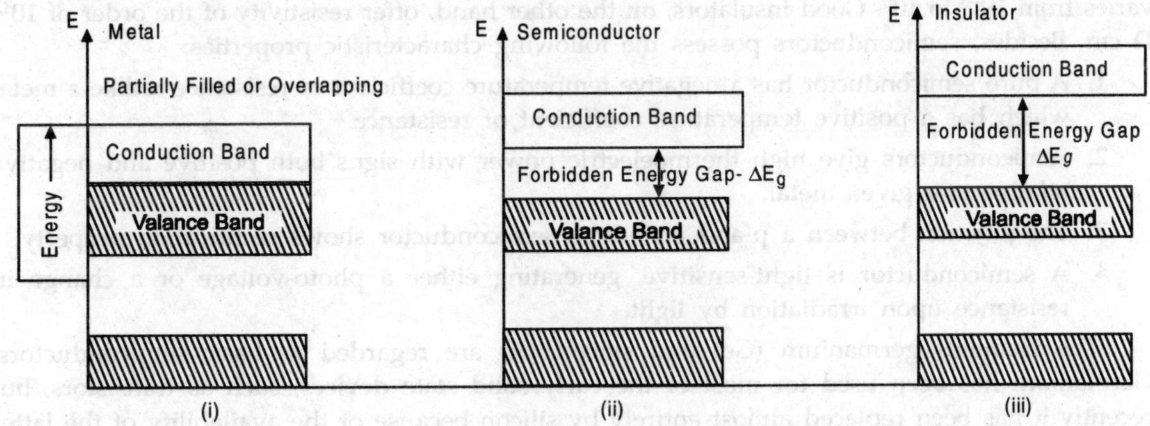

FIGURE 15

The electrical conductivity of a solid depends on two factors: (*i*) whether the conduction band is completely empty or is partially filled with electrons, and (*ii*) on the forbidden energy gap.

In the electrical conductors (such as all metals) the conduction band is partially filled [Figure 15 (*i*)] or the conduction band and the valence band overlap each other [Figure 15(*ii*)], hence the forbidden energy gap ΔE is zero. Thus without any additional energy the electrons easily move from the valence band to the conduction band. This is why the conductivity of conductors is more.

In the semiconductors (such as silicon, germanium etc.), the forbidden energy gap ΔE between the valence band and conduction band is of the order of few electron-volt. For example for silicon and germanium ΔE ≈ 1 *e*V [Figure 15(*ii*)] Thus even at room temperature, from the energy obtained due to thermal agitations the electrons reach from the valence band to the conduction band and weak electric conduction becomes possible.

In insulators (such as diamond, ebonite, mica etc.) the valence band is completely filled and the conduction band is completely empty and the forbidden energy gap between them is nearly 7*e*V (for diamond) or more than it [Figure 15 (*iii*)]. Since this amount of energy can not be obtained by the thermal agitations at room temperatures, hence electric conduction does not become possible. Only at high temperatures, some electrons from the valence band can reach the conduction band and conduction can become possible. Thus only at high temperatures these substances behave like semi-conductors. Actually all the pure semi-conductors at T = 0 K are insulators and at very high temperatures all the insulators can be assumed to be semiconductors.

6.2 SEMICONDUCTORS

We have seen that the band structure of a semiconductor is similar to that of an insulator except that in the former case the band gap is smaller. The band gap of semiconductors ordinarily varies from 0.2 *e*V to 2.5 *e*V while the band gap of a typical insulator like diamond is about 6 *e*V. As a result, unlike an insulator, a semiconductor exhibits a moderate electrical conductivity at room temperature. Generally, the term semiconductor applies to a class of materials having conductivity intermediate between a conductor and an insulator. At room temperature, the resistivity of a good conductor is about 10^{-6} Ω cm, where as that of semiconductor varies from 10^{-3} to 10^{6}. Good insulators, on the other hand, offer resistivity of the order of 10^{12} Ω cm. Besides, semiconductors possess the following characteristic properties:

1. A pure semiconductor has a negative temperature coefficient or resistance unlike a metal which has a positive temperature coefficient of resistance.
2. Semiconductors give high thermoelectric power with signs both positive and negative relative to a given metal.
3. The junction between a p-and an n-type semiconductor shows rectification property.
4. A semiconductor is light-sensitive, generating either a photo-voltage or a change in resistance upon irradiation by light.

The elements germanium (Ge) and silicon (Si) are regarded as basic semiconductors. Germanium has been used for most of the early solid state devices, such as transistors, but recently it has been replaced almost entirely by silicon because of the availability of the latter in abundance. Besides, the complex integrated circuits (IC) of present-day electronics are manufactured from silicon.

In addition to the *elemental semiconductors*, there are *compound semiconductors* which are successfully used for the fabrication of electronic devices. Important compound semiconductors are cadmium sulphide (CdS), lead sulphide (PbS), lead telluride (PbTe), indium antimonide (InSb), gallium arsenide (GaAs), indium phosphide (InP) etc. Of these, CdS has been used in light meters; PbS and PbTe have been used in infrared detectors. GaAs has been applied in the manufacture of transistors, solid state lasers, and some special high frequency devices.

Some of the compound semiconductors form alloys which have important semi conducting properties. They are known as alloy semiconductors. Among the alloy semiconductors, gallium indium arsenide ($Ga_x In_{1-x} As$) is used in high frequency and optical devices, mercury cadmium telluride ($H_{g1-x} Cd_xTe$) is used for the fabrication of efficient infrared detectors, and gallium arsenide phosphide ($Ga As_x P_{1-x}$) is used for the manufacture of light emitting diodes. Table 1 gives a list of several semiconductors with their applications.

Table 1 Some semiconductors and their applications

Semiconductors	Applications
Ge	Diodes and early transistors.
Si	Diodes, transistors, IC's etc.
*pb*Te	Infrared detectors.
GaAs	High frequency transistors, lasers, and some special devices.
$Ga_xIn_{1-x}As$	High frequency and optical devices.
$Hg_{1-x}cd_xTe$	Infrared detectors.
$GaAs_xP_{1-x}$	Light-emitting diodes (LED's).

Pure or Intrinsic Semiconductor

When the conductivity of a semiconductor is solely determined by the thermally generated carriers the semiconductor is termed a *pure or intrinsic semiconductor*.

When a pure semiconductor is kept at 0°K its valence band is completely filled and the conduction band is completely empty because the thermal energy of the electrons is zero. Therefore, *at 0°K a pure semiconductor behaves as an insulator*. On the other hand, when a pure semiconductor is kept at room temperature some of the valence-band electrons gain sufficient energy, jump into the conduction band, and become free (Figure 16).

The vacancies formed in the valence band of the semiconductor when some of the valence-band electrons jumps into the conduction band are termed holes. A *hole* carries a positive charge equal in magnitude to the electron charge. The number of thermally generated electrons always equals the number of holes. Thus, if n_i and p_i represent respectively the concentrations of thermally generated electrons and holes, then $n_i = p_i$. The quantity n_i (or p_i) is referred to as the *intrinsic carrier concentration*. The generation of the electrons in the conduction band and the holes in the valence band is shown in Figure 16.

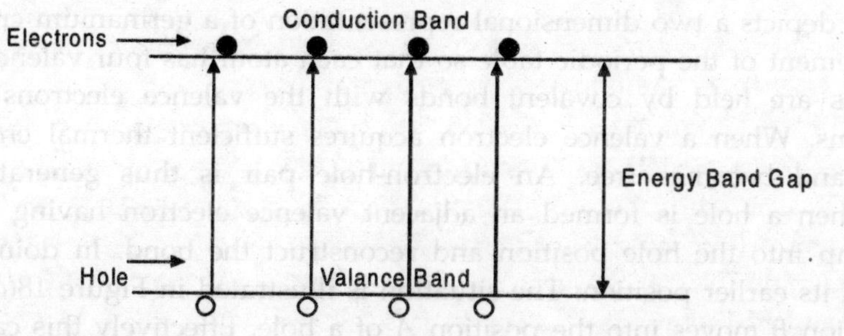

FIGURE 16 **Generation of electrons and holes in a pure semiconductor.**

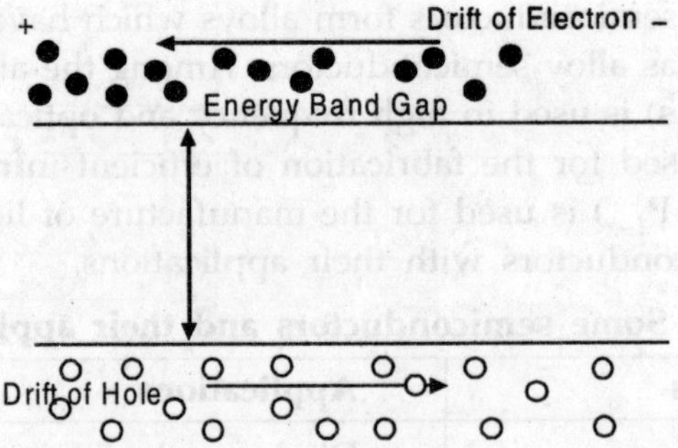

FIGURE 17 **Drift of electrons and holes in a biased semiconductor.**

The electrons in the conduction band and the holes in the valence band are free and move within the crystal at random due to their thermal energy. An external voltage applied to the semiconductor superimposes upon the random thermal motion of the electrons and the holes a drift velocity. This gives rise to a current flow. Thus, when a semiconductor is connected to a battery the current is contributed by the free electrons in the conduction band and the free holes in the valence band. The electrons move toward the positive electrode while the holes move toward the negative electrode of the battery, as shown in Figure 17. The currents due to the opposite movements of these two charge carriers aid since a hole carries positive charge.

In the case of covalent semiconductors like *Ge* and *Si*, the generation of electrons and noles and their movement are readily understood by considering the crystal structure, as explained in Figure 18.

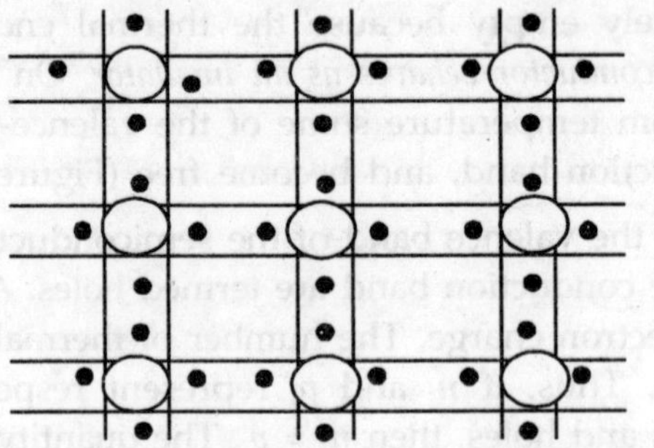

FIGURE 18(a) **A two dimensional representation of Ge crystal.**

Figure 18(*a*) depicts a two dimensional representation of a germanium crystal. Germanium is a group IV element of the periodic table so that each atom has four valence electrons. These valence electrons are held by covalent bonds with the valence electrons of four adjacent germanium atoms. When a valence electron acquires sufficient thermal energy it breaks its covalent bond and becomes free. An electron-hole pair is thus generated, as shown in Figure 18(*b*). When a hole is formed an adjacent valence electron having sufficient thermal energy may jump into the hole position and reconstruct the bond. In doing so, the electron creates a hole in its earlier position. The situation is illustrated in Figure 18(*c*) where a valence electron at position *B* moves into the position *A* of a hole. Effectively this causes a movement of the hole from position *A* to position *B*. That is, the movement of the hole takes place in the direction opposite to that of the valence electron.

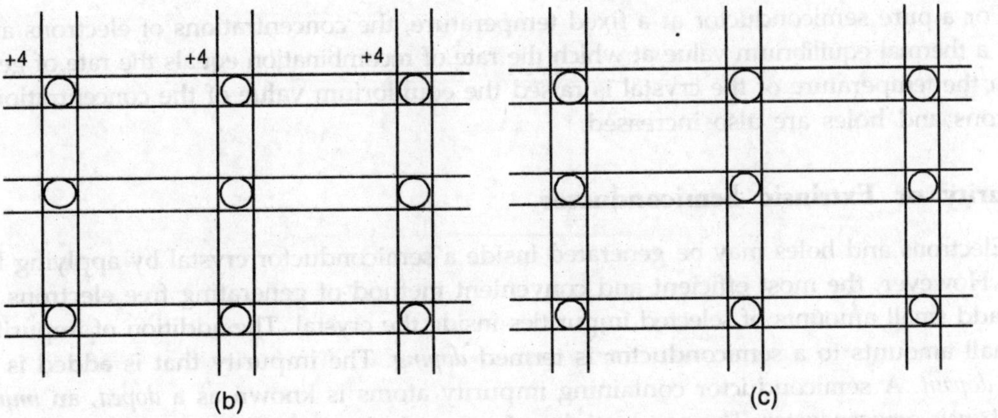

FIGURE 18(b) **Generation of an electron and a hole by breaking or a covalent bond (*c*) Movement of a hole in a crystal**

Recombination of Electrons and Holes

It is very likely that at every instant of time some of the free electrons while moving at random inside a semiconductor crystal encounter holes and combine with them. The process is termed *recombination*. When electrons and holes combine they are lost. The rate of recombination is roughly proportional to the product of the concentrations of the electrons and the holes. Note that the minimum energy required for the, generation of an electron-hole pair is the band-gap energy. This amount of energy is ordinarily obtained from thermal sources. Similarly, when an electron jumps from the conduction band into the valence band and combines with a hole (Figure 19) the minimum energy released in the form of electromagnetic radiation is again equal to the band-gap energy. The frequency of the emitted electromagnetic radiation, v, obeys the relationship

$$E_g = h_v$$

where E_g is the band-gap, and h is Planck's constant. The wave length of the radiation is

$$\lambda = \frac{c}{v} = \frac{ch}{E_g},$$

where c is the velocity of light in free space.

It should be remembered that while the free electrons and holes are lost by recombination new pairs of electrons and holes are generated due to thermal energy.

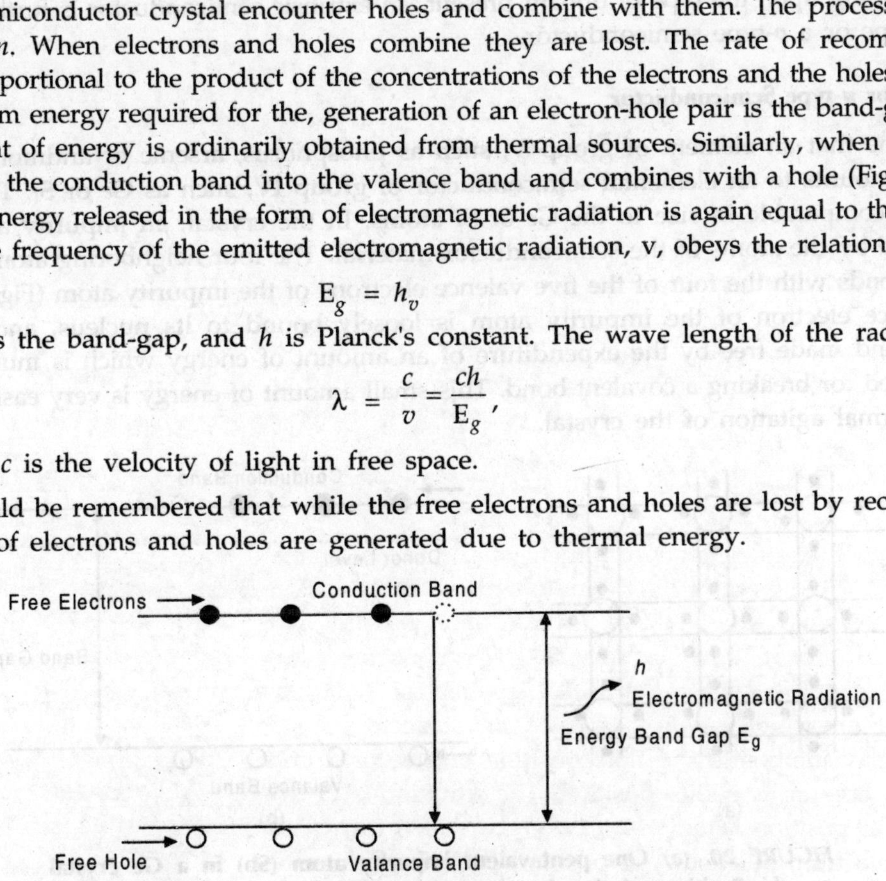

FIGURE 19 **Recombination of an electron and a hole.**

For a pure semiconductor at a fixed temperature, the concentrations of electrons and holes reach a thermal equilibrium value at which the rate of recombination equals the rate of generation. When the temperature of the crystal is raised the equilibrium value of the concentrations of the electrons and holes are also increased.

Impurity or Extrinsic Semiconductor

Electrons and holes may be generated inside a semiconductor crystal by applying heat and light. However, the most efficient and convenient method of generating free electrons or holes is to add small amounts of selected impurities inside the crystal. The addition of impurity atoms in small amounts to a semiconductor is termed *doping*. The impurity that is added is referred to as *dopant*. A semiconductor containing impurity atoms is known as a *doped*, an *impurity*, or an *extrinsic semiconductor*. The conductivity of an extrinsic semiconductor is determined by the excess electrons or holes due to the impurity atoms. Impurities frequently employed for *Ge* and *Si* are the elements of group III and V of the periodic table. They are commonly boron, gallium, indium and aluminium from group III, and phosphorus, antimony and arsenic from group V. In a compound semiconductor like I*nSb*, Indium (T*n*) is a group III element and antimony (*Sb*) is a group V element. In this case, the impurities generally employed are tellurium (T*e*), a group VI element and zinc (Z*n*), a group II element.

Depending upon the type of dopant present, an extrinsic semiconductor is further classified as an *n*-type or a *p*-type semiconductor.

(*i*) Donor or *n*-type Semiconductor

Assume that an element of group V, such as phosphorus, arsenic or antimony, is added in small amounts to an elemental semiconductor of group IV, such as *Ge* or *Si*. The impurity atoms are comparable in size to the *Ge* or *Si* atoms. In the crystal, an impurity atom will be surrounded by the atoms of the semiconductor material. The four neighboring atoms will form covalent bonds with the four of the five valence electrons of the impurity atom (Figure 20). The fifth valence electron of the impurity atom is loosely bound to its nucleus, and so can be detached and made free by the expenditure of an amount of energy which is much less than that required for breaking a covalent bond. This small amount of energy is very easily provided by the thermal agitation of the crystal.

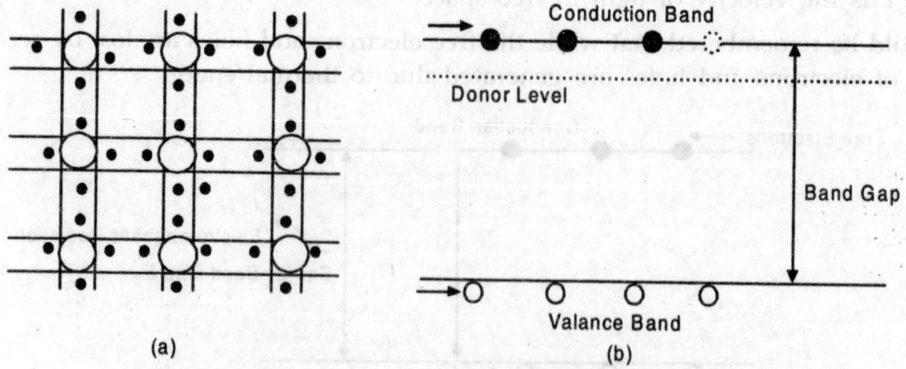

FIGURE 20. **(*a*) One pentavalent impurity atom (Sb) in a Ge crystal (*b*) Position of the donor level in the energy-band diagram.**

The energy state corresponding to the fifth valence electron is in the forbidden gap and is just below the conduction band. This energy, level is indicated by the dashed line in Figure 20, and is called the *donor level*. When the fifth valence electron is separated and transferred to the conduction band the impurity atom becomes positively charged immobile ion. Each impurity atom in this way donates a free electron to the semiconductor. Such impurity elements are called *donors*. The semiconductor containing donor type impurity is called an *n-type semiconductor* since the current carriers produced are negative charges (electrons). The free electrons produced in this way are called *excess electrons*. In a III-V compound semiconductor like I*nSb*, a group VI element like T*e*, replacing the group V element *Sb*, acts as a donor.

The electrons are *majority carriers* and thermally generated holes are the *minority carriers* in *n*-type semiconductors. In an *n*-type semiconductor, if *n* and *p* represent respectively the electron and hole concentrations, it can be shown that $np = n_i p_i = n_i^2$, where n_i and p_i are the intrinsic values.

Note that for each impurity atom there is a free electron in the conduction band but there is no corresponding positive hole in the valence band. As n_i is a constant at a particular temperature, the product *np* is also a constant at a given temperature. This result can be understood in the following way. The rate of recombination is proportional to the electron concentration *n* and also to the hole concentration *p, i.e.*, to the product *np*. As the rate of recombination is fixed at a given temperature, the product *np* must also be so. When the concentration of electrons is increased over the intrinsic value by the addition of donors, the concentration of holes falls below its intrinsic value, making the product *np* a constant, equal to n_i^2.

(*ii*) Acceptor or p-type Semiconductor

Suppose an element of group III, such as aluminum, is introduced in small amounts into the crystal structure of an elemental semiconductor, such as G*e* or S*i*. For each atom of group III element introduced into the crystal structure there remains an unfilled bond indicated by the dashed line in Figure 21. In other words, an *unfilled energy state* is created due to the introduction of a group III impurity atom. This additional energy state is located just above the valence band of the semiconductor. This is shown by the dashed line in Figure 22 and is termed the *acceptor level*. When an electron in the valence band acquires sufficient energy and jumps into the unfilled energy state, the impurity atom becomes a negatively charged ion. The energy required to make such a transfer is very small and is easily provided by the thermal agitation of the crystal. When an electron from the valence' band comes to the new energy state to fill the incomplete bond, it leaves behind a hole in the valence band. The negatively charged ion being immobile does not contribute to current.

Impurity atoms that cause generation of holes inside a semiconductor crystal are termed *acceptors* since they accept electrons and become negatively charged ions. The semiconductor containing acceptor type of impurity is called a *p-type semiconductor* because the current carriers are positive charges (holes). Holes formed in this manner are often referred to as *excess holes*. In a III-V semiconductor like InSb, a group II element like Z*n*, replacing the group III element in, acts as an acceptor.

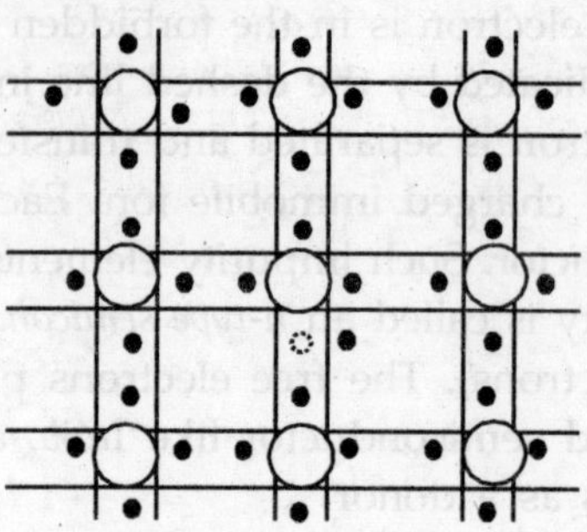

FIGURE 21. **One trivalent impurity atom in a germanium crystal.**

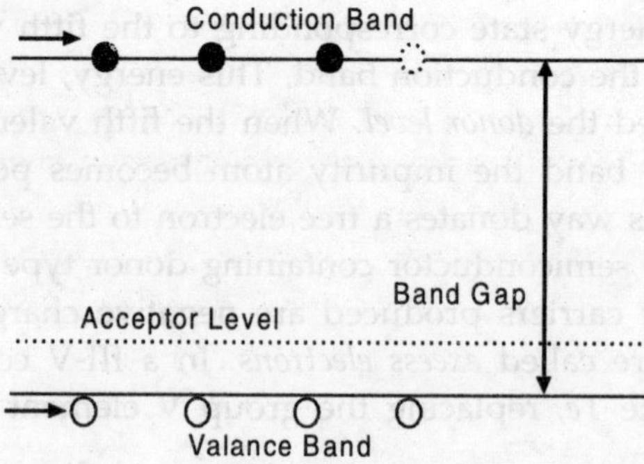

FIGURE 22. **Location of an acceptor level in the energy band diagram.**

In a p-type semiconductor, the *holes are the majority carriers* and thermally generated *electrons are the minority carriers*. Like an n-type semiconductor a p-type semiconductor also satisfies the relation $pn = p_i n_i = n_i^2$.

In the case of p-type semiconductors also for each impurity atom there is a free hole in the valence band but there is no corresponding generation of free electron in the conduction band. Thus, $p > p_i$ but $n < n_i$.

Effect of Temperature on Impurity Semiconductors

Assuming complete ionization of all donor atoms if the temperature of an n-type semiconductor is increased the generation of electron-hole pairs will increase. Therefore, with the increase in temperature a situation will arise when the concentration of thermally generated electrons in the conduction band will be much larger than the concentration of donor electrons. Under this condition the concentrations of the holes and the electrons will become almost equal, and the semiconductor will behave almost intrinsically. The same type of argument can be put forward for the *p-type* material. Thus we conclude that as the temperature of an impurity semiconductor increases the semiconductor becomes almost intrinsic.

Fermi Level in Semiconductors

The Fermi level is a quantity of great importance in semiconductor electronics. We will therefore consider physically the position of the Fermi level in the band diagram of pure and impure semiconductors.

Fermi level in Intrinsic Semiconductor

The energy-band diagram of an intrinsic semiconductor is shown in Figure 23(*a*). Here E_c and E_v. represent respectively the lowest energy level in the conduction band and the highest energy level in valence band. E_g is the band-gap energy. On the right of this band diagram is shown the Fermi-Dirac distribution function [Figure 23 (*b*)]. At absolute zero (T = 0° K) the probability of finding an electron in the conduction band and a hole in the valence band is zero, and the nature of the distribution function is as shown by the curve marked T = 0° K. At room temperature (T = 300° K) some of the electrons from the valence band will move into the conduction band and will occupy the states near the bottom of the band. This gives a non-zero value for the distribution function, or in other words, the tail of the distribution function curve extends into the conduction band. Similarly, an equal number of holes are present near the top of the valence band, and the probability of occupancy there decreases from unity. At 300° K the distribution function will therefore be as shown in Figure 22(*b*) by the curve labelled T = 300° K. Since the Fermi level, E_F is defined as that energy level at which the probability of occupancy becomes half we find that *the Fermi level for the intrinsic semiconductor is near the middle of the forbidden gap*.

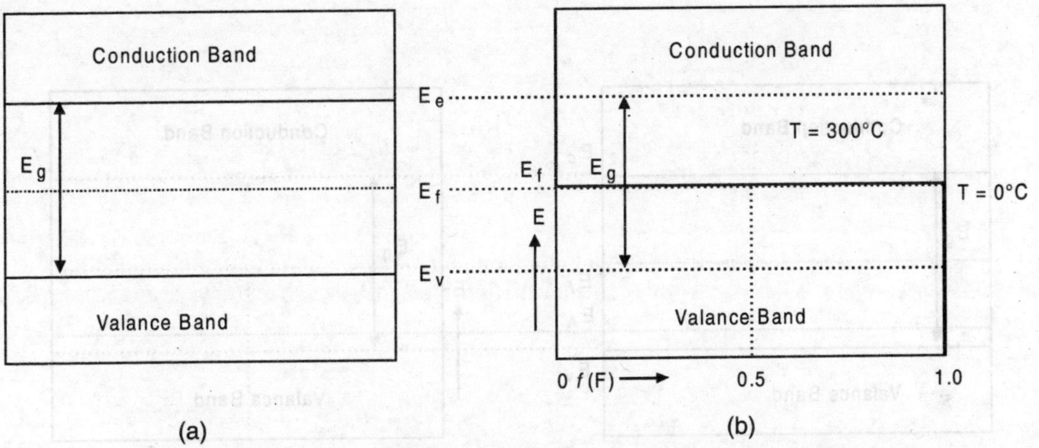

FIGURE 23. (a) **Energy-band diagram of an intrinsic semiconductor.**
(b) **Fermi-Dirac distribution function at 0° K and at 300° K.**

Fermi Level in Impurity Semiconductor

The band diagrams for *n-* and *p-type* semiconductor are shown in Figures 24 and 25, respectively. E_D and E_A represent the energy levels corresponding to the donor and the acceptor impurities respectively. In the case of an *n*-type semiconductor, if we assume that all the donor atoms are ionized then the donor electrons will occupy the states near the bottom in the conduction band.

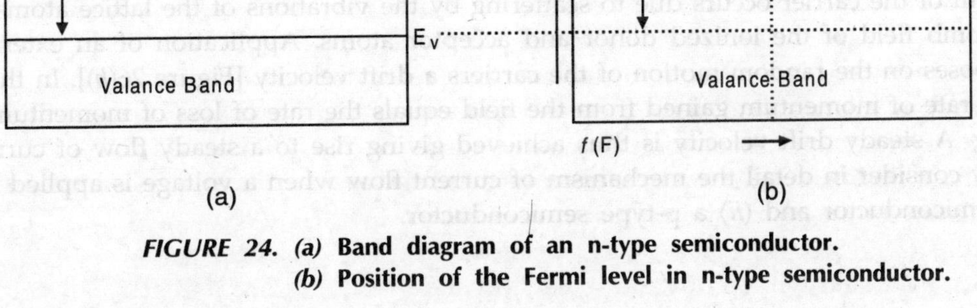

FIGURE 24. (a) **Band diagram of an n-type semiconductor.**
(b) **Position of the Fermi level in n-type semiconductor.**

Obviously then a smaller number of electrons from the valence band will be able to cross the gap by thermal agitation. As a result, the number of holes in the valence band is decreased. Since the Fermi level gives the energy state for which the probability of occupancy is half, E_F for an *n*-type semiconductor is closer to the conduction band [Figure 24 (b)]. Thus if an intrinsic semiconductor is doped with a donor type impurity the Fermi level will move toward the conduction band. If the doping is very large the Fermi level may move into the conduction band.

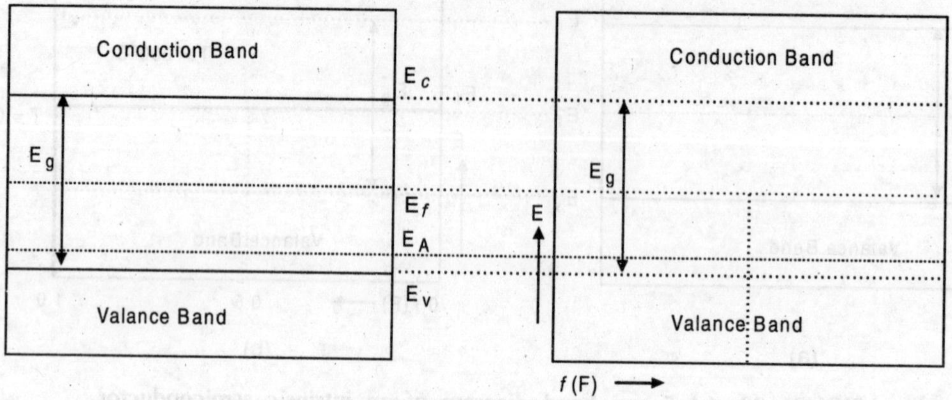

***FIGURE 25 (a)* Band diagram of a p-type semiconductor**
***(b)* Location of the Fermi level in p-type semiconductor.**

The same kind of reasoning, when applied to a *p-type* material, leads to the result that for a *p-type* semiconductor the position of the Fermi level is located near the valence band (Figure 25). If an intrinsic semiconductor is doped with acceptor type impurity the Fermi level will move toward the valence band, and for sufficiently large doping it may move into the valence band. With the increase in temperature of the semiconductor, since both *n-* and *p-type* semiconductors become essentially intrinsic, we may conclude that *with temperature rise the Fermi level E_F moves toward the centre of the forbidden gap for both types of semiconductors.*

Mechanism or Current Conduction in Semiconductors

When no external field is applied to a semiconductor each carrier (electron or hole) moves in a random way owing to its thermal energy [Figure 26(*a*)]. The frequent change in direction of the path of the carrier occurs due to scattering by the vibrations of the lattice atoms and by the Coulomb field of the ionized donor and acceptor atoms. Application of an external field superimposes on the random motion of the carriers a drift velocity [Figure 26(*b*)]. In the steady state, the rate of momentum gained from the field equals the rate of loss of momentum due to scattering. A steady drift velocity is thus achieved giving rise to a steady flow of current. We shall now consider in detail the mechanism of current flow when a voltage is applied to (*i*) an n-type semiconductor and (*ii*) a p-type semiconductor.

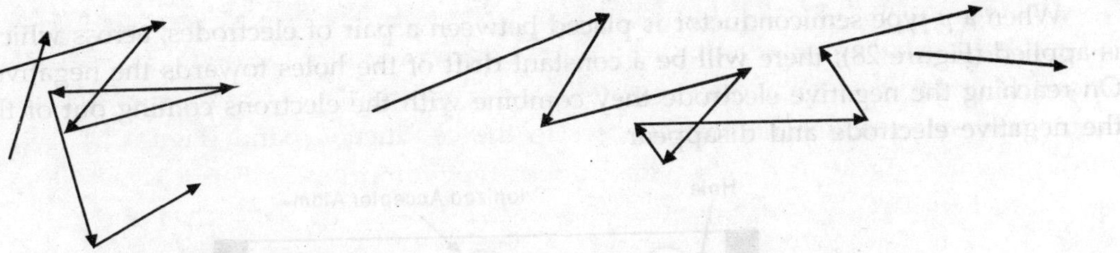

FIGURE 26 **(a) Random motion of a carrier in a semiconductor**
(b) Random motion with superimposed drift.

(i) Current Flow in a Biased *n*-type Semiconductor

An n-type semiconductor has (*i*) electrons as majority carriers, (*ii*) an (almost) equal number of immobile positive ions and (*iii*) holes as minority carriers. Assume that the hole contribution to current is negligible.

Consider an *n*-type semiconductor placed between a pair of electrodes across which a voltage is applied (Figure 27). Due to the field produced by the voltage there will be a steady drift of the free electrons towards the positive electrode. The electrons reaching the positive terminal disappear at the metal of the electrode and the immobile positive ions in the vicinity of the negative electrode remain unneutralized due to the drift of the free electrons. These ions immediately attract electrons from the negative electrode. Thus a continuous flow of electrons from one terminal of the voltage source to the other terminal via the semiconductor 'takes place. The rate of flow of electrons from the negative electrode to the semiconductor and from the semiconductor to the positive electrode is determined by the applied voltage and the conductivity of the semiconductor which account for the current flow.

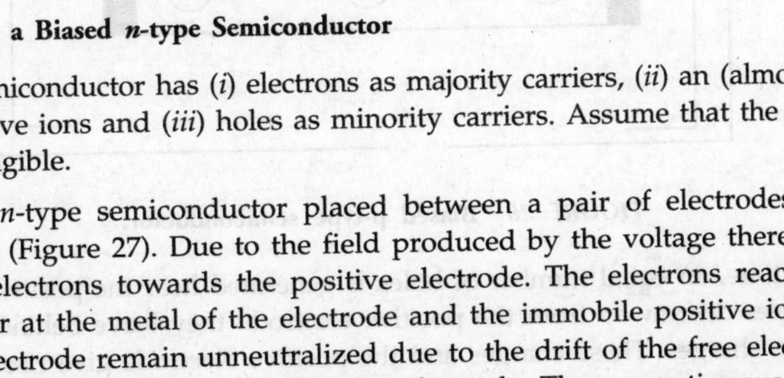

FIGURE 27 **Biased n-type semiconductor.**

(ii) Current Flow in a Biased *p*-type Semiconductor

A *p*-type semiconductor has (*i*) holes as majority carriers, (*ii*) an (almost) equal number of immobile negative ions, and (*iii*) electrons as minority carriers. In this case the electron current is considered negligible.

When a p-type semiconductor is placed between a pair of electrodes, across which a voltage is applied (Figure 28), there will be a constant drift of the holes towards the negative electrode. On reaching the negative electrode they combine with the electrons coming out of the metal of the negative electrode and disappear.

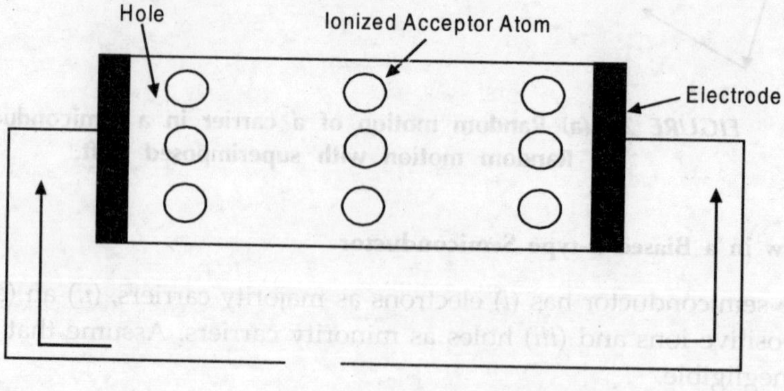

FIGURE 28 **Biased p-type semiconductor.**

At the same time, an equal number of holes is generated near the positive electrode. This is because as the holes drift away from the positive electrode they leave behind the unneutralised immobile negative charges. These charges and the positive electrode give rise to an electric field which causes the ionizing electrons to leave the acceptor atoms and come to the positive, electrode where they are lost. By loosing an electron in the above process the acceptor atom attempts to steal an electron from an adjacent bond. A hole is thus formed. The generation of holes takes place at the rate equal to the rate at which they disappear at the negative electrode. This rate is again determined by the magnitude of the applied voltage and the conductivity of the semiconductor.

Note that the current inside the p-type semiconductor is due to the motion of the holes whereas in the external circuit it is due to the motion of electrons.

Lifetime of Minority Carriers

If in an extrinsic semiconductor the concentration of minority carriers is increased over the thermal equilibrium value, the number of these excess carriers' decreases with time due to recombination.

Let n_o and P_o be the thermal equilibrium concentrations of electrons and holes, respectively. If Δn and Δp denote the small changes from the thermal equilibrium concentrations, the recombination rate R', being proportional to the product np, is given by

$$R' = Cnp = C\,(n_o + \Delta n)\,(P_o + \Delta p),$$

where C is the proportionality constant. In equilibrium, $\Delta n = \Delta p = 0$, and we have the thermal generation and recombination rate given by Cn_op_o. The recombination rate for the excess carriers R is obtained by subtracting Cn_op_o from R', *i.e.*,

$$R = C[(n_o + \Delta n)\,(p_o + \Delta p) - n_op_o)] \approx C\,(n_o\,\Delta p + p_o\,\Delta n).$$

In the above approximation, the product of the two small quantities $\Delta n \Delta p$ has been neglected. In an n-type semiconductor' $n_o\Delta p >> p_o\Delta\, n$, since $n_o >> p_o$ and Δn is taken to be nearly equal to Δp. Hence

$$R \approx Cn_o\Delta p.$$

On the other hand, for a *p*-type semiconductor $p_o \Delta n >> n_o \Delta p$, and $R \approx C p_o \Delta n$

Thus we find that the recombination rate is proportional to the concentration of excess minority carriers. For an *n*-type semiconductor we can write

$$R = -\frac{d\Delta p}{dt} = \frac{Dp}{\tau_p},$$

where $\tau_p = 1/Cn_o$. The quantity τp has the dimension of time and is referred to as the lifetime of an excess minority carrier, *i.e.*, hole. Solving the differential equation for Δ^p we get

$$\Delta p(t) = \Delta p(0)\, exp(-t/\tau_p),$$

where $\Delta p(0)$ is the initial change in hole concentration. Thus τ_p denotes the time constant of the exponential decay of Δp. That is, Δp falls to $1/exp(1)$ of its initial value in time τ_p.

We can similarly define the lifetime of an excess electron τ_n in a p-type semi-conductor, where $\tau_n = 1/Cp_0$

Mobility, Current Density and Conductivity

Mobility of a carrier is defined as the average drift velocity per unit electric field. If v is the drift velocity in meter/sec. of a carrier in an electric field of ε volt/meter then the mobility (μ) of the carrier is given by

$$\mu = \frac{v}{\varepsilon}\ m^2\ volt^{-1}\ \sec^{-1} \qquad (1)$$

Current density is defined as the current per unit cross-sectional area, i.e., the charge crossing unit cross-sectional area per second. J is the symbol generally used for the current density and is expressed in ampere per square meter.

Conductivity is the current density per unit applied electric field. If *J* is the current density due to an applied electric field ε, the conductivity is given by $\sigma = j/\varepsilon$. If *J* is expressed in amp/m^2 and ε in *v/m*, σ is expressed in *mho/m*.

Consider a cylindrical semiconductor of length L and area of cross section A. Suppose *v* is the drift velocity of the carriers due to an electric field ε applied along the length of the cylinder (Figure 29). Therefore, the time taken by a carrier to travel the distance L is given by

$$t = \frac{L}{v} \qquad (2)$$

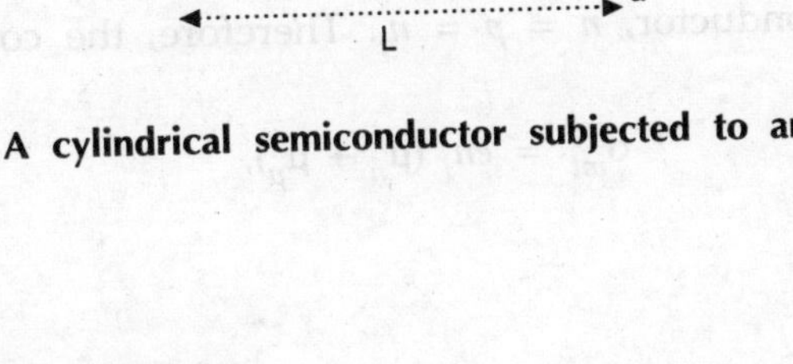

FIGURE 29 **A cylindrical semiconductor subjected to an electric field.**

If n_c *denotes* the carrier concentration then the total number of carriers contained in the cylinder is n_cAL. The total amount of charge inside the cylinder is, therefore, n_cAL e, where e is the charge of a carrier. This amount of charge passes through the cross section *aa'* in time t. Hence the charge crossing *aa'* per second is $\frac{n_c \text{AL}e}{t}$. This, by definition, is the current flowing through the semiconductor, *i.e.*,

$$\text{I} = \frac{n_c \text{AL}e}{t} \tag{3}$$

Therefore the current density J is given by

$$J = \frac{\text{I}}{\text{A}} = \frac{n_c \text{L}e}{t} \tag{4}$$

Using (2) we get

$$\text{J} = n_c ev \tag{5}$$

If n_c is expressed in m^{-3}, e in coulomb, and v in m sec^{-1}, J is expressed in amp. m^{-2}. Since $v = \mu\varepsilon$, where μ is the mobility, we can write

$$\text{J} = n_c ev = n_c e\mu\varepsilon. \tag{6}$$

Again, if σ is the conductivity of the semiconductor, then

$$\text{J} = \sigma\varepsilon. \tag{7}$$

Comparing equations (6) and (7) we get for σ

$$\sigma = n_c e\mu. \tag{8}$$

When n_c is in m^{-3}, e is in coulomb, and μ is in m^2. volt^{-1}. sec^{-1}., σ is expressed in *mho* m^{-1}.

Expressions for Conductivity in Semiconductors

If the conductivity of a semiconductor due to the electrons is denoted by σ_n then from equation (8) we can write

$$\sigma_n = ne\mu_n \tag{9}$$

where n is the electron concentration and μn is the electron mobility.

Similarly, the conductivity of a semiconductor due to the holes is

$$\sigma_p = pe\mu_p \tag{10}$$

where p is the hole Generally concentration and μ_p is the hole mobility.

Generally, $\mu_n > \mu_p$.

In the presence of au electric field, the currents due to the electrons and, the holes will be in the same direction since they carry opposite charges. Hence the overall conductivity of the semiconductor containing electrons and holes is

$$\sigma = \sigma_n + \sigma_p = e(n\mu_n + p\mu_p). \tag{11}$$

For an intrinsic semiconductor, $n = p = n_i$. Therefore, the conductivity of the intrinsic semiconductor is

$$\sigma_{int} = en_i\,(\mu_n + \mu_p). \tag{12}$$

For an n-type semiconductor, $n >> p$ hence

$$\sigma_{\text{n-type}} \approx en\mu_n. \tag{13}$$

For a p-type semiconductor, $p >> n$ and the product $p\mu_p$ is larger than $n\mu_n$ for heavily doped semiconductors. In that case

$$\sigma_{\text{p-type}} \approx ep\mu_p. \tag{14}$$

Expression for Mobility

When an external electric field is applied to a semiconductor the charge carriers (electrons or holes) experience forces not only due to this applied field but also due to the field produced by the internal crystal potential. The effect of the internal crystal field may be shown to modify the mass of the carriers in such a way that the carriers respond to the externally applied field with this modified mass, obeying the laws of mechanics. This modified mass is called the *effective mass* of the carrier, and is generally different from the free electron mass.

When a field ε is applied, the rate at which momentum is gained from the field by a carrier of charge e is given by the force $e\varepsilon$, according to Newton's second law of motion. The rate of loss of momentum due to scattering in most cases can be written in the form m^*v/τ where m^* is the effective mass of the carrier, v is its drift velocity and τ is a quantity called the relaxation time. The relaxation time is different from the life time defined earlier. The relaxation time is typically about 10^{-13} s while the life time is about 10^{-4} s. The relaxation time has the significance that when the exciting force is removed, the momentum falls to 1/exp (1) of its initial value in time τ. In the steady state the rate of gain pf momentum from the field is equal to the rate of loss of momentum due to scattering, *i.e,*

$$e\varepsilon = \frac{m^* v}{\tau}. \tag{15}$$

The mobility of the carrier is therefore given by

$$\mu = \frac{v}{\varepsilon}\frac{e\tau}{m^*}. \tag{16a}$$

τ is a function of energy and so in the expression for mobility an average value of or has to be used. A rigorous treatment shows that the mobility is given by

$$\mu = \frac{e\langle\tau\rangle}{m^*} \tag{16b}$$

where $\langle \tau \rangle$ represents the average value of τ involving the distribution function of the carriers. If e is expressed in coulomb, τ in s, and m^* in kg, μ is expressed in $m^2\ v^{-1}\ S^{-1}$.

Generally, the effective mass of the holes is greater than the effective mass of the electrons. The holes are therefore less mobile than the electrons, *i.e.,* $\mu p > \mu_n$.

Diffusion

Whenever there is a concentration gradient in a system of particles, the particles tend to move from a region of higher concentration to a region of lower concentration in a way to destroy the non uniformity in concentration. This movement of particles is referred to as the phenomenon of *diffusion.*

In a semiconductor a non uniform concentration of charge carriers (electrons or holes) may exist. Let us suppose that the concentration p of holes varies with distance x in the semiconductor, the concentration gradient being dp/dx. Suppose that dp/dx it negative, *i.e.*, the hole concentration is greater for lower values of x. The holes move at random owing to their thermal energy, but there is a diffusive flow of holes along the positive x-direction. This flow of holes constitutes diffusion current in the positive x-direction. It may be mentioned that diffusion is not a result of mutual repulsion among charge particles of like sign, but is the result of a statistical phenomenon. The diffusion hole-current density J_p is proportional to the concentration gradient dp/dx and is given by

$$J_p = -eD_p \frac{d_p}{d_x}, \tag{16c}$$

where e is the magnitude of the electron charge, and D_p is termed the diffusion constant, diffusion coefficient, or diffusivity of holes. The negative sign in Eq. (16c) is required since dp/dx is negative and J_p is positive in the positive x-direction.

A similar' equation can be written for the diffusion electron-current density J_n with p and D_p replaced respectively by n and D_n, and the minus sign replaced by a plus sign in Eq. (16*c*).

If there is an electric field ε and a concentration gradient of holes in the x-direction, the total hole current is the sum of the drift current $pe\mu_p\varepsilon$ and the diffusion current given by Eq. (16*c*). That is,

$$J_p = pe\mu_p\varepsilon - eD_p \frac{d_p}{d_x}. \tag{16d}$$

Similarly, for electrons the total current density is

$$J_n = ne\mu_n\varepsilon + eD_n \frac{dn}{dx}. \tag{16e}$$

Einstein Relation

The quantities D and μ are not independent; they are related by the following equation, known as the Einstein relation:

$$\frac{D_p}{\mu_p} = \frac{D_n}{\mu_n} = \frac{kT}{e}, \tag{16f}$$

where k is the Boltzmann constant and T is the absolute temperature. At room temperature (T = 300°K), substituting $k = 1.38 \times 10^{-23}$ Joules/°K and $e = 1.6 \times 10^{-19}$ coulomb in Eq. (16*f*), we get

$$\frac{D_p}{\mu_p} = \frac{D_n}{\mu_n} = 0.026 \text{ V}. \tag{16g}$$

The diffusion coefficient is expressed in $m^2\ s^{-1}$, or $cm^2\ s^{-1}$. For electrons in germanium at 300° K, we have $\mu_n = 3800\ cm^2/\text{volt sec}$. Therefore, from Eq. (16*g*) we get $D_n = 99\ cm^2/\text{sec}$ at 300° K.

Junction Diodes

A junction between a *p*-and an *n*-type semiconductor is most important in electronic applications since this forms the basis of semiconductor devices like diodes, transistors etc. In this section, we shall describe the physical principle and the operation of a p-n junction. The various device applications of *p-n* junctions are also given.

p-n Junction

When acceptor type impurities are introduced into one half and donors into the other half of a single-crystal semiconductor a *p-n* junction is formed, as shown in Figure 30. This is a two terminal device and is called a *junction diode*.

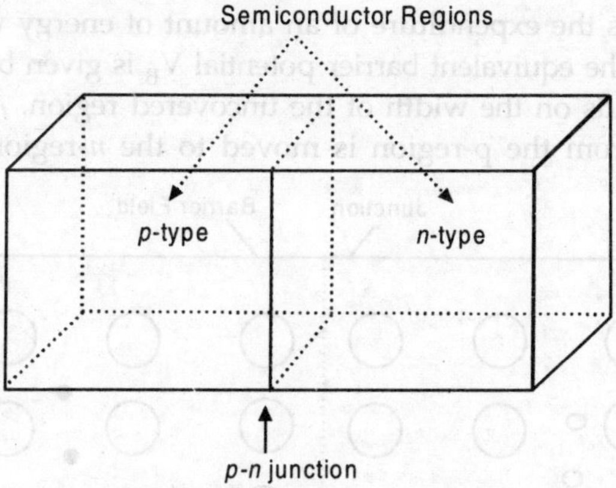

***FIGURE 30* A *p-n* Junction.**

The *p-n* junction may be (*i*) a step-graded junction or (*ii*) a linearly-graded junction.

In a step-graded junction the acceptor or the donor impurity density in the semiconductor is constant up to the junction. This type of junction is formed by placing a small sphere of a trivalent impurity say indium, on a wafer of *n*-type germanium and heating the combination to a high temperature for a short duration. During the process indium dissolves into the germanium and changes the conductivity of the germanium from *n*- to *p*-type over a portion of the semiconductor and forms a *p-n* junction.

In a linearly-graded junction, the impurity density varies almost linearly with distance away from the junction. This type of junction is formed by drawing a single crystal from a melt of germanium which at the start of the process contained impurity of one kind. During the drawing process impurity of the other type is added in sufficient quantity to change the conductivity type of the melt.

An Unbiased *p-n* Junction

A schematic representation of a *p-n* junction is shown in Figure 31. Here the circled charges represent the ionized acceptor and donor atoms.

For the sake of simplicity, let us assume for the moment that there are only holes in the p-type side and electrons in the n-type side. Due to the existence of density gradient across the junction electrons will diffuse through the junction to the left and holes to the right. After crossing the junction they combine with each other leaving the immobile ions in the neighborhood of the junction un-neutralized. These are called the uncovered charges. The variation of concentration of (*i*) immobile ions (*ii*) mobile carriers and (*iii*) uncovered charges are shown in

Figures 31(b), (c) and (d), respectively. The positive and negative charges which are uncovered produce an electric field across the junction. This field is directed from the n-side to the p-side and is termed the barrier field. This field opposes the further diffusive motion of electrons and holes across the junction. Equilibrium is established at a barrier field that is just sufficient to stop further diffusion of electrons and holes. Under this condition there is no movement of carriers across the junction. Since the vicinity of the junction is depleted of mobile charges it is called the depletion or the space charge region. The thickness of this region is around 0.5 μm. Due to the existence of a barrier field across the junction transfer of an electron from the n-side to the p-side requires the expenditure of an amount of energy which is termed the barrier energy (E_B) [Figure 31(e)]. The equivalent barrier potential V_B is given by $E_B = eV_B$. The magnitude of the barrier energy depends on the width of the uncovered region. An equal amount of energy is involved when a hole from the p-region is moved to the *n*-region.

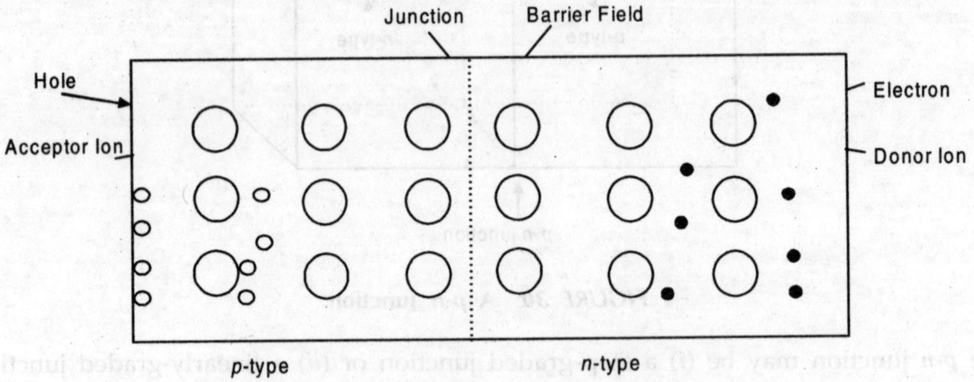

(a) Diagram showing the positions of carriers.

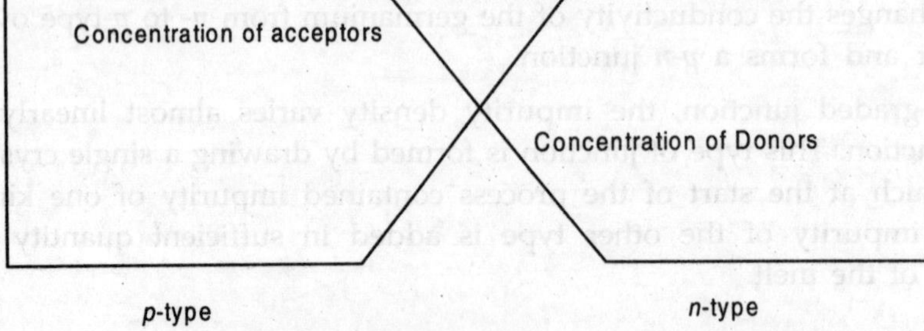

(b) Variation of concentration of immobile ions.

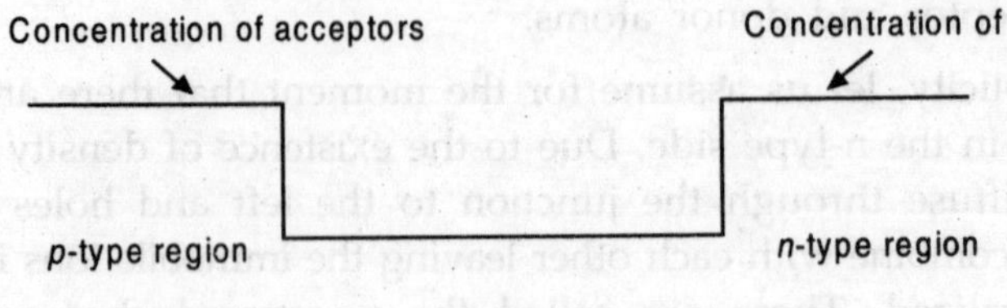

(c) Variation of concentration of mobile carriers.

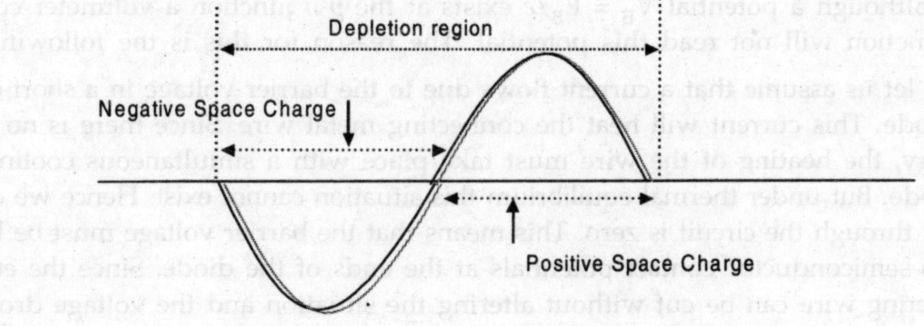

(*d*) Uncovered charge concentration.

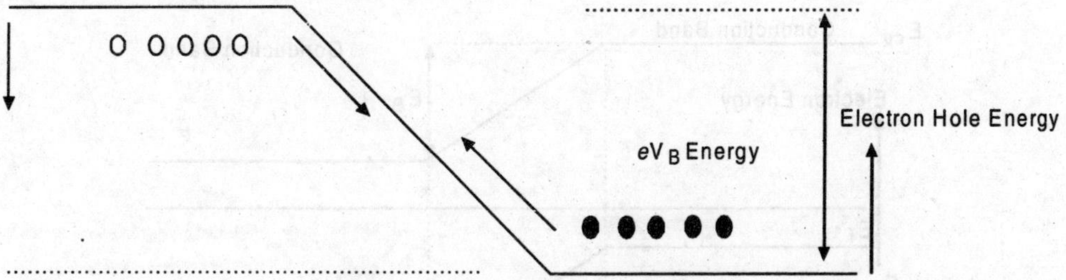

(e) Variation of electron and hole energy.

FIGURE 31 **Schematic representation of a *p-n* junction.**

Now consider the actual situation where the *p*-region contains electrons as minority carriers and the *n*-region contains holes as minority carriers. When the *p-n* junction is unbiased the barrier field acts in such a way that the electrons from the *p*-type side and the holes from the *n*-type side can easily cross the junction. The movement of the minority carriers therefore constitutes a current flow. Under equilibrium condition this current is exactly counterbalanced by an equal and opposite flow of majority carriers that gain from the thermal sources the energy that is sufficient to cross the barrier.

Energy-Band Diagram of an Unbiased *p-n* Junction

We know that for an *n*-type semiconductor the Fermi level lies near the conduction band edge E_c and for the *p*-type semiconductor it is near the valence band edge E_v. When a *p-n* junction is formed, under equilibrium condition the Fermi level attains a constant value throughout the sample, as shown in Figure 32. Clearly, then the conduction band edge E_{cp} of the *p*-type will not be at the same level as E_{cn} of the n-type. Similarly, E_{vp} will not be in line with E_{vn}. From Figure 32, we find

$$E_B = E_{cp} - E_{cn} = E_{vp} - E_{vn} = eV_B'$$

where E_B is the barrier energy, *e* is the electronic charge and V_B is the barrier potential.

Note that although a potential $V_B = E_B/e$ exists at the *p-n* junction a voltmeter connected across a *p-n* junction will not read this potential. The reason for this is the following:

If possible, let us assume that a current flows due to the barrier voltage in a short-circuited *p-n* junction diode. This current will heat the connecting metal wire. Since there is no external source of energy, the heating of the wire must take place with a simultaneous cooling of the *p-n* junction diode. But under thermal equilibrium this situation cannot exist. Hence we conclude that the current through the circuit is zero. This means that the barrier voltage must be balanced by the metal-to-semiconductor contact potentials at the ends of the diode. Since the current is zero the connecting wire can be cut without altering the situation and the voltage drop across the cut must be zero. Thus a voltmeter connected to the terminals of a diode show's zero voltage reading.

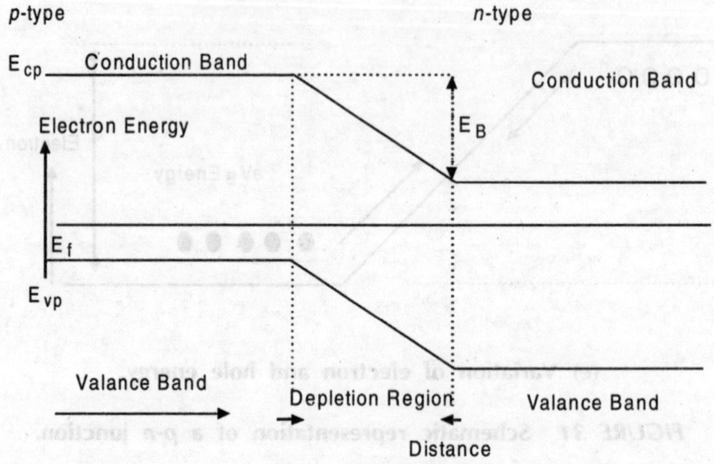

FIGURE 32 **Energy band diagram of a *p-n* junction.**

A Biased *p-n* Junction

When the positive terminal of a battery is connected to the *p*-type side and the negative terminal to the *n*-type side of a *p-n* junction, the junction allows a large current to flow through it. In this case, the *p-n* junction is said to be forward-biased. When the terminals of the battery are reversed, *i.e.*, the positive terminal is connected to the *n*-type side and the negative terminal to the *p*-type side, the junction allows a very small current to flow through it. Under this condition the *p-n* junction is called reverse biased.

The behavior of a *p-n* junction makes it suitable for use as a rectifier. We now consider qualitatively the mechanism involved when the junction is (*i*) forward-biased and (*ii*) reverse biased.

(*i*) *Forward-Biased p-n Junction:* A forward-biased *p-n* junction and its symbolic representation are shown in Figures 33(a) and (b), respectively. The forward biasing voltage V exerts a force on the holes of the *p*-type side and on the electrons of the *n*-type side.

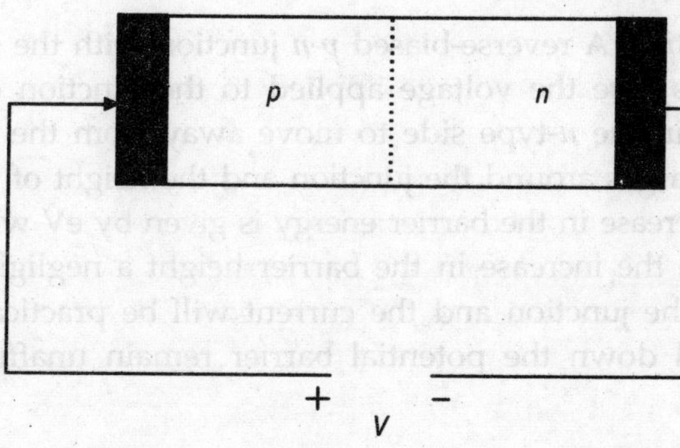

(a) A forward-biased *p-n* junction diode.

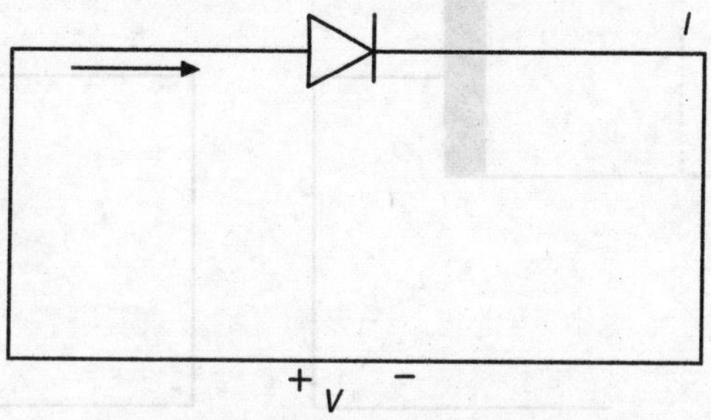

(b) Its symbolic representation.

FIGURE 33

This force causes them to move toward the junction. This reduces the width of the uncovered charges and the height of the barrier, *i.e.*, the barrier energy (Figure 34). The magnitude by which the barrier energy is reduced is given by eV, where V is the applied voltage. Due to the reduction in the barrier height the current flow due to the majority carriers, *i.e.*, holes from *p*-type side to *n*-type side and electrons from *n*-type side to *p*-type side increases. On the other hand, the minority carrier current flowing in the opposite direction to that of the majority carrier current is not affected by the forward bias. This is because the minority carrier current depends only on the temperature.

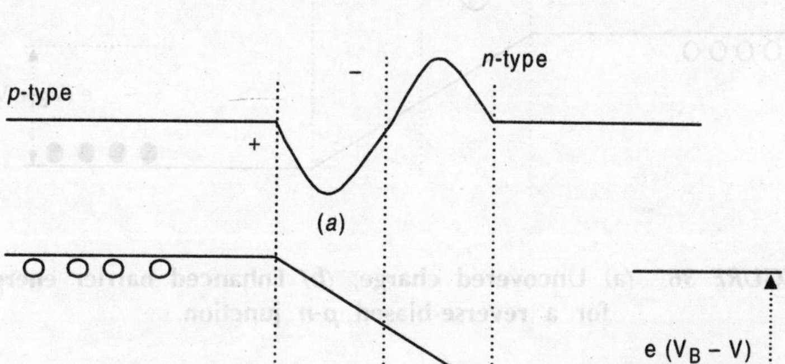

FIGURE 34 **(*a*) Uncovered charges (*b*) Reduced barrier energy for a forward-biased *p-n* junction.**

(*ii*) *Reverse-Biased p-n Junction*: A reverse-biased *p-n* junction with the symbolic representation is shown in Figure 35. In this case the voltage applied to the junction causes the holes in the *p*-type side and the electrons in the *n*-type side to move away from the junction. This increases the width of the uncovered charges around the junction and the height of the barrier [Figure 36(a) and (b)]. The amount of the increase in the barrier energy is given by eV where V is the magnitude of the applied voltage. Due to the increase in the barrier height a negligible number of majority carriers will be able to cross the junction and the current will be practically zero. However, the minority carriers which travel down the potential barrier remain unaffected and give a amah current.

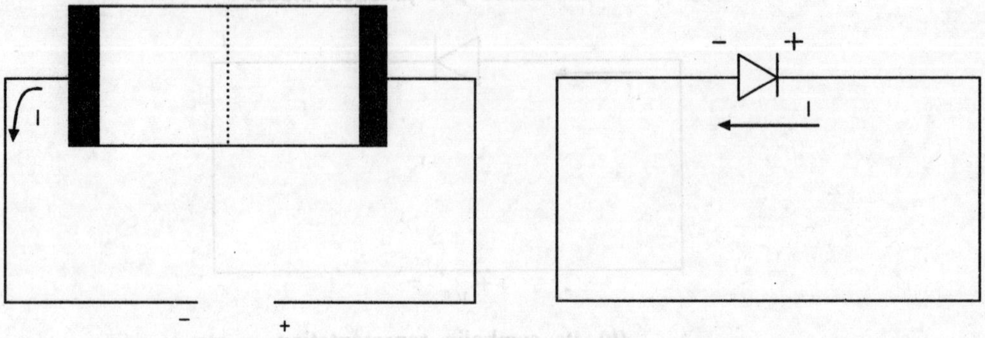

FIGURE 35 ***(a)*** **A reverse-biased *p-n* junction (*b*) Symbolic representation**

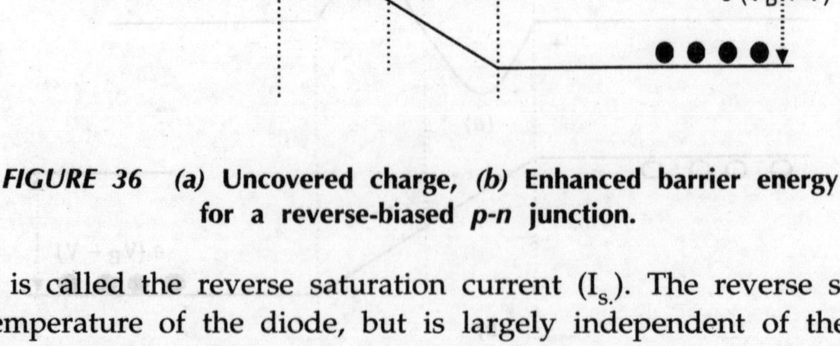

FIGURE 36 ***(a)*** **Uncovered charge, (*b*) Enhanced barrier energy for a reverse-biased *p-n* junction.**

This current is called the reverse saturation current (I_s). The reverse saturation current increases with temperature of the diode, but is largely independent of the reverse voltage applied. A rise of temperature enhances the minority carrier concentration leading to an increase of the reverse saturation current.

The Volt-Ampere Characteristic of a *p-n* Junction

It can be shown that the total current flowing, through a *p-n* junction due to the application of a voltage V across the junction is given by

$$I - I_S \left[\exp\left(\frac{eV}{\eta kT}\right) - 1\right] \quad (1)$$

where

I_S = reverse saturation current,

e = charge of an electron (= 1.6×10^{-19} coulombs),

k = Boltzmann's constant (= 1.38×10^{-23} Joules° K^{-1}),

T = temperature in °K, and η is a numerical constant depending upon the material of the diode. For germanium $\eta = 1$, and for silicon $\eta \approx 2$.

When V is positive the junction is forward-biased and when V is negative the junction is reverse-biased. At room temperature (T = 300° K), we get from equation (1)

$$I - I_S \left[\exp\left(\frac{39\,V}{\eta}\right) - 1\right] \quad (2)$$

Note that the voltage V in equation (2) refers to the voltage drop across the junction. However, this is nearly equal to the voltage applied to the diode since the voltage drops in the bulk *p* and *n* regions are very small. A typical plot of equation (2) is shown in Figure 37.

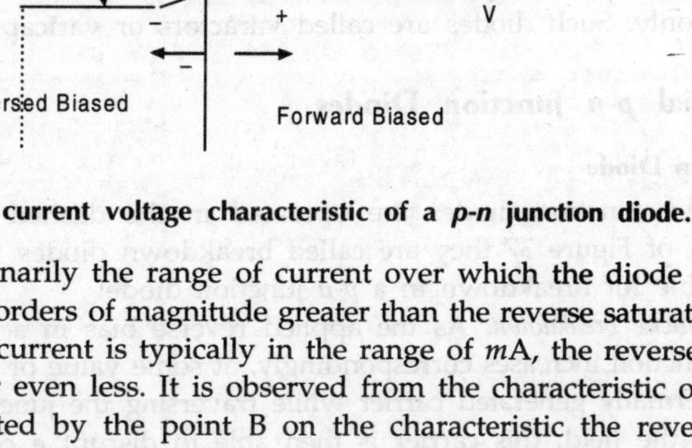

FIGURE 37 **A typical current voltage characteristic of a *p-n* junction diode.**

It should be noted that ordinarily the range of current over which the diode is operated in the forward direction is many orders of magnitude greater than the reverse saturation current. For example, while the forward current is typically in the range of *m*A, the reverse saturation current is in, the range of ηA or even less. It is observed from the characteristic of Figure 37 that at a reverse voltage indicated by the point B on the characteristic the reverse current increases suddenly. The p-n junction diode when operated in the dashed portion of the characteristic is termed a breakdown diode. We shall discuss the breakdown diodes later in this section.

The static or dc resistance (r_{dc}) of a diode is defined as

$$r_{dc} = \frac{V}{I} \quad (3)$$

The *dc* resistance of a diode varies widely with V and I.

The dynamic or *a-c* resistance (r_{ac}) of a diode is defined by

$$r_{dc} = \frac{dV}{dI} \tag{4}$$

When the diode is sufficiently forward-biased, the term unity within parentheses in equation (2) may be neglected. Making this simplification and differentiating equation (2), we get for the diode dynamic resistance in the forward direction at room temperature

$$r_{dc} = \frac{\eta\,1000}{39\,I} = \frac{\eta\,26}{I}, \tag{5}$$

where I is expressed in milliamperes and r_{ac} in ohms. Thus, for $\eta = 1$ and for a diode current of 26 *m*A the dynamic resistance is 1 *ohm*. The slope of the characteristic displayed in Figure 37 shows that the dynamic resistance in the reverse direction is very large.

A noteworthy feature of the diode characteristic is that if the forward bias is below a value V_y the current is very small. Beyond V_y the current rises very sharply. The voltage V_y is referred to as the cutin or offset or threshold voltage of the diode. Typically, for germanium diodes $V_y \approx 0.2$ V and for silicon diodes $V_y \approx 0.6$ V.

Junction Capacitance

It is observed from Figure 31 that the depletion region around a *p-n* junction contains immobile positive charges on the *n*-type side and immobile negative charges on the *p*-type side. Also we note from Figure 36 that application of a reverse voltage to a *p-n* junction causes these uncovered charges to increase. This may be considered as a capacitive effect and the junction is assumed to exhibit a capacitance, called the junction capacitance. This capacitance is not a constant, but may be shown to decrease with the increase in reverse voltage. This property of the *p-n* junction finds a number of applications. They are: (*i*) in voltage tuning of an LC resonant circuit, (*ii*) in self-balancing bridge circuits and (*iii*) in-special type of amplifiers, called parametric amplifiers, in the above applications, *p-n* junction diodes are made to give voltage-variable capacitance only. Such diodes are called varactors or varicaps.

Some Special *p-n* Junction Diodes

(1) Breakdown Diode

When *p-n* junction diodes are operated in the dashed portion of the reverse-voltage characteristic of Figure 37 they are called breakdown diodes. The following two mechanisms are responsible for breakdown in a *p-n* junction diode:

(*i*) *Avalanche breakdown:* As the applied reverse bias in a *p-n* junction increases, the field across the junction increases correspondingly, at some value of the bias, the field becomes large so that a thermally generated carrier while traversing the junction acquires a large amount of energy from the field, this carrier is then able to disrupt a covalent bond and create a new electron-hole pair while colliding with an immobile ion. The new carriers again pick up sufficient energy from the applied field and collide with other immobile ions generating further electron-hole pairs. This process being cumulative in nature results in generation of an avalanche of carriers within a very short time. This mechanism of carrier generation is referred to as avalanche multiplication. The result of the process is the flow of a large amount of current at some value of the reverse bias, as shown by the dashed portion of the characteristic of Figure 37.

(ii) *Zener breakdown:* Zener breakdown takes place when the reverse-bias field across a *p-n* junction is such that the field may exert a strong force on a bound electron to tear it out from a covalent bond. Thus, a large number of electron-hole pairs will be generated through a direct rupture of the covalent bonds. The electron-hole pairs so generated increase the reverse current. Note that in zener breakdown carrier generation is not due to the collisions of carriers with the immobile ions as it happens in the case of avalanche multiplication.

In spite of the two distinct mechanisms breakdown diodes are ordinarily called zener diodes. The symbol for a zener diode is shown in Figure 38. The voltage-current characteristic of a zener diode is given by dashed portion of the characteristic of Figure 37. The zener characteristic is almost parallel to the current axis indicating that the voltage across the diode remains almost constant although the current through it may vary.

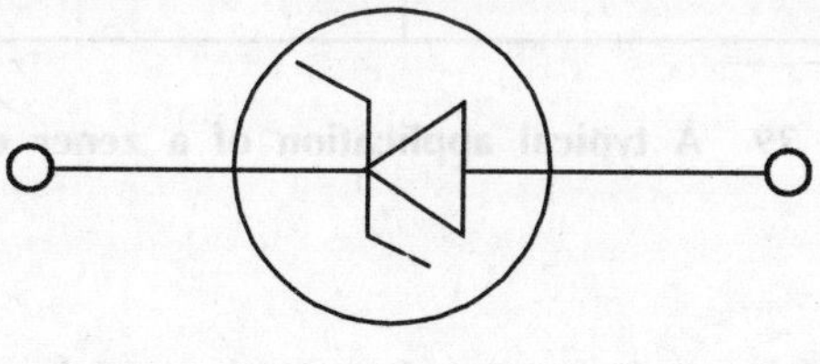

FIGURE 38 **Symbol used for a zener diode.**

The voltage across a zener diode thus serves as a reference and the diode is referred to as *a reference diode*. A typical application of a zener diode as a reference diode is given in Figure 39. The voltage V and the resistance R are determined so that the diode current is within the specified limit and the diode is operated in the breakdown region. The voltage V_o across the load resistance R_L is a constant here although the supply voltage V and the load resistance R_L may vary. The upper limit of the diode current is determined by the power dissipation rating of the diode.

In Figure 39, if *I* be the current drawn from the source, and I_z and I_L be the currents through the zener diode and the load resistance respectively, then Kirchhoff's current and voltage laws give

$$I = I_z + I_L, \qquad (6)$$

And

$$V_o = V - IR. \qquad (7)$$

Also

$$V_o = I_L R_L. \qquad (8)$$

Let the supply voltage V remain constant and the load resistance R_L be varied. Since the zener voltage V_o tends to remain constant, equation (7) gives $\delta I = 0$. We then obtain from equation (6)

$$\delta I = \delta I_z + \delta I_L = 0,$$

or,

$$\delta I_z = -\delta I_L.$$

Thus, when the load resistance is increased but the supply voltage is fixed, the current *h* decreases and the current I_z increases by an equal amount so that the total current *I* remains constant. Now suppose that the load resistance R_L is fixed and the supply voltage V is varied. Since V_o tends to remain constant, we obtain from equation (7)

$$\delta V = R\delta I.$$

We obtain from equation (8) $\delta I_L = 0$, and from equation (6) $\delta I = \delta I_z$ that is, when the supply voltage is altered but the load resistance is kept fixed, the total current I and the zener current I_z change equally to keep the load current I_L constant.

Note that though the zener voltage V_o is fixed, the zener current I_z may vary. Hence the DC resistance of the zener diode, *i.e.*, V_o/I_z is not a constant. The zener dynamic resistance r_z is given by the inverse of the slope of the characteristic, *i.e.*, by $\delta V_o/\delta I_z$. As the zener characteristic is nearly parallel to the current axis, $\delta V_o = 0$, *i.e.*, $r_Z \approx 0$. In practice, r_z has a finite value, but it must be small for a good zener diode.

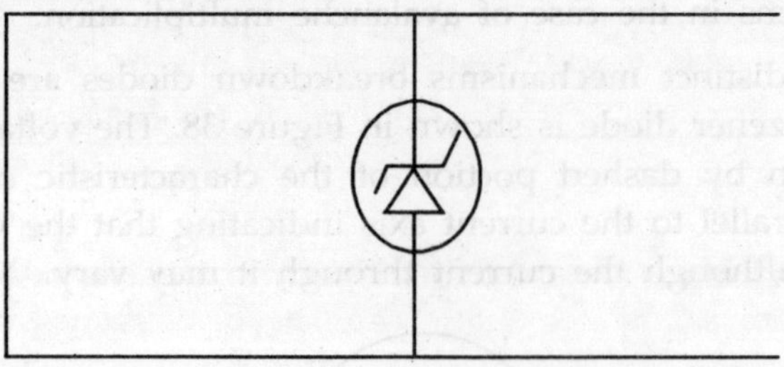

FIGURE 39 **A typical application of a zener diode.**

(2) Tunnel Diode

When the concentration of impurity atoms is very large (about 10^{18} or 10^{19} cm^{-3}) in both *p*-and *n*-regions, the barrier-width of a *p-n* junction diode becomes very small (≈ 100Å). The typical volt-ampere characteristic of such a diode is shown in Figure 40.

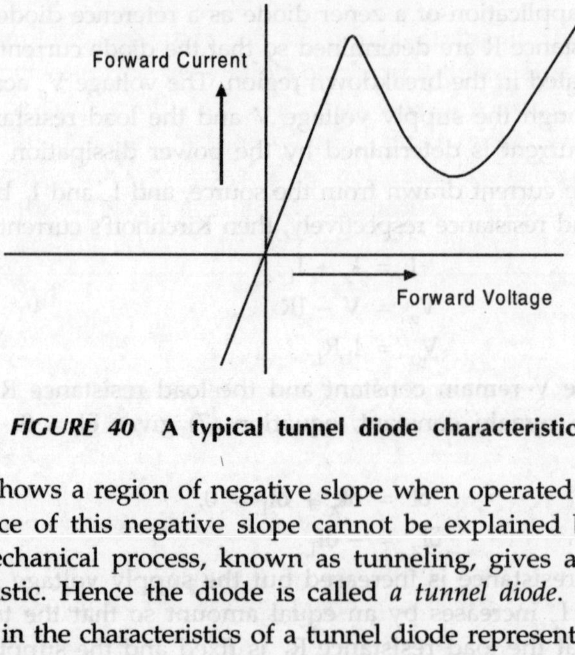

FIGURE 40 **A typical tunnel diode characteristic.**

The characteristic shows a region of negative slope when operated in the forward direction. Note that the occurrence of this negative slope cannot be explained by the mechanism given earlier. A quantum mechanical process, known as tunneling, gives a satisfactory explanation of the above characteristic. Hence the diode is called *a tunnel diode.*

The negative slope in the characteristics of a tunnel diode represents the negative differential resistance. The differential resistance at any point in the volt-ampere characteristic is given by the reciprocal of the slope of the characteristic at that point. If the slope is negative, evidently the differential resistance is negative.

The negative slope, *i.e.*, the negative differential resistance exhibited by a tunnel diode can be utilized in the construction of amplifiers, oscillators or switching devices. Since the tunneling process is very fast tunnel diode devices can work at very high frequencies (about 10 GHz).

(3) Photo Diode

When light is allowed to fall on a reverse-biased p-n junction diode additional electron-hole pairs are created in both *p*-and *n*-regions. This produces a very small change in majority carrier concentration and a very large change in minority carrier concentration. These additional minority carriers enhance the reverse current since they fall down the barrier potential. It is found that the current through the diode varies almost linearly with the light flux. The diodes designed to operate on this principle are called photo diodes. Such diodes are used in light detection, in light operated switches, reading of computer punched cards and tapes etc.

(4) Light-Emitting Diode (LED)

Just as the generation of an electron-hole pair requires the expenditure of energy, so the recombination of an electron with a hole involves the release of energy. In the case of certain semiconductors, such as *GaAs*, when an electron from the conduction band falls into the valence band the energy that is released appears in the form of infrared radiation. In the case of an alloy semiconductor gallium arsenide phosphide ($GaAs_xP_{1-X}$) the emitted radiation is red light. A *p-n* junction diode constructed with such semiconductors is called a light emitting diode (LED) when the diode is forward biased the electrons move into the p-side and encounter a large number of holes. The electron-hole recombination probability is thus increased causing emission of radiation. Diodes built with $GaAs_xP_{1-x}$ are used in the manufacture of signal lamps and displays. The infrared LED is a potential light source for optical-fibre communication.

Under some conditions the emitted light is coherent. Such a diode is referred to as an injection junction laser.

(5) Solar Cell

The solar cell is basically a *p-n* junction diode that converts sunlight directly to electricity with large conversion efficiency. A physical explanation is given below.

Minority carriers are injected in both p-and n-regions of a p-n junction diode when it is exposed to light. When the diode *i*5 unbiased and left open-circuited the current due to these injected minority carriers must be balanced by an equal and opposite current flow due to the minority carriers since the net current through an open-circuited diode is zero. But the flow of additional majority carriers is possible only when the barrier voltage across the junction is reduced. That is, when a *p-n* junction diode is exposed to light a voltage just equal to the amount by which the battier voltage is reduced, is produced across the junction. This voltage gives rise to a current when the diode is connected to an external circuit as long as the junction is exposed to light. The magnitude of the current is proportional to the light intensity.

At present solar cells are used extensively in satellites and space vehicles as most important long duration power supply. Solar cells are constructed with silicon, gallium arsenide, cadmium sulphide and with many other semiconductors and in various device configurations.

Example 1. *The saturation current density of a p-n junction germanium diode is 250 mA/m² at 300° K. Find the voltage that would have to be applied across the junction to cause a forward current density of 10⁵ A/m² to flow.*

Solution. We have

$$I = I_s\left[\exp\left(\frac{eV}{kT}\right)-1\right].$$

Dividing by the area of the diode we get the expression for current density,

$$J = J_s\left[\exp\left(\frac{eV}{kT}\right)-1\right],$$

where J_s is the saturation current density.

Given

$$J_S = 250 \text{ mA/m}^2 \text{ and } J = 10^5 \text{ A/m}^2.$$

$$\therefore \quad \exp\left(\frac{eV}{kT}\right)-1 = \frac{J}{J_S} = \frac{10^5}{250\times 10^{-3}} = 4\times 10^5$$

or, $$\frac{eV}{kT} \approx \log_e 4\times 10^5 = 12.9$$

or, $$V = \frac{12.9\times 1.38\times 10^{-23}\times 300}{1.6\times 10^{-19}}$$

$$= 0.33\text{V}.$$

Example 2. *Find the static and the dynamic resistances of a p-n junction germanium diode if the temperature is 27°C and $I_s = 1\mu A$ for an applied forward bias of 0.2 V.*

Solution. The forward current through the diode is

$$I = I_s\left[\exp\left(\frac{eV}{kT}\right)-1\right].$$

Substituting the values we get

$$I = 1 - 10^{-6}\left[\exp\left(\frac{1.6\times 10^{-19}\times 0.2}{1.38\times 10^{-23}\times 300}\right)-1\right] \text{ A}$$

$$= 2.27 \text{ mA}$$

The static resistance,

$$r_{dc}\ \frac{V}{I} = \frac{V}{I}\ \frac{0.2}{2.27\times 10^{-3}} = 88\ \Omega$$

The dynamic resistance,

$$r_{ac} \approx \frac{26}{I} = \frac{26}{2.27} = 11.4\ \Omega.$$

Example 3. *Refer to Figure 39. Assume that the supply voltage V is 9 volts and that the zener voltage V_o is 6 volts. If the maximum zener current that can safely flow be 20 mA, determine the value of the series resistance R. If a load resistance R_L of 1 kΩ be connected across the zener diode, calculate the load current and the zener current. Also calculate the minimum value of R_L that can be employed.*

Solution. The zener current is a maximum when the load resistance R_L is infinite. Hence

$$R = \frac{V - V_0}{I} = \frac{9-6}{20\times10^{-3}} = 150\Omega$$

The current through the load resistance $R_L = 1\ k\Omega$ is

$$I_L = \frac{V_0}{R_L} = \frac{6}{10^3} A = 6mA.$$

The zener current is

$$I_z = I - I_L = 20 - 6 = 14 \text{ mA}.$$

As R_L is decreased, the current through it increases, and in the limiting case I_L becomes equal to I. *i.e.*, 20 mA hence the minimum value of R_L is

$$(R_L) \min = \frac{6 \text{ volts}}{20\ mA} = 300\ \Omega$$

6.3 THE HALL EFFECT

When a current-carrying conductor is placed in a magnetic field, a potential difference is generated in a direction perpendicular to both the current and the magnetic field. This phenomenon, first observed by Edwin Hall (1855–1938) in 1879, is known as the Hall effect. It arises from the deflection of charge carriers to one side of the conductor as a result of the magnetic force they experience. The Hall effect gives information regarding the sign of the charge carriers and their density; it can also be used to measure the magnitude of magnetic fields.

The arrangement for observing the Hall effect consists of a flat conductor carrying a current *I* in the *x* direction, as shown in Figure 41. A uniform magnetic field B is applied in the *y* direction. If the charge carries are electrons moving in the negative *x* direction with a drift velocity v_d, they experience an upward magnetic force $F_B = qv_d \times B$, are deflected upward, and accumulate at the upper edge of the flat conductor, leaving an excess of positive charge at the lower edge (Figure 42(*b*)). This accumulation of charge at the edges establishes an electric field in the conductor and increases until the electric force on carriers remaining in the bulk of the conductor balances the magnetic force acting on the carriers. When this equilibrium condition is reached, the electrons are no longer deflected upward. A sensitive voltmeter or potentiometer connected across the sample, as shown in Figures 42(*b*) and (*c*), can measure the potential difference—known as the Hall voltage ΔV_H– generated across the conductor.

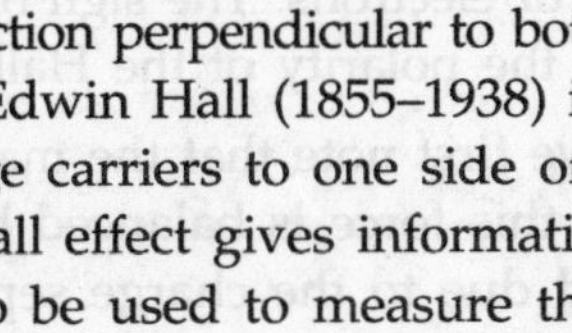

FIGURE 41

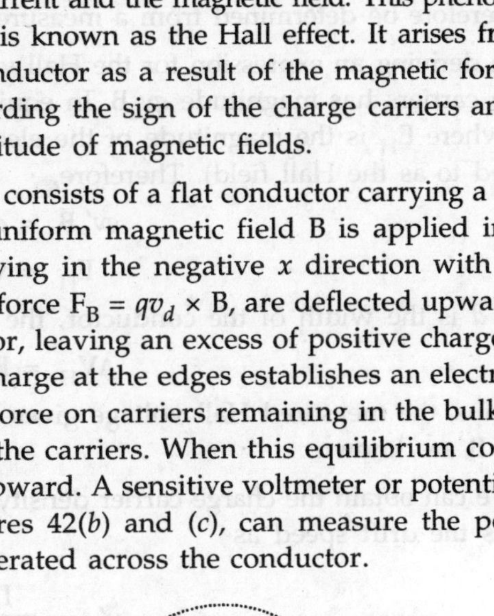

FIGURE 42(a)

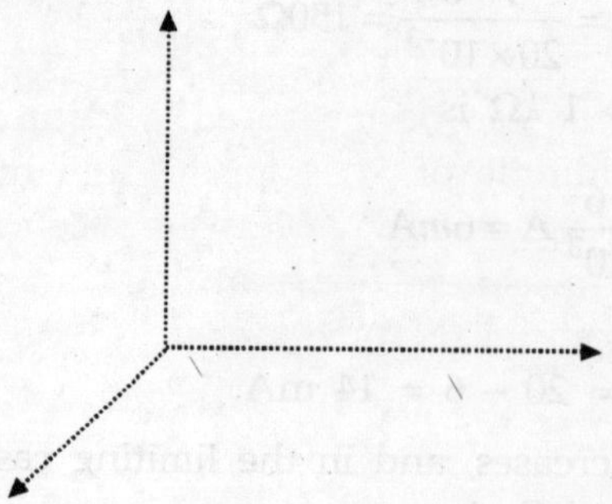

FIGURE 42(b)

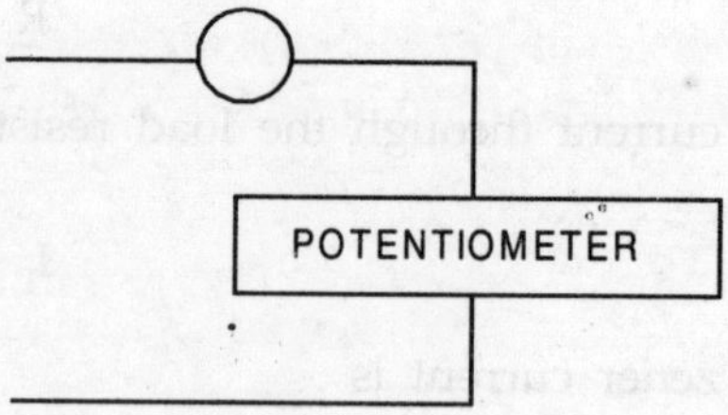

FIGURE 42(c)

If the charge carriers are positive and hence move in the positive x direction (for rightward current), as shown in Figures 42(a) and (c), they also experience an upward magnetic force $qv_d \times B$. This builds-up positive charge on the upper edge and leaves an excess of negative charge on the lower edge. Hence, the sign of the Hall voltage generated in the sample is opposite the sign of the Hall voltage resulting from the deflection of electrons. The sign of the charge carriers can therefore be determined from a measurement of the polarity of the Hall voltage.

In deriving an expression for the Hall voltage, we first note that the magnetic force exerted on the carriers has magnitude qv_dB. In equilibrium, this force is balanced by the electric force qE_H, where E_H is the magnitude of the electric field due to the charge separation (sometimes referred to as the Hall field). Therefore,

$$qv'_dB = qE_H$$

$$E_H = v'_dB$$

If d is the width of the conductor, the Hall voltage is

$$\Delta V_H = E_Hd = v'_dBd \tag{1}$$

Thus, the measured Hall voltage gives a value for the drift speed of the charge carriers if d and B are known.

We can obtain the charge carrier density n by measuring the current in the sample. We can express the drift speed as

$$v'_d = \frac{I}{nqA} \tag{2}$$

Where A is the cross-sectional area of the conductor. Substituting equation 2 into equation 1 we obtain

$$\Delta V_H = \frac{IBd}{nqA} \tag{3}$$

Because A = td, where t is the thickness of the conductor, we can also express equation 3

$$\Delta V_H = \frac{IB}{nqt} = \frac{R_H IB}{t} \tag{4}$$

Where $R_H = 1/nq$ is the Hall coefficient. This relationship shows that a properly calibrated conductor can be used to measure the magnitude of an unknown magnetic field.

Because all quantities in equation 4 other than nq can be measured, a value for the Hall coefficient is readily obtainable. The sign and magnitude of R_H give the sign of the charge carriers and their number density. In most metals, the charge carriers are electrons, and the charge-carrier density determined from Hall-effect measurements is in good agreement with calculated values for such metals as lithium (Li), sodium (Na), copper (Cu), and silver (Ag), whose atoms each give up one electron to act as a current carrier. In this case, n is approximately equal to the number of conducting electrons per unit volume. However, this classical model is not valid for metals such as iron (Fe), bismuth (Bi), and cadmium (Cd) or for semiconductors. These discrepancies can be explained only by using a model based on the quantum nature of solids.

An interesting medical application related to the Hall effect is the electromagnetic blood flow meter, first developed in the 1950s and continually improved since then. Imagine that we replace the conductor in Figures 42(b) and (c) with an artery carrying blood. The blood contains charged ions that experience electric and magnetic forces like the charge carriers in the conductor. The speed of flow of these ions can be related to the volume rate of flow of blood. Solving equation 1 for the speed v'_d of the ions in the blood we obtain

$$v_d = \frac{\Delta V_H}{Bd}$$

Thus, by measuring the voltage across the artery, the diameter of the artery, and the applied magnetic field, the speed of the blood can be calculated.

Example 4. *A rectangular copper strip 1.5 cm wide and 0.10 cm thick carries a current of 5.0 A. Find the Hall voltage for a 1.2- T magnetic field applied in a direction perpendicular to the strip. Also discuss the problem when the strip is of a semiconducting material.*

Solution. If we assume that one electron per atom is available for conduction, we can take the charge carrier density to be 8.49×10^{28} electrons/m^3 substituting this value and the given data into equation 4 gives

$$\Delta V_H = \frac{IB}{nqt} = \frac{(5.0\,\text{A})(1.2\,\text{T})}{(8.49\times10^{28}\ m^{-3})(1.6\times10^{-19}\,\text{C})(0.0010\,m)}$$

$$\Delta V_H = 0.44\ \mu\text{V}$$

Such an extremely small Hall voltage is expected in good conductors. (Note that the width of the conductor is not needed in this calculation,)

In semiconductors, n is much smaller than it is in metals that contribute one electron per atom to the current; hence, the Hall voltage is usually larger because it varies as the inverse of n currents on the order of 0.1 mA are generally used for such materials. Consider a piece of silicon that has the same dimensions as the copper strip in this example and whose value for n is 1.0×10^{20} electrons/m^3. Taking B = 1.2 T and I = 0.10 mA, we find that $\Delta V_H = 7.5\ m$V. A potential difference of this magnitude is readily measured.

Elementary Theory of Hall Effect

Consider a slab of material subject to an external field E_x along x-axis and a magnetic field H_z along z-axis as a result of the applied electric field, a current density I_x will flow in the direction of E_x. For the moment let us suppose that current is carried by electrons of charge $-e$. Under the influence of magnetic field the electrons as discussed above will be subjected to a Lorentz force such that lower surface collects a negative charge and upper surface a positive charge. Ultimately a stationary state is obtained in which current along y-direction vanishes and a field E_y is set up. If the charge carriers were positive the upper surface would become negative and lower surface positive *i.e.*, E_y would be reversed. Thus measurement of Hall voltage gives the information about the sign of charge carriers.

Now electric force on an electron having charge-e is $-eE$

Force due to magnetic field H is $(-e/c)\ v \wedge H$

Hence, total force on an electron is

$$F = -e\left[\vec{E} + \frac{1}{c}\vec{v} \wedge \vec{H}\right]$$

But in the present case

$$E_y = -e\left[E_y - \frac{1}{c}v_x H_z\right] \tag{5}$$

Hall Voltage and Hall Coefficient

In the steady state, the force due to accumulation of electrons becomes equal to the magnetic force and so the flow of electrons stops *i.e.*, $E_y = 0$. Substituting this condition in equation 5, we get

$$0 = -e\left[E_y - \frac{1}{c}v_x H_z\right] \text{ or } E_y = ? \tag{6}$$

The E_y is called Hall voltage, where v_x is average drift velocity.

Further, current density may be expressed in terms of number of electrons n per unit volume

$$I_y = -n\,e\,v_x \quad \text{or} \quad v_x = \frac{I_x}{ne} \quad \text{[negative sign]}$$

Substituting this value of v_x in equation 6, we get

$$E_y = \frac{1}{c}\frac{I_x}{ne}.H_z \text{ [negative sign]}$$

or

$$\frac{E_y}{I_x.H_z} = R_H = -\frac{1}{nec} = R_H = -\frac{1}{nec} \quad \text{(7) [space between '}n\ e\ c\text{']}$$

where R_H is known as Hall coefficient. This expression for Hall coefficient is in *e.s.u.* and in *e.m.u.* it will be

$$R_H = -\frac{1}{ne}$$

The Hall coefficient is determined essentially by the sign of charge carriers. It is to be noted that the sign of the Hall coefficient is the same as the sign of the carrier. Thus it is negative if the conduction is by electrons. If our calculation has been based upon holes, the sign of the charge entering the Lorentz force would have been positive and corresponding Hall coefficient would have been positive.

Mobility and Hall Angle

When current carrying particles acquire a velocity per unit electric field; the velocity is known as mobility and is represented by μ *i.e.,*

$$\mu = \frac{v_x}{E_x} \quad \text{or} \quad v_x = \mu E_x \tag{8}$$

The value of v_x in equation 6 given by $E_y = \frac{1}{c} v_x H_z$ we get

$$E_y = \frac{1}{c}\mu E_x . H_z . \tag{9}$$

Equating the two values of E_y from equations 7 and 9, we get

$$R_H \cdot I_x \cdot H_z = \frac{1}{c}\mu E_x . H_z . \quad \text{or} \quad \mu = \frac{R_H \cdot I_x}{E_x} . c$$

But since $(I_x/E_x) = \sigma$, the electrical conductivity,

$$\therefore \quad \mu = R_H . \sigma\, c \text{ in } e.s.u. \tag{10}$$

and

$$\mu = R_H . \sigma\, c \text{ in e.m.u.}$$

also

$$\mu = R_H . \sigma \tag{11}$$

$$= \frac{E_y}{I_x . H_z} . \sigma \quad \text{from (7)}$$

or

$$\mu = \frac{E_y}{\sigma E_x \cdot H_z} . \sigma = \frac{E_y}{E_x} \frac{1}{H_z} = \phi = \frac{1}{H_z} = \phi = \frac{E_y}{E_x} \text{ is called}$$

Hall angle.

$$\therefore \quad \phi = \mu H_z \tag{12}$$

Importance of the Hall effect:

1. The sign of the current carrying charge is determined.
2. The number of charge carriers per unit volume can be calculated from the Hall coefficient R_H.
3. The mobility is measured directly.
4. It can be used to determine the electronic structure of the substance *i.e.,* whether these are metals, semiconductors or insulators.

5. The knowledge of Hall voltage developed enables us to measure high unknown magnetic field provided we know the Hall constant for the slab used for it.
6. It gives the concept of negative mass as suggested by Bleaney

Experimental Determination of Hall Coefficient:

We have

$$E_y = \frac{1}{c} \cdot \frac{I_x}{ne} H_z$$

If i be the total current then

$$I_x = \frac{i}{bxd}$$

where b and d are the breadth and thickness of the specimen.

Thus,

$$E_y = \frac{i}{nebd} H_z$$

or

$$E_y\, d = V_H = -\frac{1}{c} \cdot \frac{1}{neb} H_z$$

$$= R_H \cdot \frac{1}{b} . H_z$$

or

$$R_H = \frac{b}{i} \cdot \frac{V_H}{H_z} \tag{13}$$

where V_H is the Hall voltage.

Thus the value of Hall coefficient can be calculated by knowing i and H_z and by measuring experimentally the Hall voltage, V_H developed across C and D which is measured either by sensitive galvanometer or better by means of a calibrated potentiometer. For accuracy of the measurement the length of the specimen should be large as compared to its thickness.

Sources of Error and their Elimination

1. It is usually found that even in the absence of a magnetic field there is a potential difference between C and D owing to their alignment being imperfect. This is eliminated by reversing the magnetic field and measuring the potential difference again between C and D.
2. Errors may also be introduced if a temperature gradient is present. These are eliminated by reversing the current and taking another pair of readings with the magnetic field normal and reversed.
3. Averaging these four readings eliminates all errors except that due to Ettingshausen effect. However this is so small that it can be neglected.

Comparison between observed Hall coefficient R_H and those calculated on free electron theory can be done from the following table. $R_H \times 10^{11}$ (volt-cm./amp-oersted) at room temperature.

Metal	Observed	Calculated
Li	− 17.0	− 13.1
Na	− 25.0	− 24.4
K	− 42.0	− 47.0
Cs	− 78.0	− 73.0
Zn	+ 3.3	− 4.6
Cd	+ 6.0	− 6.5

Observed values of the Hall coefficient for several metals are compared in the above table with values directly from the concentration of valence electrons. For monovalent metals, the agreement between the observed and calculated values is quite satisfactory; however the sign of the effect in zinc and cadmium is opposite to that predicted for electrons. This situation represented a famous unsolved problem until clarified by band theory.

Uses of Hall Effect

1. Determination of Semiconductor Type

For an *n*-type semiconductor the Hall coefficient is negative whereas for a *p*-type semiconductor it is positive. Thus, the sign of the Hall coefficient can be used to determine whether a given semiconductor is *n*-or *p*-type.

2. Determination of Carrier Concentration

By measuring the Hall coefficient the carrier concentration of a semiconductor can be determined from the relations.

$$n = \left|\frac{1}{eR_H}\right| \text{ and } p = \frac{1}{R_H e}$$

3. Determination of Mobility

If the conduction is due to one type of carriers, e.g., electrons we have

$$\sigma = ne\mu, \tag{14}$$

where μ is the mobility of the electrons.

Also,

$$\mu = \sigma \left| R_H \right|, \tag{15}$$

i.e., knowing σ and determining R_H, the mobility μ can be determined. The mobility as determined from equation 15 is, therefore, called the Hall mobility to distinguish it from the actual drift mobility. However, since the factor *r* is not much different from unity the two mobilities are nearly the same.

4. Measurement of Magnetic Flux Density

Since the Hall voltage V_H is proportional to the magnetic flux density B for a given current I through a sample, the Hall effect can be used as the basis for the design of a magnetic flux density meter.

5. Measurement of Power in an Electromagnetic Wave

In an electromagnetic wave in free space the magnetic field H and the electric field ε are at right angles. Thus, if a semiconductor sample is placed parallel to ε it will drive a current I in the semiconductor. The semiconductor is subjected simultaneously to a transverse magnetic field H producing a Hall voltage across the sample. The Hall voltage will be proportional to the product ε H, *i.e.*, to the magnitude of the Poynting vector of the electromagnetic wave. Thus, the Hall effect can be used to determine the power flow in an electromagnetic wave.

6. Hall Effect Multiplier

If the magnetic flux density B is produced by passing a current I' through an air core coil, B will be proportional to I'. The hall voltage is thus proportional to the product of I and I'. This forms the basis of a multiplier.

Advantages of Semiconductor Devices

The present age is the age of semiconductors. In fact, most of the devices in electronics to day are built from semiconductors. The vacuum devices like diode, triode etc., have now a days been replaced by semiconductor devices like p-n junction diodes, transistors etc. The advantages of semiconductor devices over vacuum tubes are the following:

1. As the name indicates vacuum tubes require vacuum, but semiconductor devices do not require vacuum.
2. Vacuum tubes have filaments, and so require filament power. Semiconductor devices do not have filaments and hence no filament power is required.
3. Semiconductor devices are smaller in size, lighter in weight, and are mechanically very rugged.
4. Operation of semiconductor devices requires low voltage power supply. The power consumption is also smaller than that of the corresponding vacuum tubes.
5. The semiconductor devices require very small warm up time and, therefore, operate immediately after the supply voltages are switched on.
6. The performance of semiconductor devices is more reliable than that of vacuum tube devices.
7. The operating life of a semiconductor devices is appreciably longer that that of its equivalent vacuum tube.

Example 5. *The following data are given for intrinsic germanium at 300°K:*

$$n_1 = 2.4 \times 10^{19}\ m^{-3},$$

$$\mu_n = 0.39\ m^2 v^{-1} s^{-1},$$

and

$$\mu_p = 0.19\ m^2 v^{-1} s^{-1}.$$

Calculate the conductivity of intrinsic germanium.

Solution. We know

$$\sigma_1 = n_1 e(\mu_n + \mu_p)$$

Substituting the values we get

$$\sigma_1 = 2.4 \times 10^{19} \times 1.6 \times 10^{-19} \times (0.39 + 0.19)$$

$$= 2.22\ \mho/m$$

[The symbol $\mho$ stands for reciprocal of ohm (Ω) and is called mho]

Example 6. *Determine the number density of donor atoms which has to be added to an intrinsic Ge semiconductor to produce an n-type semiconductor of conductivity 5/cm. Given that the mobility of electrons in the n-type semiconductor is 3850 cm²/volt, sec.*

Solution. Assuming that the concentration of donor atoms (N_d) is equal to the concentration of free electrons (n), we have

$$\sigma_n = ne\mu_n = N_d e\mu_n$$

Given $\sigma_n = 5\mho/\text{cm}$, $\mu_n = 3850\ \text{cm}^2/\text{volt sec}$,

and $e = 1.6 \times 10^{-19}$ coulombs

$$\therefore \quad n = N_d = \sigma_n/e\mu_n = \frac{5}{1.6 \times 10^{-19} \times 3850} = 0.81 \times 10^{16}/\text{cm}^3$$

Example 7. *The bandgap of an alloy semiconductor gallium arsenide phosphide is 1.98 eV. Calculate the wavelength of radiation that is emitted when electrons and holes in this material recombine directly. Mention the colour of the emitted radiation.*

Solution. We have

$$\lambda = \frac{ch}{E_g}$$

Here $E_g = 1.98\ eV = 1.98 \times 1.6 \times 10^{-19}$ joules

Also $h = 6.6 \times 10^{-34}$ joule sec, and $c = 3 \times 10^8\ m/\text{sec}$.

Hence

$$\lambda = \frac{3 \times 10^8 \times 6.6 \times 10^{-34}}{1.98 \times 1.6 \times 10^{-19}}\ m = 6.25 \times 10^{-7}\ m$$

$$= 6250\ \text{A}$$

The colour of the emitted radiation is red.

Example 8. *An n-type germanium sample has a donor density of $10^{21}\ m^{-3}$. It is arranged in a Hall effect experiment having B = 0.5 webers/m² and J = 500 amps/m². Find the Hall voltage if the sample is 3 mm thick.*

Solution. We have

$$V_H = R_H \, BdJ = \left| \frac{Dd\,J}{ne} \right| .$$

Substituting the values we get

$$V_H = \frac{0.5 \times 0.3 \times 10^{-2} \times 500}{10^{21} \times 1.6 \times 10^{-19}}$$

$$= 4.7 \ mV \text{ (approximately).}$$

Emergence of Nanotechnology

Nanotechnology is new, but research on nanometer scale is not new at all. The study of biological systems and the engineering of many materials such as colloidal dispersions, metallic quantum dots, and catalysts have been in the nanometer regime for centuries. What has changed recently is an explosion in our ability to image, engineer and manipulate systems in the nanometer scale. What is really new about nanotechnology is the combination of our ability to see and manipulate matter on the nanoscale and our understanding of atomic scale interactions.

Although study on materials in the nanometer scale can be traced back for centuries, the current fever of nanotechnology is at least partly driven by the ever shrinking of devices in the semiconductor industry and supported by the availability of characterization and manipulation techniques at the nanometer level. The continued decrease in device dimensions has followed the well-known Moore's law predicted in 1965.

The continued size shrinkage of transistors will sooner or later meet with the limitations of the materials' fundamentals. For example, the widening of the band gap of semiconductors occurs when the size of the materials reaches de Broglie's wavelength.

Promising applications of nanotechnology in the practice of medicine, often referred to as nanomedicine, have attracted a lot of attention and have become a fast growing field. One of the attractive applications in nanomedicine is the creation of nanoscale devices for improved therapy and diagnostics. Such nanoscale devices are known as nanorobots or more simply as nanobots. These nanobots have the potential to serve as vehicles for delivery of therapeutic agents, detectors or guardians against early disease and perhaps repair of metabolic or genetic defects.

Studies in nanotechnology are not limited to miniaturization of devices.

Materials in nanometer scale may exhibit unique physical properties and have been explored for various applications. For example, gold nanoparticles have found many potential applications using surface chemistry and its uniform size. Au nanoparticles can function as carrier vehicles to accommodate multiple functionalities through attaching various functional organic molecules or bio-components. Bandgap engineered quantum devices, such as lasers and heterojunction bipolar transistors, have been developed with unusual electronic transport and optical effects. The discovery of synthetic materials, such as carbon fullerenes, carbon nanotubes,22 and ordered mesoporous materials,23 has further fueled the research in nanotechnology and nanomaterials.

The invention and development of Scanning Tunneling Microscopy (STM) in the early 1980's and subsequently other Scanning Probe Microscopy (SPM) such as Atomic Force Microscopy (AFM) have opened up new possibilities for the characterization, measurement and manipulation of nanostructures and nanomaterials. Combining with other Well-developed characterization and measurement techniques such as Transmission Electron Microscopy (TEM), it is possible to study and manipulate the nanostructures and nanomaterials to a great detail and often down to the atomic level. Nanotechnology is already all around us if you know where to look. This technology is not new, it is the combination of existing technologies and our new found ability to observe and manipulate at the atomic scale, this makes nanotechnology so compelling from scientific, business and political viewpoints.

Bottom-up and Top-down Approaches

Obviously there are two approaches to the synthesis of nanomaterials and the fabrication of nanostructures: top-down and bottom-up. Attrition or milling is a typical top-down method in making nanoparticles, whereas the colloidal dispersion is a good example of bottom-up approach in the synthesis of nanoparticles. Lithography may be considered as a hybrid approach, since the growth of thin films is bottom-up whereas etching is top-down, while nanolithography and nanomanipulation are commonly a bottom-up approach. Both approaches play very important roles in modern industry and most likely in nanotechnology as well. There are advantages and disadvantages in both approaches.

Among others, the biggest problem with top-down approach is the imperfection of the surface structure. It is well known that the conventional top-down techniques such as lithography can cause significant crystallographic damage to the processed and additional defects may be introduced even during the etching steps. For example, nanowires made by lithography is not smooth and may contain a lot of impurities and structural defects on surface. Such imperfections would have a significant impact on physical properties and surface chemistry of nanostructures and nanomaterials, since the surface over volume ratio in nanostructures and nanomaterials is very large. The surface imperfection would result in a reduced conductivity due to inelastic surface scattering, which in turn would lead to the generation of excessive heat and thus impose extra challenges to the device design and fabrication. Regardless of the surface imperfections and other defects that top-down approaches may introduce, they will continue to play an important role in the synthesis and fabrication of nanostructures and nanomaterials.

Bottom-up approach is often emphasized in nanotechnology literature, though bottom-up is nothing new in materials synthesis. Typical material synthesis is to build atom by atom on a very large scale, and has been in industrial use for over a century. Examples include the production of salt and nitrate in chemical industry, the growth of single crystals and deposition of films in electronic industry. For most materials, there is no difference in physical properties of materials regardless of the synthesis routes, provided that chemical composition, crystallinity, and microstructure of the material in question are identical. Of course, different synthesis and processing approaches often result in appreciable differences in chemical composition, crystallinity, and microstructure of the material due to kinetic reasons. Consequently, the material exhibits different physical properties.

Bottom-up approach refers to the build-up of a material from the bottom: atom-by-atom, molecule-by-molecule, or cluster-by-cluster. In organic chemistry andlor polymer science, we know polymers are synthesized by connecting individual monomers together. In crystal growth, growth species, such as atoms, ions and molecules, after impinging onto the growth surface, assemble into crystal structure one after another.

Although the bottom-up approach is nothing new, it plays an important role in the fabrication and processing of nanostructures and nanomaterials. There are several reasons for this. When structures fall into a nanometer scale, there is little choice for a top-down approach. All the tools we have possessed are too big to deal with such tiny subjects.

Bottom-up approach also promises a better chance to obtain nanostructures with less defects, more homogeneous chemical composition, and better short and long range ordering. This is because the bottom-up approach is driven mainly by the reduction of Gibbs free energy, so that nanostructures and nanomaterials such produced are in a state closer to a thermodynamic equilibrium state. On the contrary, top-down approach most likely introduces internal stress, in addition to surface defects and contaminations.

Physical Chemistry of Solid Surfaces

Nanostructures and nanomaterials possess a large fraction of surface atoms per unit volume. The ratio of surface atoms to interior atoms changes dramatically if one successively divides a macroscopic object into smaller parts. For example, for a cube of iron of 1 cm3, the percentage of surface atoms would be only when the cube is divided into smaller cubes with an edge of 10 nm, the percentage of the surface atoms would increase to 10%. In a cube of iron of 1 nm3, every atom would be a surface atom.

The total surface energy increases with the overall surface area, which is in turn strongly dependent on the dimension of material. It should be noted that the specific surface area and the total surface energy are negligible when cubes are large, but become significant for very small particles. When the particles change from centimeter size to nanometer size, the surface area and the surface energy increase seven orders of magnitude. Due to the vast surface area, all nanostructured materials possess a huge surface energy and, thus, are thermodynamically unstable or metastable. One of the great challenges in fabrication and processing of nanomaterials is to overcome the surface energy, and to prevent the nanostructures or nanomaterials from growth in size, driven by the reduction of overall surface energy. In order to produce and stabilize nanostructures and nanomaterials, it is essential to have a good understanding of surface energy and surface physical chemistry of solid surfaces. In this chapter, the origin of the surface energy will be reviewed first, followed with detailed discussion of the possible mechanisms for a system or material to reduce the overall surface energy. Then attention will be focused on the chemical potentials as a function of surface curvature and its implications. Finally, two mechanisms to prevent the agglomeration of nanomaterials will be discussed.

Surface Energy

Atoms or molecules on a solid surface possess fewer nearest neighbors or coordination numbers, and thus have dangling or unsatisfied bonds exposed to the surface. Because of the dangling bonds on the surface, surface atoms or molecules are under an inwardly directed force and the bond distance between the surface atoms or molecules and the sub-surface atoms or molecules, is smaller than that between interior atoms or molecules.

When solid particles are very small, such a decrease in bond length between the surface atoms and interior atoms becomes significant and the lattice constants of the entire solid particles show an appreciable reduction. The extra energy possessed by the surface atoms is described as surface energy, surface free energy or surface tension. Surface energy by definition, is the energy required to create a unit area of "new" surface.

On the newly created surfaces, each atom is located in an asymmetric environment and will move towards the interior due to breaking of bonds at the surface. An extra force is required to pull the surface atoms back to its original position.

Such a surface is ideal and also called singular surface. For each atom on such a singular surface, the energy required to get it back to its original position will be equal to the number of broken bonds, *Nb*, multiplying by half of the bond strength, E. Therefore, the surface energy is dependent on the surface atomic density, the number of atoms per unit area on the new surface.

There is a strong tendency for a solid or a liquid to minimize the total surface energy. There are a variety of mechanisms to reduce the overall surface energy. The various mechanisms can be grouped into atomic or surface level, individual structures and the overall system. For a given surface with a fixed surface area, the surface energy can be reduced through

(*i*) surface relaxation, the surface atoms or ions shift inwardly which occur more readily in liquid phase than in solid surface due to rigid structure in solids.

(*ii*) surface restructuring through combining surface dangling bonds into strained new chemical bonds.

(*iii*) surface adsorption through chemical or physical adsorption of terminal chemical species onto the surface by forming chemical bonds or weak attraction forces such as electrostatic or van der Waals forces.

(*iv*) composition segregation or impurity enrichment on the surface through solid-state diffusion.

Zero-Dimensional Nanostructures: Nanoparticles

Nanoparticles can be defined as particles ranging from a couple of tens to several hundreds nanometers in diameter. In nanotechnology, a particle is defined as a small object that behaves as a whole unit in terms of its transport and properties. It is further classified according to size: In terms of diameter, fine particles cover a range between 100 and 2500 nanometers, while ultrafine particles, on the other hand, are sized between 1 and 100 nanometers. Similarly to ultrafine particles, nanoparticles are sized between 1 and 100 nanometers, though the size limitation can be restricted to two dimensions. Nanoparticles may or may not exhibit size-related intensive properties that differ significantly from those observed in fine particles or bulk materials.

The fabrication of nanoparticles, a small size is not the only requirement, for any practical application, the processing conditions need to be controlled in such a way that resulting nanoparticles have the following characteristics: (*i*) identical size of all particles (also called monosized or with uniform size distribution), (*ii*) identical shape or morphology, (*iii*) identical chemical composition and crystal structure that are desired among different particles and within individual particles, such as core and surface composition must be the same, and (*iv*) individually dispersed or monodispersed, *i.e.*, no agglomeration. If agglomeration does occur, nanoparticles should be readily redispersible.

Properties

Nanoparticles are of great scientific interest as they are effectively a bridge between bulk materials and atomic or molecular structures. A bulk material should have constant physical properties regardless of its size, but at the nano-scale this is often not the case. Size-dependent properties are observed due to which the properties of materials change as their size approaches the nanoscale and as the percentage of atoms at the surface of a material becomes significant. For bulk materials larger than one micrometer the percentage of atoms at the surface is minuscule relative to the total number of atoms of the material.

Nanoparticles exhibit a number of special properties relative to bulk material. For example, the bending of bulk copper (wire, ribbon, etc.) occurs with movement of copper atoms/clusters at about the 50 nm scale. Copper nanoparticles smaller than 50 nm are considered super hard materials that do not exhibit the same malleability and ductility as bulk copper. The change in properties is not always desirable. Ferroelectric materials smaller than 10 nm can switch their magnetization direction using room temperature thermal energy, thus making them useless for memory storage.

Moreover nanoparticles have been found to impart some extra properties to various day to day products. Nano Zinc Oxide particles have been found to have superior UV blocking properties compared to its bulk substitute. This is one of the reasons why it is often used in the sunscreen lotions. Clay nanoparticles when incorporated into polymer matrices increase re-inforcement, leading to stronger plastics, verified by a higher glass transition temperature and other mechanical property tests. These nanoparticles are hard, and impart their properties to the polymer (plastic). Nanoparticles have also been attached to textile fibers in order to create smart and functional clothing.

Some of the Nanoparticles are:

Quantum dot

A quantum dot is a nano-particulate semiconductor, whose excitons are confined in all three spatial dimensions. As a result, they have properties that are between those of bulk semiconductors and those of discrete molecules Colloidal quantum dots irradiated with a UV light. Different sized quantum dots emit different color light due to quantum confinement. Researchers have studied quantum dots in transistors, solar cells, LEDs, and diode lasers. They have also investigated quantum dots as agents for medical imaging and hope to use them as qubits. Some quantum dots are commercially available.

Quantum confinement in semiconductors

In an unconfined (bulk) semiconductor, an electron-hole pair is typically bound within a characteristic length called the Bohr exciton radius. If the electron and hole are constrained further, then the semiconductor's properties change. This effect is a form of quantum confinement, and it is a key feature in many emerging electronic structures.

Other quantum confined semiconductors include:

1. quantum wires, which confine electrons or holes in two spatial dimensions and allow free propagation in the third.
2. quantum wells, which confine electrons or holes in one dimension and allow free propagation in two dimensions.

Colloidal gold

Colloidal gold, or Aqueous colloidal gold also known as "nanogold", is a suspension (or colloid) of sub-micrometre-sized particles of gold in a fluid—usually water. The liquid is usually either an intense red colour (for particles less than 100 nm), or a dirty yellowish colour (for larger particles). The nanoparticles themselves can come in a variety of shapes. Spheres, rods, cubes, and caps are some of the more frequently observed ones.

Known since ancient times, the synthesis of colloidal gold was originally used as a method of staining glass. Modern scientific evaluation of colloidal gold did not begin until Michael Faraday's work of the 1850s. Due to the unique optical, electronic, and molecular-recognition properties of gold nanoparticles, they are the subject of substantial research, with applications in a wide variety of areas, including electronics, nanotechnology, and the synthesis of novel materials with unique properties.

Colloidal silver

Colloidal silver is a liquid suspension of microscopic particles of silver. A colloid is technically defined as particles which remain suspended without forming an ionic, or dissolved solution. The broader commercial definition of "colloidal silver" includes products that contain various concentrations of ionic silver, silver colloids, ionic silver compounds or silver proteins in purified water. Colloidal silver with concentrations of 30 parts per million (ppm) or less are typically manufactured using an electrolysis process, whereas colloidal silver with higher concentrations of 50 ppm or more are usually either silver compounds such as silver chloride and silver iodide or are solutions that have been bound with a protein to disperse the particles. Silver ions are reported to kill bacteria via the oligodynamic effect by inhibiting the expression of enzymes and other proteins essential to ATP production.

Nanoscale iron particles

Environmental contaminants cover U.S. grounds. These contaminants include polychlorinated biphenyls (PCBs), chlorinated organic solvents, and organochlorine pesticides. The EPA reported that more than 1500 sites are polluted and in the last decade, less than a third of them have been cleaned up. The current clean up procedure for these contaminants are costly and limited. Much environmental research has been dedicated to less costly alternatives. One alternative that has proven to be successful in laboratory research as well as field test studies is the use of nanoscale iron particles.

Nanoscale iron particles

Nanoscale iron particles are small in size, usually 1–100 nm, making them highly reactive because of their large surface areas. Furthermore, they are easily transportable through ground water. This attribute allows in situ treatment. The nanoparticle-water slurry can be injected into the contaminated area and stay there for long periods of time.

Platinum nanoparticles

Platinum nanoparticles are usually in the form of a suspension or colloid[1] of sub-micrometre-sized particles of platinum in a fluid, usually water. A colloid is technically defined as particles which remain suspended without forming an ionic, or dissolved solution. The broader commercial definition of "colloidal platinum" includes products that contain various concentrations of ionic platinum, platinum colloids, ionic platinum compounds or platinum nanoparticles in purified water.

The platinum nanoparticle sizes range between 2–3 nanometres (nm). Trillions of platinum nanoparticles are suspended in the brownish red or black colored colloidal solution. Nanoparticles come in wide variety of shapes including spheres, rods, cubes, and caps.

Due to the antioxidant properties of the platinum nanoparticles, they are the subject of substantial research with applications in a wide variety of areas, including nanotechnology, medicine and the synthesis of novel materials with unique properties.

One-Dimensional Nanostructures: Nanowires and Nanorods

One-dimensional nanostructures have been called by a variety of names including: whiskers, fibers or fibrils, nanowires and nanorods. Nanowires are extremely narrow threads (less than 50 nm wide). A nanowire is a wire of diameter of the order of a nanometer (10^{-9} meters). Alternatively, nanowires can be defined as structures that have a lateral size constrained to tens of nanometers or less and an unconstrained longitudinal size. At these scales, quantum mechanical effects are important—hence such wires are also known as "quantum wires". Many different types of nanowires exist, including metallic (e.g., Ni, Pt, Au), semiconducting (e.g., Si, InP, GaN, etc.), and insulating (e.g., SiO_2,TiO_2). Molecular nanowires are composed of repeating molecular units either organic (e.g., DNA) or inorganic (e.g., $Mo_6S_{9-x}I_x$). The nanowires could be used, in the near future, to link tiny components into extremely small circuits. Using nanotechnology, such components could be created out of chemical compounds.They have potential to be used in nanoscale electrical devices. The vision is of electronic chips so small and cheap that they could be used in almost any way. In biology, they could form the heart of extremely sensitive biosensors, identifying molecules associated with disease or the binding of chemicals to a drug target. Nanorod in nanotechnology, nanorods are one morphology of nanoscale objects. Each of their dimensions range from 1–100 nm. They may be synthesized from metals or semiconducting materials. Standard aspect ratios (length divided by width) are 3–5. Nanorods are produced by direct chemical synthesis. A combination of ligands act as shape control agents and bond to different facets of the nanorod with different strengths. This allows different faces of the nanorod to grow at different rates, producing an elongated object. The applications of nanorods are diverse, ranging from display technologies (the reflectivity of the rods can be changed by changing their orientation with an applied electric field) to microelectromechanical systems (MEMS).

In many cases, nanotubules and nanocables are also considered one-dimensional structures. Although whiskers and nanorods are in general considered to be shorter than fibers and nanowires, the definition is often a little arbitrary. In addition, one-dimensional structures with diameters ranging from several nanometers to several hundred microns were referred to as whiskers and fibers, whereas nanowires and nanorods with diameters not exceeding a few hundred nanometers are used predominantly in the recent literature. Many fundamental understandings and techniques of growth of one-dimensional nanostructures are based on the early work on the growth of whiskers and fibers, *albeit* with less emphasis on nanometer scale. Many techniques have been developed in the synthesis and formation of one-dimensional nanostructured materials. These techniques can be generally grouped into four categories:

(1) Spontaneous growth:
  (*a*) Evaporation (or dissolution)-condensation
  (*b*) Vapor (or solution)-liquid-solid (VLS or SLS) growth
  (*c*) Stress-induced recrystallization

(2) Template-based synthesis:
  (*a*) Electroplating and electrophoretic deposition
  (*b*) Colloid dispersion, melt, or solution filling
  (*c*) Conversion with chemical reaction

(3) Electrospinning

(4) Lithography

Spontaneous growth, template-based synthesis and electrospinning are considered as a bottom-up approach, whereas lithography is a top-down technique. Spontaneous growth commonly results in the formation of single crystal nanowires or nanorods along a preferential crystal growth direction depending on the crystal structures and surface properties of the nanowires materials. Template-based synthesis mostly produces polycrystalline or even amorphous products.

Two-Dimensional Nanostructures: Thin Films

Deposition of thin films has been a subject of intensive study for almost a century, and many methods have been developed and improved. Many such techniques have been developed and widely used in industries, which in turn provides a great driving force for further development and improvement of the deposition techniques. Film growth methods can be generally divided into two groups: vapor-phase deposition and liquid-based growth. The former includes, for example, evaporation, molecular beam epitaxy (MBE), sputtering, chemical vapor deposition (CVD), and atomic layer deposition (ALD). Examples of the latter are electrochemical deposition, chemical solution deposition (CSD), Langmuir-Blodgett films and self-assembled monolayers (SAMs).

The film deposition involves predominantly heterogeneous processes including heterogeneous chemical reactions, evaporation, adsorption and desorption on growth surfaces, heterogeneous nucleation and surface growth. In addition, most film deposition and characterization processes are conducted under a vacuum. Therefore, in this chapter, before discussing the details of various methods for thin film deposition and growth, a brief discussion will be devoted to the fundamentals of heterogeneous nucleation followed by a general introduction to vacuum science and technology.

Other aspects of heterogeneous processes and relevant vacuum issues will be incorporated into the various deposition methods where the subject is relevant.

Fundamentals of Film Growth

Growth of thin films, as all phase transformation, involves the processes of nucleation and growth on the substrate or growth surfaces. The nucleation process plays a very important role in determining the crystallinity and microstructure of the resultant films. For the deposition of thin films with thickness in the nanometer region, the initial nucleation process is even more important. The size and the shape of the initial nuclei are assumed to be solely dependent on the change of volume of Gibbs free energy, due to super saturation, and the combined effect of surface and interface energies governed by Young's equation. No other interactions between the film or nuclei and the substrate were taken into consideration. In practice, the interaction between film and substrate plays a very important role in determining the initial nucleation and the film growth. Many experimental observations revealed that there are three basic nucleation modes:

(1) Island or Volmer-Weber growth,

(2) Layer or Frank-van der Merwe growth, and

(3) Island-layer or Stranski-Krastonov growth.

Special Nanomaterials

There are a number of important nanomaterials their syntheses are unique and important in nanotechnology Examples of such nanomaterials are carbon fullerenes and nanotubes, ordered mesoporous materials, organicinorganic hybrids, intercalation compounds, and oxide-metal core-shell structures. In addition, bulk materials with nanosized building blocks, such as nanograined ceramics and nanocomposites. Most of these nanomaterials are unique, do not exist in nature, and are truly "man-made" relatively recently, therefore, a brief introduction to the materials, such as their peculiar structures and properties, has also been included in the discussion.

Fullerene

Fullerenes were discovered in 1985 by Robert Curl, Harold Kroto and Richard Smalley at the University of Sussex and Rice University, and are named after Richard Buckminster Fuller. Fullerenes are a family of carbon allotropes, molecules composed entirely of carbon, in the form of a hollow sphere, ellipsoid, tube, or plane. Spherical fullerenes are also called buckyballs, and cylindrical ones are called carbon nanotubes or buckytubes. Graphene is an example of a planar fullerene sheet. Fullerenes are similar in structure to graphite, which is composed of stacked sheets of linked hexagonal rings, but may also contain pentagonal (or sometimes heptagonal) rings that would prevent a sheet from being planar.

Carbon Fullerenes

Carbon is a unique material, and can be a good metallic conductor in the form of graphite, a wide band gap semiconductor in the form of diamond, or a polymer when reacted with hydrogen. Carbon provides examples of materials showing the entire range of intrinsic nanometer scaled structures from fbllerenes, which are zero-dimensional nanoparticles, to carbon nanotubes, one-dimensional nanowires to graphite, a two-dimensional layered anisotropic material, to fullerene solids, a three-dimensional bulk materials with the fullerene molecules as the fundamental building block of the crystalline phase.

Carbon fullerenes

Carbon fullerene commonly refers to a molecule with 60 carbon atoms, C_{60}, and with an icosahedral symmetry, but also includes larger molecular weight fullerenes C_n ($n > 60$). Examples of larger molecular weight fullerenes are C_{70}, C_{76}, C_{78}, C_{80}, and higher mass fullerenes, which possess different geometric structure. C_{70} has a rugby ball-shaped symmetry.

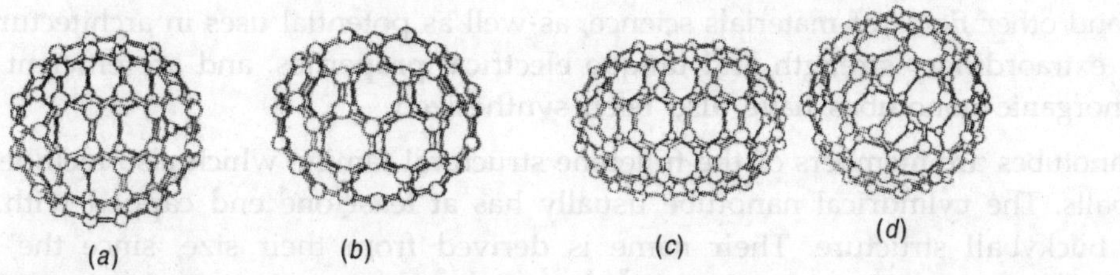

***FIGURE 43** (a) The icosahedral C_{60} molecule. (b) The C70 molecule as a rugby ball-shaped molecule. (c) The C_{80}, molecule as an extended rugby ball-shaped molecule. (d) The Cs0 molecule as an icosahedron.*

Figure shows the structure and geometry of some fullerene molecule. The name of fullerene was given to this family of carbon molecules because of the resemblance of these molecules to the geodesic dome designed and built by R. Buckminster Fuller, whereas the name of Buckminster fullerene or buckyball was specifically given to the C_{60} molecules, which are the most widely studied in the fullerene family and deserve a little more discussion on its structure and properties.

The 60 carbon atoms in C_{60} are located at the vertices of a regular truncated icosahedron and every carbon site on C_{60} is equivalent to every other site. The average nearest neighbor C-C distant in C_{60} (1.44) is almost identical to that in graphite (1.42 A). Each carbon atom in C_{60} is trigonally bonded to other carbon atoms, the same as that in graphite, and most of the faces on the regular truncated icosahedron are hexagons. There are 20 hexagonal faces and 12 additional pentagonal faces in each C_{60} molecule, which has a molecule diameter of 7.10 A.3 > 10 Although each carbon atom in C_{60} is equivalent to every other carbon atom, the three bonds emanating from each atom are not equivalent. Each carbon atom has four valence electrons for the formation of three chemical bonds, and there will be two single bonds and one double bond. The hexagonal faces consist of alternating single and double bonds, whereas the pentagonal faces are defined by single bonds. In addition, the length of single bonds is 1.46 A, longer than the average bond length, 1.44 A, while the double bonds are shorter, 1.40 A. The structures of other fullerene molecules can be considered as a modification of C_{60} by varying the number of hexagonal faces as far as the Euler's theorem is not violated, which states that a closed surface consisting of hexagons and pentagons has exactly 12 pentagons and an arbitrary number of hexagons. For example, C_{70},, structure can be envisioned by adding a belt of five hexagons around the equatorial plane of the C_{60} molecule normal to one of the five-fold axis.

Fullerenes are usually synthesized by using an arc discharge between graphite electrodes in approximately 200torr of He gas, first demonstrated in 1990 by Kratschmer and coworkers. The heat generated at the contact point between the electrodes evaporates carbon to form soot and fullerenes, which condense on the water-cooled walls of the reactor. This discharge produces carbon soot that can contain up to ~15% fullerenes: C_{60} (~13%) and C_{70} (~2%). The fullerenes are next separated from the soot, according to their mass, by use of liquid chromatography and using a solvent such as a toluene. However, there is no definite understanding of the growth mechanism of the fullerenes. Fullerene chemistry has been a very active research field, because of the uniqueness of the C_{60} molecule and its ability to have a variety of chemical reaction.

Carbon nanotubes (CNTs) are allotropes of carbon with a nanostructure that can have a length-to-diameter ratio greater than 1,000,000. These cylindrical carbon molecules have novel properties that make them potentially useful in many applications in nanotechnology, electronics, optics and other fields of materials science, as well as potential uses in architectural fields. They exhibit extraordinary strength and unique electrical properties, and are efficient conductors of heat. Inorganic nanotubes have also been synthesized.

Nanotubes are members of the fullerene structural family, which also includes the spherical buckyballs. The cylindrical nanotube usually has at least one end capped with a hemisphere of the buckyball structure. Their name is derived from their size, since the diameter of a nanotube is in the order of a few nanometers (approximately 1/50,000th of the width of a human hair), while they can be up to several millimeters in length. Nanotubes are categorized as single-walled nanotubes (SWNTs) and multi-walled nanotubes (MWNTs).

The nature of the bonding of a nanotube is described by applied quantum chemistry, specifically, orbital hybridization. The chemical bonding of nanotubes is composed entirely of sp^2 bonds, similar to those of graphite. This bonding structure, which is stronger than the sp^3 bonds found in diamonds, provides the molecules with their unique strength. Nanotubes naturally align themselves into "ropes" held together by Van der Waals forces. Under high pressure, nanotubes can merge together, trading some sp^2 bonds for sp^3 bonds, giving the possibility of producing strong, unlimited-length wires through high-pressure nanotube linking.

Fullerene Chemistry

Fullerene chemistry is a field of organic chemistry devoted to the chemical properties of fullerenes. Research in this field is driven by the need to functionalize fullerenes and tune their properties. For example fullerene is notoriously insoluble and adding a suitable group can enhance solubility. By adding a polymerizable group, a fullerene polymer can be obtained. Functionalized fullerenes are divided into two classes: exohedral with substituents outside the cage and endohedral fullerenes with trapped molecules inside the cage.

Structural

- clothes: water proof tear-resistant cloth fibers.
- combat jackets: MIT is working on combat jackets that use carbon nanotubes as ultrastrong fibers and to monitor the condition of the wearer. Cambridge university has developed the fibres and given a license to a company.
- concrete: In concrete, they increase the tensile strength, and halt crack propagation.
- polyethylene: Researchers have found that adding them to polyethylene increases the polymer's elastic modulus by 30%.
- sports equipment: Stronger and lighter tennis rackets, bike parts, golf balls, golf clubs, golf shaft and baseball bats.
- space elevator: This will be possible only if tensile strengths of more than about 70 GPa can be achieved. Monoatomic oxygen in the Earth's upper atmosphere would erode carbon nanotubes at some altitudes, so a space elevator constructed of nanotubes would need to be protected (by some kind of coating). Carbon nanotubes in other applications would generally not need such surface protection.

- ultrahigh-speed flywheels: The high strength/weight ratio enables very high speeds to be achieved.
- Bridges: For instance in suspension bridges (where they will be able to replace steel).

Electromagnetic

- artificial muscles
- buckypaper - a thin sheet made from nanotubes that are 250 times stronger than steel and 10 times lighter that could be used as a heat sink for chipboards, a backlight for LCD screens or as a faraday cage to protect electrical devices/aeroplanes.
- chemical nanowires: Carbon nanotubes additionally can also be used to produce nanowires of other chemicals, such as gold or zinc oxide. These nanowires in turn can be used to cast nanotubes of other chemicals, such as gallium nitride. These can have very different properties from CNTs—for example, gallium nitride nanotubes are hydrophilic, while CNTs are hydrophobic, giving them possible uses in organic chemistry that CNTs could not be used for.
- computer circuits: A nanotube formed by joining nanotubes of two different diameters end to end can act as a diode, suggesting the possibility of constructing electronic computer circuits entirely out of nanotubes. Because of their good thermal properties, CNTs can also be used to dissipate heat from tiny computer chips. The longest electricity conducting circuit is a fraction of an inch long. (Source: June 2006 National Geographic).
- conductive films: A 2005 paper in Science notes that drawing transparent high strength swathes of SWNT is a functional production technique (Zhang et al., vol. 309, p. 1215). Additionally, Eikos Inc of Franklin, Massachusetts and Unidym Inc.[2] of Silicon Valley, California are developing transparent, electrically conductive films of carbon nanotubes to replace indium tin oxide (ITO) in LCDs, touch screens, and photovoltaic devices. Carbon nanotube films are substantially more mechanically robust than ITO films, making them ideal for high reliability touch screens and flexible displays. Nanotube films show promise for use in displays for computers, cell phones, PDAs, and ATMs.
- electric motor brushes: Conductive carbon nanotubes have been used for several years in brushes for commercial electric motors. They replace traditional carbon black, which is mostly impure spherical carbon fullerenes. The nanotubes improve electrical and thermal conductivity because they stretch through the plastic matrix of the brush. This permits the carbon filler to be reduced from 30% down to 3.6%, so that more matrix is present in the brush. Nanotube composite motor brushes are better-lubricated (from the matrix), cooler-running (both from better lubrication and superior thermal conductivity), less brittle (more matrix, and fiber reinforcement), stronger and more accurately moldable (more matrix). Since brushes are a critical failure point in electric motors, and also don't need much material, they became economical before almost any other application.
- light bulb filament: alternative to tungsten filaments in incandescent lamps.
- magnets: MWNTs coated with magnetite
- optical ignition: A layer of 29% iron enriched SWNT is placed on top of a layer of explosive material such as PETN, and can be ignited with a regular camera flash.

- solar cells: GE's carbon nanotube diode has a photovoltaic effect. Nanotubes can replace ITO in some solar cells to act as a transparent conductive film in solar cells to allow light to pass to the active layers and generate photocurrent.
- superconductor: Nanotubes have been shown to be superconducting at low temperatures.
- ultracapacitors: MIT is researching the use of nanotubes bound to the charge plates of capacitors in order to dramatically increase the surface area and therefore energy storage ability.[3]
- displays: One use for nanotubes that has already been developed is as extremely fine electron guns, which could be used as miniature cathode ray tubes in thin high-brightness low-energy low-weight displays. This type of display would consist of a group of many tiny CRTs, each providing the electrons to hit the phosphor of one pixel, instead of having one giant CRT whose electrons are aimed using electric and magnetic fields. These displays are known as field emission displays (FEDs).
- transistor: developed at Delft, IBM, and NEC.

Chemical

- air pollution filter: Future applications of nanotube membranes include filtering carbon dioxide from power plant emissions.[4]
- biotech container: Nanotubes can be opened and filled with materials such as biological molecules, raising the possibility of applications in biotechnology.
- hydrogen storage: Research is currently being undertaken into the potential use of carbon nanotubes for hydrogen storage. They have the potential to store between 4.2 and 65% hydrogen by weight. This is an important area of research, since if they can be mass produced economically there is potential to contain the same quantity of energy as a 50L gasoline tank in 13.2L of nanotubes. See also, Hydrogen Economy.[5]
- water filter: Recently nanotube membranes have been developed for use in filtration. This technique can purportedly reduce desalination costs by 75%. The tubes are so thin that small particles (like water molecules) can pass through them, while larger particles (such as the chloride ions in salt) are blocked.

Mechanical

- oscillator: fastest known oscillators (> 50 GHz).
- nanotube membrane: Liquid flows up to five orders of magnitude faster than predicted by classical fluid dynamics.
- slick surface: slicker than Teflon and waterproof.

In Electrical Circuits

Carbon nanotubes have many properties—from their unique dimensions to an unusual current conduction mechanism that make them ideal components of electrical circuits. Currently, there is no reliable way to arrange carbon nanotubes into a circuit.

The major hurdles that must be jumped for carbon nanotubes to find prominent places in circuits relate to fabrication difficulties. The production of electrical circuits with carbon nanotubes are very different from the traditional IC fabrication process. The IC fabrication process is somewhat like sculpture - films are deposited onto a wafer and pattern-etched away. Because carbon nanotubes are fundamentally different from films, carbon nanotube circuits can so far not be mass produced.

Researchers sometimes resort to manipulating nanotubes one-by-one with the tip of an atomic force microscope in a painstaking, time-consuming process. Perhaps the best hope is that carbon nanotubes can be grown through a chemical vapor deposition process from patterned catalyst material on a wafer, which serve as growth sites and allow designers to position one end of the nanotube. During the deposition process, an electric field can be applied to direct the growth of the nanotubes, which tend to grow along the field lines from negative to positive polarity. Another way for the self assembly of the carbon nanotube transistors consist in using chemical or biological techniques to place the nanotubes from solution to determinate place on a substrate.

Even if nanotubes could be precisely positioned, there remains the problem that, to this date, engineers have been unable to control the types of nanotubes-metallic, semiconducting, single-walled, multi-walled-produced. A chemical engineering solution is needed if nanotubes are to become feasible for commercial circuits.

Metallic and Semiconducting Nanotubes

Most experimentally observed CNTs are multi-walled structures with outer most shell diameters exceeding 10 nm. Since current conduction in a MWCNT is known to be mostly confined to the outermost single-walled nanotube and since band gap of a SWCNT varies inversely with its diameter, MWCNTs are metallic in nature. SWCNTs can be either metallic or semiconducting depending on the way the roll-up of the graphene sheet occurs - an aspect termed as Chirality, and if all the roll-up types are realized with equal probability, 1/3 of the SWCNTs end up being metallic and 2/3 semiconducting. Thus, when CNTs are fabricated either by arc growth, laser ablation or Chemical Vapor Deposition (CVD), a mixture of metallic and semiconducting nanotubes is formed.This carbon nanotubes can be used as light emitting semiconductors.

Carbon Nanotube Interconnects

Metallic CNTs have aroused a lot of research interest in their applicability as Very-large-scale integration (VLSI) interconnects of the future because of their desirable properties of high thermal stability, high thermal conductivity and large current carrying capacity. An isolated CNT can carry current densities in excess of 1000 MA/sq-cm without any signs of damage even at an elevated temperature of 250 degrees C, thereby eliminating electromigration reliability concerns that plague Cu interconnects. Recent modeling work comparing the performance, power dissipation and thermal/reliability aspects of CNT interconnect to scaled copper interconnects have shown that CNT bundle interconnects can potentially offer advantages over copper[6] Additionally, the concept of hybrid CNT/Cu interconnects-employing CNT vias in tandem with copper interconnects has been shown to offer advantages from a reliability/

thermal-management perspective. More information on state-of-the-art of CNT interconnects (including their fabrication) can be found in the literature.

Carbon Nanotube Transistors

Semiconducting CNTs have been used to fabricate field effect transistors (CNTFETs), which show promise due to their superior electrical characteristics over silicon based MOSFETs. Since the electron mean free path in SWCNTs can exceed 1 micrometer, long channel CNTFETs exhibit near-ballistic transport characteristics, resulting in high speed devices. In fact, CNT devices are projected to be operational in the frequency range of hundreds of GHz. Recent work detailing the advantages and disadvantages of various forms of CNTFETs have also shown that the tunneling based CNTFET offers better characteristics compared to other CNTFET structures. This device has been found to be superior in terms of subthreshold slope—a very important property for low power applications.

Challenges in Electronic Design and Design Automation

Although CNT devices and interconnects have been separately shown to be promising in their own respects, there have been few efforts to successfully combine them in a realistic circuit. Most CNTFET structures employ the silicon substrate as a back gate. Applying different back gate voltages might become a concern when designing large circuits out of these devices. Several top-gated structures have also been demonstrated, which can alleviate this concern. Recently, a fully integrated logic circuit built on a single nanotube has been reported. However, this circuit also employs a back-gate. Additionally, there are still several process related challenges that need to be addressed before CNT-based devices and interconnects can enter mainstream VLSI process. This makes it an exciting and open field for research. Problems like purification, separation of carbon nanotubes, control over nanotube length, chirality and desired alignment, low thermal budget as well as high contact resistance are yet to be fully resolved. Although these are serious technological challenges, innovative ideas have been proposed to build practical transistors out of nano-networks. Since lack of control on chirality produces a mix of metallic as well as semi-conducting CNTs from any fabrication process and it is difficult to control the growth direction of the CNTs, random arrays of SWCNTs (that are easily produced) have been proposed to build thin film transistors. This idea can be further exploited to build practical CNT based transistors and circuits without the need for precise growth and assembly.

As Fiber and Film

One application for nanotubes that is currently being researched is high tensile strength fibers. Two methods are currently being tested for the manufacture of such fibers. A French team has developed a liquid spun system that involves pulling a fiber of nanotubes from a bath which yields a product that is approximately 60% nanotubes[citation needed]. The other method, which is simpler but produces weaker fibers uses traditional melt-drawn polymer fiber techniques with nanotubes mixed in the polymer. After drawing, the fibers can have the polymer component burned out of them leaving only the nanotube or they can be left as they are.

Ray Baughman's group from the NanoTech Institute at University of Texas at Dallas produced the current toughest material known as of mid-2003 by spinning fibers of single wall carbon nanotubes with polyvinyl alcohol. Beating the previous contender, spider silk, by a factor of four, the fibers require 600 J/g to break. In comparison, the bullet-resistant fiber Kevlar is 27-33 J/g. In mid-2005, Baughman and co-workers from Australia's Commonwealth Scientific and Industrial Research Organization developed a method for producing transparent carbon nanotube sheets 1/1000th the thickness of a human hair capable of supporting 50,000 times their own mass. In August 2005, Ray Baughman's team managed to develop a fast method to manufacture up to seven meters per minute of nanotube tape.[8] Once washed with ethanol, the ribbon is only 50 nanometers thick; a square kilometer of the material would only weigh 30 kilograms.

In 2004, Alan Windle's group of scientists at the Cambridge-MIT Institute developed a way to make carbon nanotube fiber continuously at the speed of several centimetres per second just as nanotubes are produced. One thread of carbon nanotubes was more than 100 metres long. The resulting fibers are electrically conductive and as strong as ordinary textile threads.

Allotropes of Carbon

Eight allotropes of carbon: (*a*) Diamond, (*b*) Graphite, (*c*) Lonsdaleite, (*d*) C_{60} (Buckminsterfullerene or buckyball), (*e*) C_{540}, (*f*) C_{70}, (*g*) Amorphous carbon, and (*h*) single-walled carbon nanotube or buckytube.

This is a list of the allotropes of carbon.

- 1 Diamond
- 2 Graphite
  - o 2.1 Graphene
- 3 Amorphous carbon
- 4 Buckminsterfullerenes
  - o 4.1 Buckyballs
  - o 4.2 Carbon nanotubes
  - o 4.3 Carbon nanobuds
- 5 Aggregated diamond nanorods
- 6 Glassy carbon
- 7 Carbon nanofoam
- 8 Lonsdaleite (hexagonal diamond)
- 9 Linear Acetylenic Carbon (LAC)
- 10 Variability of carbon
- 11 Other possible forms
- 12 References
- 13 External links

Diamond

Diamond is one of the best known allotropes of carbon, whose hardness and high dispersion of light make it useful for industrial applications and jewelry. Diamond is the hardest known natural mineral, which makes it an excellent abrasive and makes it hold polish and luster extremely well. No known naturally occurring substance can scratch, let alone cut, a diamond.

The market for industrial-grade diamonds operates much differently from its gem-grade counterpart. Industrial diamonds are valued mostly for their hardness and heat conductivity, making many of the gemological characteristics of diamond, including clarity and color, mostly irrelevant. This helps explain why 80% of mined diamonds (equal to about 100 million carats or 20,000 kg annually), unsuitable for use as gemstones and known as *bort*, are destined for industrial use. In addition to mined diamonds, synthetic diamonds found industrial applications almost immediately after their invention in the 1950s; another 400 million carats (80,000 kg) of synthetic diamonds are produced annually for industrial use-nearly four times the mass of natural diamonds mined over the same period.

The dominant industrial use of diamond is in cutting, drilling (drill bits), grinding (diamond edged cutters), and polishing. Most uses of diamonds in these technologies do not require large diamonds; in fact, most diamonds that are gem-quality can find an industrial use. Diamonds are embedded in drill tips or saw blades, or ground into a powder for use in grinding and polishing applications. Specialized applications include use in laboratories as containment for high pressure experiments (see diamond anvil), high-performance bearings, and limited use in specialized windows.

With the continuing advances being made in the production of synthetic diamond, future applications are beginning to become feasible. Garnering much excitement is the possible use of diamond as a semiconductor suitable to build microchips from, or the use of diamond as a heat sink in electronics. Significant research efforts in Japan, Europe, and the United States are under way to capitalize on the potential offered by diamond's unique material properties, combined with increased quality and quantity of supply starting to become available from synthetic diamond manufacturers.

Each carbon atom in a diamond is covalently bonded to four other carbons in a tetrahedron. These tetrahedrons together form a 3-dimensional network of puckered six-membered rings of atoms. This stable network of covalent bonds and the three dimensional arrangement of bonds is the reason that diamond is so strong.

Graphite

Graphite (named by Abraham Gottlob Werner in 1789, from the Greek γρα'φετν (graphein, "to draw/write", for its use in pencils) is one of the most common allotropes of carbon. Unlike diamond, graphite is an electrical conductor, and can be used, for instance, as the material in the electrodes of an electrical arc lamp. Graphite holds the distinction of being the most stable form of carbon under standard conditions. Therefore, it is used in thermochemistry as the standard state for defining the heat of formation of carbon compounds.

Graphite is able to conduct electricity, due to delocalization of the pi bond electrons above and below the planes of the carbon atoms. These electrons are free to move, so are able to conduct electricity. However, the electricity is only conducted along the plane of the layers. In diamond all four outer electrons of each carbon atom are 'localised' between the atoms in covalent bonding. The movement of electrons is restricted and diamond does not conduct an electric current. In graphite, each carbon atom uses only 3 of its 4 outer energy level electrons in covalently bonding to three other carbon atoms in a plane. Each carbon atom contributes one electron to a delocalised system of electrons that is also a part of the chemical bonding. The delocalised electrons are free to move throughout the plane. For this reason, graphite conducts electricity along the planes of carbon atoms, but does not conduct in a direction at right angles to the plane.

Graphite powder is used as a dry lubricant. Although it might be thought that this industrially important property is due entirely to the loose interlamellar coupling between sheets in the structure, in fact in a vacuum environment (such as in technologies for use in space), graphite was found to be a very poor lubricant. This fact led to the discovery that graphite's lubricity is due to adsorbed air and water between the layers, unlike other layered dry lubricants such as molybdenum disulfide. Recent studies suggest that an effect called superlubricity can also account for this effect.

When a large number of crystallographic defects bind these planes together, graphite loses its lubrication properties and becomes what is known as pyrolytic carbon, a useful material in blood-contacting implants such as prosthetic heart valves.

Natural and crystalline graphites are not often used in pure form as structural materials due to their shear-planes, brittleness and inconsistent mechanical properties.

In its pure glassy (isotropic) synthetic forms, pyrolytic graphite and carbon fiber graphite is an extremely strong, heat-resistant (to 3000 °C) material, used in reentry shields for missile nosecones, solid rocket engines, high temperature reactors, brake shoes and electric motor brushes.

Intumescent or expandable graphites are used in fire seals, fitted around the perimeter of a fire door. During a fire the graphite intumesces (expands and chars) to resist fire penetration and prevent the spread of fumes. A typical start expansion temperature (SET) is between 150 and 300 degrees Celsius.

Density: its specific gravity is 2.3 which makes it lighter than diamond.

Effect of heat: it is the most stable allotrope of carbon. At a temperature of 2500 degree Celsius, it can be transformed into diamond. At about 700 degree Celsius it burns in pure oxygen forming carbon dioxide.

Chemical activity: it is slightly more reactive than diamond. This is because the reactants are able to penetrate between the hexagonal layers of carbon atoms in graphite. It is unaffected by ordinary solvents, dilute acids, or fused alkalis. However, chromic acid oxidises it to carbon dioxide.

Graphene

A single layer of graphite, once thought to be impossible, is called graphene and has extraordinary electrical, thermal, and physical properties. It can be produced by epitaxy (vapor deposition) on an insulating surface or by mechanical exfoliation (repeated peeling). Its applications may include replacing silicon in high-performance electronic devices.

Amorphous Carbon

Amorphous carbon is the name used for carbon that does not have any crystalline structure. As with all glassy materials, some short-range order can be observed, but there is no long-range pattern of atomic positions.

While entirely amorphous carbon can be made, natural amorphous carbon (such as soot) actually contains microscopic crystals of graphite, [1] sometimes diamond [2]. On the macroscopic scale, amorphous carbon has no definite structure as it conisists of small irregular cystals, but on the nanomicroscopic scale, we can see it is made of regularly arranged carbon atoms.

Coal and soot are both informally called amorphous carbon. However, both are products of pyrolysis (the process of decomposing a substance by the action of heat), which does not produce true amorphous carbon under normal conditions. The coal industry divides coal up into various grades depending on the amount of carbon present in the sample compared to the amount of impurities. The highest grade, anthracite, is about 90 percent carbon and 10% other elements. Bituminous coal is about 75–90 percent carbon, and lignite is the name for coal that is around 55 percent carbon.

Buckminsterfullerenes

The *buckminsterfullerenes*, or usually just *fullerenes* for short, were discovered in 1985 by a team of scientists from Rice University and the University of Sussex, three of whom were awarded the 1996 Nobel Prize in Chemistry. They are named for the resemblance of their alliotropic structure to the geodesic structures devised by the scientist and architect Richard Buckminster "Bucky" Fuller. Fullerenes are molecules of varying sizes composed entirely of carbon, which take the form of a hollow sphere, ellipsoid, or tube.

As of the early twenty-first century, the chemical and physical properties of fullerenes are still under heavy study, in both pure and applied research labs. In April 2003, fullerenes were under study for potential medicinal use—binding specific antibiotics to the structure to target resistant bacteria and even target certain cancer cells such as melanoma.

Buckyballs--Spherical fullerenes are also called buckyballs. They have the formula C_{60}.

Carbon Nanotubes

Carbon nanotubes, also called buckytubes, are cylindrical carbon molecules with novel properties that make them potentially useful in a wide variety of applications (e.g., nano-electronics, optics, materials applications, etc.). They exhibit extraordinary strength, unique electrical properties, and are efficient conductors of heat. Inorganic nanotubes have also been synthesized. A nanotube (also known as a buckytube) is a member of the fullerene structural family, which also includes buckyballs. Whereas buckyballs are spherical in shape, a nanotube is cylindrical, with at least one end typically capped with a hemisphere of the buckyball structure. Their name is derived from their size, since the diameter of a nanotube is on the order of a few nanometers (approximately 50,000 times smaller than the width of a human hair), while they can be up to several centimeters in length. There are two main types of nanotubes: single-walled nanotubes (SWNTs) and multi-walled nanotubes (MWNTs).

Carbon Nanobuds

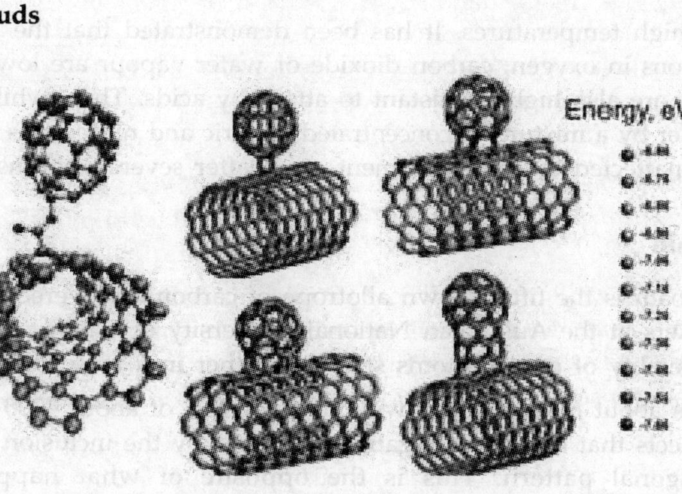

FIGURES 44 **Computer models of stable NanoBud structures.**

Carbon NanoBuds are a newly discovered allotrope of carbon in which fullerene like "buds" are covalently attached to the outer sidewalls of the carbon nanotubes. This hybrid material has useful properties of both fullerenes and carbon nanotubes. In particular, they have been found to be exceptionally good field emitters.

Aggregated Diamond Nanorods

Aggregated diamond nanorods, or ADNRs, are an allotrope of carbon believed to be the least compressible material known to humankind, as measured by its isothermal bulk modulus; aggregated diamond nanorods have a modulus of 491 gigapascals (GPa), while a conventional diamond has a modulus of 442 GPa. ADNRs are also 0.3% denser than regular diamond. The ADNR material is also harder than type IIa diamond and ultrahard fullerite.

Glassy Carbon

Glassy carbon is a class of non-graphitizing carbon which is widely used as an electrode material in electrochemistry, as well as for high temperature crucibles and as a component of some prosthetic devices. It was first produced by workers at the laboratories of The General Electric Company, UK, in the early 1960s, using cellulose as the starting material. A short time later, Japanese workers produced a similar material from phenolic resin.

It was first produced by Bernard Redfern in the mid 1950's at the laboratories of the Carborundum Company, Trafford Park, Manchester, UK. He set out to develop a polymer matrix to mirror a diamond structure and discovered a resole (phenolic) resin that would, with special preparation, set without a catalyst. Using this resin the first glassy carbon was produced. Patents were filed some of which were withdrawn in the interests of national security. Original research samples of resin and product exist.

The preparation of glassy carbon involves subjecting the organic precursors to a series of heat treatments at temperatures up to 3000°C. Unlike many non-graphitizing carbons, they are impermeable to gases and are chemically extremely inert, especially those which have been prepared at very high temperatures. It has been demonstrated that the rates of oxidation of certain glassy carbons in oxygen, carbon dioxide or water vapour are lower than those of any other carbon. They are also highly resistant to attack by acids. Thus, while normal graphite is reduced to a powder by a mixture of concentrated sulfuric and nitric acids at room temperature, glassy carbon is unaffected by such treatment, even after several months.

Carbon Nanofoam

Carbon nanofoam is the fifth known allotrope of carbon discovered in 1997 by Andrei V. Rode and co-workers at the Australian National University in Canberra. It consists of a low-density cluster-assembly of carbon atoms strung together in a loose three-dimensional web.

Each cluster is about 6 nanometers wide and consists of about 4000 carbon atoms linked in graphite-like sheets that are given negative curvature by the inclusion of heptagons among the regular hexagonal pattern. This is the opposite of what happens in the case of buckminsterfullerenes, in which carbon sheets are given positive curvature by the inclusion of pentagons.

The large-scale structure of carbon nanofoam is similar to that of an aerogel, but with 1% of the density of previously produced carbon aerogels—only a few times the density of air at sea level. Unlike carbon aerogels, carbon nanofoam is a poor electrical conductor.

Lonsdaleite (Hexagonal Diamond)

Lonsdaleite is a hexagonal allotrope of the carbon allotrope diamond, believed to form when meteoric graphite falls to Earth. The great heat and stress of the impact transforms the graphite into diamond, but retains graphite's hexagonal crystal lattice.

Lonsdaleite was first identified from the Canyon Diablo meteorite at Barringer Crater (also known as Meteor Crater) in Arizona. It was first discovered in 1967. Lonsdaleite occurs as microscopic crystals associated with diamond in the Canyon Diablo meteorite; Kenna meteorite, New Mexico; and Allan Hills (ALH) 77283, Victoria Land, Antarctica meteorite. It has also been reported from the Tunguska impact site, Russia.

Linear Acetylenic Carbon (LAC)

Chemists in the USA have recently reported (ca 1995) an allotrope of carbon consisting of long chains of carbon atoms where the alternate carbon-carbon bonds are of different lengths; and consist of C-C bonds and C = C bonds.

The same polymer was synthesized in early 1960s by group of Soviet chemists and was called carbyne (Russian: 8"D$4>). It appeared to be a semiconductor that is very sensitive to light, thus it was suggested to use it in photodiodes and similar devices.

Carbyne, or polyyne, is also another name for Linear Acetylenic Carbon (LAC) the carbon allotrope that has the chemical structure -(C:::C)n-. Carbon in this modification is linear with *sp* orbital hybridisation, and is a polymer with alternating single and triple bonds. This type of carbyne is of considerable interest to nanotechnology as its Young's modulus is forty times that of diamond.

Variability of Carbon

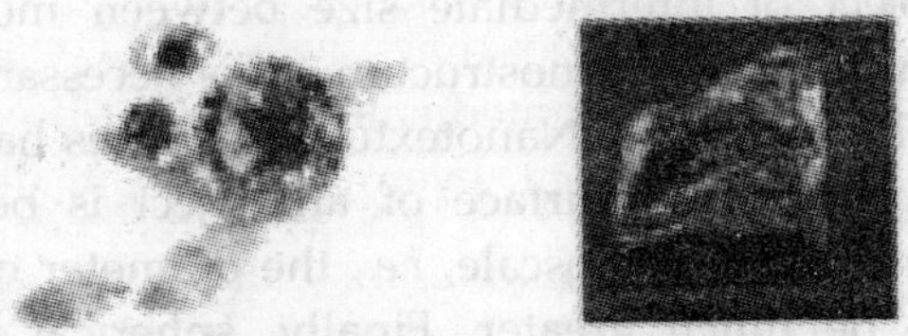

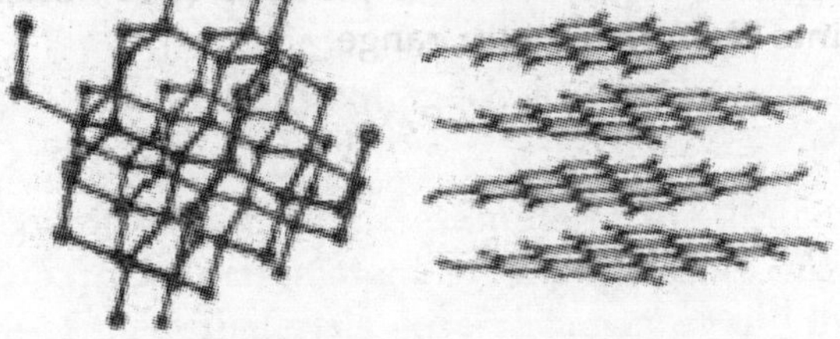

FIGURE 45

Diamond and graphite are two allotropes of carbon: pure forms of the same element that differ in structure.

The system of carbon allotropes spans an astounding range of extremes, considering that they are all merely structural formations of the same element.

Between diamond and graphite:

- Diamond crystallizes in the cubic system but graphite crystallizes in the hexagonal system.
- Diamond is hardest mineral known to man (10 on Mohs scale), but graphite is one of the softest (1–2 on Mohs scale).
- Diamond is the ultimate abrasive, but graphite is a very good lubricant.
- Diamond is an excellent electrical insulator, but graphite is a conductor of electricity.
- Diamond is an excellent thermal conductor, but some forms of graphite are used for thermal insulation (*i.e.*, heatshields and firebreaks)

Other Possible Forms

- Chaoite is a mineral believed to have been formed in meteorite impacts. It has been described as slightly harder than graphite with a reflection colour of grey to white. However, the existence of carbyne phases is disputed—see the entry on chaoite for details.
- Metallic carbon: Theoretical studies have shown that carbon (diamond) when brought at enormous pressure, there are regions in the phase diagram where it's metallic. It seems that it can also become superconducting at very low temperatures (4 kelvins).
- Hexagonite: in theory, instead of having the 6-arom rings of graphite, one sp carbon atom could be inserted between each of the 6 sp^2 atoms.
- Prismane C8 is another possible metastable form.

Nanostructure

A nanostructure is an object of intermediate size between molecular and microscopic (micrometer-sized) structures. In describing nanostructures it is necessary to differentiate between the number of dimensions on the nanoscale. Nanotextured surfaces have one dimension on the nanoscale, *i.e.*, only the thickness of the surface of an object is between 0.1 and 100 nm. Nanotubes have two dimensions on the nanoscale, *i.e.*, the diameter of the tube is between 0.1 and 100 nm; its length could be much greater. Finally, spherical nanoparticles have three dimensions on the nanoscale, *i.e.*, the particle is between 0.1 and 100 nm in each spatial dimension. The terms nanoparticles and ultrafine particles (UFP) often are used synonymously although UFP can reach into the micrometre range.

List of Nanostructures

1. Nanocages
2. Nanocomposite
3. Nanofabrics
4. Nanofiber
5. Nanoflake
6. Nanoflower
7. Nanofoam
8. Nanomesh
9. Nanopillar
10. Nanopin film
11. Nanoring
12. Nanorod
13. Nanoshell
14. Quantum dot
15. Quantum heterostructure
16. Sculptured thin film

1. Nanocages

Nanocages are hollow porous gold nanoparticles ranging in size from 10 to over 150 nm. They are created by reacting silver nanoparticles with chloroauric acid ($HAuCl_4$) in boiling water. While gold nanoparticles absorb light in the visible spectrum of light (at about 550 nm), gold nanocages absorb light in the near-infrared, where biological tissues absorb the least light. Because they are also biocompatible, gold nanocages are promising as a contrast agent for optical coherence tomography, which uses light scattering in a way analogous to ultrasound to produce in-vivo images of tissue with resolution approaching a few micrometres. A contrast agent is required if this technique will be able to image cancers at an early, more treatable stage. Gold nanocages also absorb light and heat up, killing surrounding cancer cells. The Xia group at the Washington University, the original inventors of the nanocages, has functionalized nanocages with cancer-specific antibodies so they specifically attach to cancer cells.

2. Nanocomposite

Definition Nanocomposites can be defined[1] as multiphase solid materials where one of the phases has a dimension of less than 100 nanometres (nm): structures having nanometre scale dimensional repeat distances between the different phases that make up the material. In the broadest sense this definition can include porous media, colloids, gels and copolymers, but is more usually taken to mean the solid combination of nano-dimensional phases differing in properties due to dissimilarities in structure and chemistry. The mechanical, electrical, optical, electrochemical, catalytic properties of the nanocomposite will differ markedly from that of the component materials. Size limits for these effects have been proposed, < 5*nm* for catalytic activity, < 20*nm* for making a hard magnetic material soft, < 50*nm* for refractive index changes, and < 100*nm* for achieving superparamagnetism, mechanical strengthening or restricting matrix dislocation movement.

Nanocomposites are found in nature as for example in the structure of the abalone shell and bone. The use of nanoparticle rich materials long predates the understanding of the physical and chemical nature of these materials. Jose-Yacaman et al[3] investigated the origin of the depth of colour and the resistance to acids and bio-corrosion of Maya blue paint, attributing it to a nanoparticle mechanism. From the mid 1950s nanoscale organo-clays have been used to control flow of polymer solutions (*e.g.*, as paint viscosifiers) or the constitution of gels (*e.g.*, as a thickening substance in cosmetics, keeping the preparations in homogeneous form). By the 1970s polymer/clay composites were the topic of textbooks, although the term "nanocomposites" was not in common use.

In mechanical terms, nanocomposites differ from conventional composite materials due to the exceptionally high surface to volume ratio of the reinforcing phase. The reinforcing material can be made up of particles *e.g.*, exfoliated clay particles, short fibres *e.g.*, carbon nanotubes, or long fibres *e.g.*, electrospun fibres. The area of the interface between the matrix and reinforcement phases is typically an order of magnitude greater than that in conventional composite materials. The matrix material is significantly affected by proximity to the reinforcement. Ajayan *et al* [1] note that with polymer nanocomposites, properties related to local chemistry, degree of thermoset cure, polymer chain mobility, polymer chain conformation, polymer chain conformation, degree of polymer chain ordering or crystallinity can all vary significantly and continuously from being in contact with the reinforcement into the bulk of the matrix. This large amount of reinforcement surface area means that the effect of a relatively small amount of nanoscale reinforcement can have an observable effect on the macroscale properties of the composite. For example, adding carbon nanotubes improves the electrical and thermal conductivity. Other kinds of nanoparticulates may result in enhanced optical properties, dielectric properties or mechanical properties such as stiffness and strength. In general, the nano reinforcement is dispersed into the matrix during processing. The percentage by weight (called *mass fraction*) of the nanoparticulates introduced can remain very low (on the order of 0.5% to 5%) due to the low filler percolation threshold, especially for the most commonly used non-spherical, high aspect ratio fillers (*e.g.*, nanometer-thin platelets, such as clays, or nanometer—diameter cylinders, such as carbon nanotubes).

3. Nanofabrics

Nanofabrics is an emerging nanotechnology that deals with building specialized fabrics. Active Camouflage is a costume that hides the wearer by redirecting light from one side to the other.

4. Nanofiber

Nanofibers are defined as fibers with diameters on the order of 100 nanometers. They can be produced by interfacial polymerization and electrospinning. Carbon nanofibers are graphitized fibers produced by catalytic synthesis.

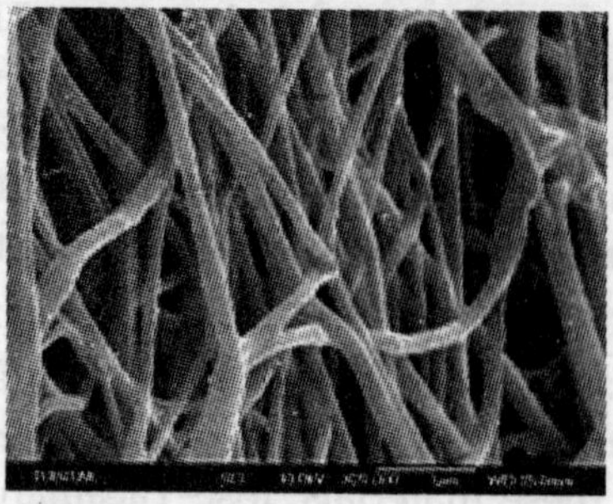

FIGURE 46 **SEM image of nanofibers by Nano FMG.**

Potential Applications

- Medical: Artificial organ components, Tissue engineering, Implant material, Drug delivery, Wound dressing, Medical textile materials.
- Protective materials: Sound absorption materials, Protective clothings against chemical and biological warfare agents, Sensor applications for detecting chemical agents.
- Energy: Batteries, Photovoltaic Cells, Polymer Electrolytes, Membrane Fuel Cells.
- Textile: Sport Apparels, Sport Shoes, Climbing, Rainwear, Outerwear Garments, Baby Diapers.
- Filtration: HVAC system filters, HEPA, ULPA High efficient filters, Air, oil, fuel filters for automotive, Filters for beverage, pharmacy, medical applications.
- In one study, combined neural stem cells with carbon nanofibers triggered neural tissue regeneration in the brains of rats that had suffered a simulated stroke.[citation needed] On their own, neither nanofibers nor stem cells could heal the rats.
- Napkins with nanofibers contain antibodies against numerous biohazards and chemicals that signal by changing color (potentially useful in identifying bacteria in kitchens).
- In wound healing nanofibers assemble at the injury site and stay put, drawing the body's own growth factors to the injury site.
- Nano FMG-Nano Fiber Membranes Group develops polymer nanofiber membranes and industrial electrospinning unit for nanofiber production.
- Donaldson develops nanofiber filter media for new air and liquid filtration applications, such as vacuum cleaners.

Other applications include industrial and high-tech applications for aerospace, capacitors, transistors, battery separators, energy storage, fuel cells and information technology.

5. Nanoflake

Nanoflake has been discovered in the Center. It is a perfect crystalline structure that also absorbe all light. Nanoflake have the potential to convert up to 30 per cent of the solar energy into electricity and that is twice the amount converted today. It can reduce the solar cell production costs because uses less of the expensive semiconducting silicon in the process due to the use of nanotechnology.

6. Nanoflower

From Wikipedia, the free encyclopedia

Jump to: navigation, search

A nanoflower, in chemistry, refers to a compound of certain elements that results in formations which in microscopic view resemble flowers or, in some cases, trees that are called nanobouquets or nanotrees. These formations are nanometers long and thick so they can only be observed using special equipment.

7. Nanofoam

From Wikipedia, the free encyclopedia

Jump to: navigation, search

Nanofoams are a class of nanostructured, porous materials, foams, containing a significant population of pores with diameters less than 100 nm. Aerogels are one example of nanofoam.

Metal Nanofoams

In 2006, Dr. Bryce Tappan at Los Alamos National Laboratory discovered a technique for producing metal nanofoams by igniting pellets of energetic metal bis(tetrazolato)amine complexes. Nanofoams of iron, cobalt, nickel, copper, silver, and palladium have been prepared through this technique. These materials exhibit densities as low as 11 mg/cm^3, and surface areas as high as 258 m^2/g. These foams are effective catalysts and are being investigated for other applications.

8. Nanomesh

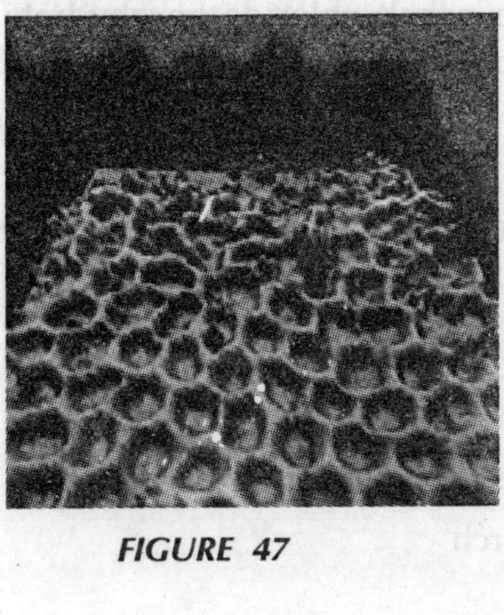

FIGURE 47

Perspective view of nanomesh, whose structure ends at the back of the figure. The distance between two pore centers is 3.2nm, and the pores are 0.05nm deep.

The nanomesh is a new inorganic nanostructured two-dimensional material, similar to graphene. It was discovered in 2003 at the University of Zurich, Switzerland.

It consists of a single layer of boron (B) and nitrogen (N) atoms, which forms by self-assembly a highly regular mesh after high-temperature exposure of a clean rhodium or ruthenium surface to borazine under ultra-high vacuum.

The nanomesh looks like an assembly of hexagonal pores at the nanometer (nm) scale. The distance between 2 pore centers is only of 3.2 nm, whereas each pore has a diameter of about 2 nm and is 0.05 nm deep. The lowest regions bind strongly to the underlying metal, while the wires (highest regions) are only bound to the surface through strong cohesive forces within the layer itself.

The boron nitride nanomesh is not only stable under vacuum, air and some liquids, but also up to temperatures of 796°C (1070 K). In addition it shows the extraordinary ability to trap molecules and metallic clusters, which have similar sizes to the nanomesh pores, forming a well-ordered array. These characteristics promise interesting applications of the nanomesh in areas like nanocatalysis, surface functionalisation, spintronics, quantum computing and data storage media like hard drives.

9. Nanopillar

Nanopillars are a metamaterial.

10. Nanopin Film

Material in nanotechnology developed in 2005 with unusual superhydrophobic properties. A droplet of water makes contact with the surface of this film and forms an almost perfect sphere with a contact angle of 178°. The film is able to do this because it is covered with nanoscale topped off pins or cones perpendicular to the surface. The surface is regarded as a composite material with mostly air and a small fraction made up by the tops of the cones. When the contact angle of the cone material is sufficiently large Cassie's law predicts large contact angle values for the composite.

This particular nanopin film is produced with borosilicate glass as the primary substrate. A solution of $CoCl_2 \bullet 6H_2O$ or cobalt chloride hexahydrate is heated at 60 °C for 24 hours in a chemical bath deposition to form a brucite type cobalt(II) hydroxide layer with composition

$$CoOH_{1.13}Cl_{0.09}(CO_3)_{0.39}.0.05H_20$$

The top coating is provided by lauric acid in a separate step. A 3 square micrometer surface now contains on average 166 such cones with cone height of around 100 nm and the cone diameter at the tip of 6.5 nm. The Cassie's law prediction for this material with the lauric acid surface area of 0.000612 and flat film contact angle of 75° is 177.8°.

11. Nanoring

From Wikipedia, the free encyclopedia

Jump to: navigation, search

A nanoring is a small ringformed crystal. The first nanoring made was a zinc oxide nanoring discovered by researchers at Georgia Institute of Technology. They are made by a spontaneous self-coiling process of nanobelts. Many layers of nanobelts are rolled together as coils, layer-by-layer.

12. Nanorod

From Wikipedia, the free encyclopedia

Jump to: navigation, search

In nanotechnology, nanorods are one morphology of nanoscale objects. Each of their dimensions range from 1–100 nm. They may be synthesized from metals or semiconducting materials. Standard aspect ratios (length divided by width) are 3–5. Nanorods are produced by direct chemical synthesis. A combination of ligands act as shape control agents and bond to different facets of the nanorod with different strengths. This allows different faces of the nanorod to grow at different rates, producing an elongated object.

The applications of nanorods are diverse, ranging from display technologies (the reflectivity of the rods can be changed by changing their orientation with an applied electric field) to microelectromechanical systems (MEMS).

13. Nanoshell

A nanoshell is composed of a spherical core of a particular compound surrounded by a shell of a few nanometer in thickness. In Europe it is also referred to as "core-shell".

One of the promising applications concerns the biological field. Research is being performed to create nanoshells with high absorptions at biologically useful wavelengths by altering the thickness of the shells. Particularly, the Near Infra Red region, which corresponds with low absorption by tissue, may be useful.

In the literature, special attention is given to gold nanoshell with a dielectric core (gold sulfide,silicon dioxide,...). Gold is a biocompatible compound, making it a useful material for medical applications.

Nanoshells are currently being investigated as:

- a treatment for cancer similar to chemotherapy but without the toxic side-effects
- inexpensive, quick analysis of "samples as small as a single molecule" and builds on Raman spectroscopy

Nanoshells featured in Alex Rider novel *Scorpia*, where they were filled with cyanide. The fictional organziation Scorpia planned to use satellite dishes to activate the nanoshells to kill thousands of children in Great Britain.

14. Quantum Dot

A quantum dot is a nano-particulate semiconductor, whose excitons are confined in all three spatial dimensions. As a result, they have properties that are between those of bulk semiconductors and those of discrete molecules. Colloidal quantum dots irradiated with a UV light. Different sized quantum dots emit different color light due to quantum confinement. Researchers have studied quantum dots in transistors, solar cells, LEDs, and diode lasers. They have also investigated quantum dots as agents for medical imaging and hope to use them as qubits. Some quantum dots are commercially available.

15. Quantum Heterostructure

Quantum heterostructure is a heterostructure in a substrate (usually a semiconductor material), with size restricting the movements of the charge carriers and forcing them into a quantum confinement, leading to formation of a set of discrete energy levels at which the carriers can exist. Quantum heterostructures have sharper density of states than structures of more conventional sizes.

Quantum heterostructures are important for fabrication of short-wavelength light-emitting diodes and diode lasers, and for other optoelectronic applications, e.g., high-efficiency photovoltaic cells.

Examples of quantum heterostructures are:

- Quantum wells
- Quantum dots
- Quantum wires

16. Sculptured Thin Film

Sculptured Thin Films (STFs) are nanostructured materials with unidirectionally varying properties that can be designed and realized in a controllable manner using variants of physical vapor deposition. The ability to virtually instantaneously change the growth direction of their columnar morphology, through simple variations in the direction of the incident vapor flux, leads to a wide spectrum of columnar forms.

Challenges in Nanotechnology

Although many of the fundamentals have long been established in different fields such as in physics, chemistry, materials science and device science and technology, and research on nanotechnology is based on these established fundamentals and technologies, researchers in the field face many new challenges that are unique to nanostructures and nanomaterials.

Challenges in nanotechnology include the integration of nanostructures and nanomaterials into or with macroscopic systems that can interface with people.

Challenges include the building and demonstration of novel tools to study at the nanometer level what is being manifested at the macro level. The small size and complexity of nanoscale structures make the development of new measurement technologies more challenging than ever. New measurement techniques need to be developed at the nanometer scale and may require new innovations in metrological technology. Measurements of physical properties of nanomaterials require extremely sensitive instrumentation, while the noise level must be kept very low. Although material properties such as electrical conductivity, dielectric constant, tensile strength, are independent of dimensions and weight of the material in question, in practice, system properties are measured experimentally. For example, electrical conductance, capacitance and tensile stress are measured and used to calculate electrical conductivity, dielectric constant and tensile strength. As the dimensions of materials shrink from centimeter or millimeter scale to nanometer scale, the system properties would change accordingly, and mostly decrease with the reducing dimensions of the sample materials. Such a decrease can easily be as much as orders of magnitude as sample size reduce from centimeter to nanometer scale. Other challenges arise in the nanometer scale, but are not found in the macro level.

The challenge will be not only to achieve reproducible and uniform distribution of dopant atoms in the nanometer scale, but also to precisely control the location of dopant atoms. To meet such a challenge, the ability to monitor and manipulate the material processing in the atomic level is crucial. Furthermore, doping itself also imposes another challenge in nanotechnology, since the self-purification of nanomaterials makes doping very difficult. For the fabrication and processing of nanomaterials and nanostructures, the following challenges must be met:

1. Overcome the huge surface energy, a result of enormous surface area or large surface to volume ratio.
2. Ensure all nanomaterials with desired size, uniform size distribution, morphology, crystallinity, chemical composition, and microstructure that altogether result in desired physical properties.
3. Prevent nanomaterials and nanostructures from coarsening through either Ostwald ripening or agglomeration as time evolutes.

EXERCISES

Short Questions for Discussion

1. What is meant by the free electron gas?
2. Which statistics is obeyed by the free electron inside the metal?
3. What type of energy levels are of a one-dimensional free electron gas: continuous or descrete?
4. How many maximum number of electrons are found in the energy level below the Fermi energy level of one-dimensional free electron gas?
5. What is meant by the Fermi energy?

   (Jabalpur 1996, Saugar 1991, Bhopal 1990, Rewa 1993, 95, Vikram 1994)
6. Define the Fermi temperature?
7. Write down the relationship between the Fermi temperature T_F and Fermi energy E_F.
8. What is meant by the valance band and conduction band?
9. What do you mean by the forbidden gap? What is the value of forbidden energy gap for metals? (Jabalpur 1993)
10. What is Hall effect ?
11. What is super-conductivity ?
12. What do you mean by the critical temperature of a metal? What is the critical temperature of mercury ?

Long Questions for Discussion

1. Explain the free electron model in metals and on this basis explain the physical significance of Fermi energy.
2. Explain the merits and demerits of free election model.
3. An electron is in motion along X-axis in a length L under a constant potential ($v = 0$). Write down the Schrodinger equation for it and solve it. Also find the eigen values of energy of electron and density of states.
4. Give the solution of Schrodinger equation for a uniform potential according to Sommerfeld's model and discuss the conclusions obtained from it.
5. Explain the meaning of density of energy states and Fermi energy. Obtain expressions for it.
6. What is meant by Fermi energy? Explain its physical significance and deduce expression for the Fermi energy for the one-dimensional motion of electron under a constant potential.
7. Obtain expression for the Fermi energy E_F of N free electrons of a system.
8. Write Schrodinger equation for a constant one-dimensional potential well and solve it. Show that the value of fourth energy state is four times the value of second energy state.
9. Explain the Kronig-Penny model fo periodic potential. Write down the Schrodinger equation and its solution on the basis of this model.
10. Explain the behaviour of electron in a periodic potential on the basis of Kronig-penny model.
11. Discuss the conclusions regarding energy bands in solids obtained from the Kronig-penny model.
12. Find the effective mass of electron on the basis of Kronig-penny model.
13. Explain the formation of energy bands in solids. On this basis explain the distinction between metals, insulators and semiconductors.
14. Draw neat energy level diagram for conductors, insulators and semiconductors.
15. Explain the formation of the barrier field across the junction of a *p-type* and an *n-type* semiconductor.
16. In the case of an unbiased p-n junction sketch the variation of space charge and barrier energy. Distinguish between a step junction and a linearly graded junction.
17. Explain how a semiconductor diode can be used as a rectifier.
18. What is the effect on the space charge width at a p-n junction when the junction is (*i*) forward-biased, and (*ii*) reverse biased. Draw the circuit diagrams of (*i*) a forward-biased p-n junction, and (*ii*) a reverse-biased p-n junction.
19. Draw the energy-band diagram of an unbiased p-n junction. Explain the terms: barrier energy, barrier potential and depletion region, as applied to a p-n junction.
20. "The barrier potential across a p-n junction diode cannot be measured by placing a voltmeter across the diode terminals"-, Explain.

21. Write down the expression for the voltage-current characteristic of a p-n junction diode. Why does the reverse saturation current vary with temperature? Does it vary with the applied bias?

22. (*a*) Distinguish between the static and the dynamic resistances of a p-n junction diode. Do they depend on temperature and bias voltage?

(*b*) Define the term 'cutin voltage' for a p-n junction diode. What are its typical values for a Ge diode and an Si diode?

23. What is junction capacitance? How does the junction capacitance vary with (*i*) the depletion layer width, and (*ii*) the applied reverse voltage?

24. What is breakdown diode? What is its use? Describe physically how breakdown in a p-n junction occurs.

25. Distinguish between avalanche breakdown and zener breakdown. Draw a simple circuit for operating a zener diode.

26. What is Hall effect? Deduce expression for the Hall coefficient of a solid and describe method for its determination experimentally. What important informations are obtained from its measurement?

27. What is tunnel diode? Draw a typical volt, ampere characteristic of a tunnel diode. Name some uses of tunnel diode.

28. Write short notes on (*i*) Photo diode, and (*ii*) Light-emitting diode.

Numericals

[If required take : $h = 6.6 \times 10^{-34}$ Js, $m_c = 9.1 \times 10^{-31}$kg. Avogadro number N = 6.02×10^{23} per gm atom. Boltzmann constant $k = 1.38 \times 10^{-23}$ J/K]

1. An electron is in between the two walls of infinite heights and separation 2.5×10^{-10} *m*. Determine the three possible quantum energies of electron.

[**Ans.** 5.98 eV, 23.93 eV,53.85eV]

2. An electron is enclosed in a box of infinite height and length 2A. Calculaete: (*i*) energy of electron in the ground state. (*ii*) density of ground energy state, (*iii*) Fermi energy to fill 10 electrons in the box.

[**Ans.** (*i*) 9.34 eV (*ii*) 6.7×10^{17} per joule, (*iii*) 2.33eV]

3. The Fermi temperature of potassium metal is 2.46×10^4 K and the depth of potential well is 4.34 eV. Determine: (*i*) Fermi energy, (*ii*) work function.

[**Hint:** Depth of potential well $E_s = E_F + \phi$] [**Ans.** (*i*) 2.12 eV (*ii*) 2.22eV]

4. The depth of potential well for the atomic wire of silver is 9.9 eV and the work function of silver is 4.5 eV. Calculate its Fermi energy and Fermi temperature.

[**Ans.** (*i*) 5.4 eV, (*ii*) 6.26×10^4 K]

5. Assuming the depth of potential well for 2cm length of one-dimensional copper wire to be 11.5 eV. Find the number of electrons present in it. (Given: work function of copper = 4.4 eV) [**Ans.** 1.74×10^8]

6. In a thin copper rod of length 10 cm and width 2cm, a current of 2μA flows along its length. The rod is placed in a magnetic field of strength 2 × 10^{-6} weber/m^2 perpendicular to its length. If Hall coefficient of copper is 0.55 × 10^{10} volt metre^3/ampere weber, find the Hall voltage produced.

[Hint: Hall voltage $V_{Hall} = R_{Hall} \frac{1}{b} H_2$] [Ans. 1.1 volt]

7. A copper strip of width 2 cm and thickness 1.0 mm is placed in a magnetic field of strength 1.5 *wb/m*^{2} such that the magnetic field is along the width of strip. 2A current flows through it. Calculate the Hall voltage across the strip. Number of electrons per unit volume of copper is 3.26 × 10^{25} m^{-3}.

[Ans. 0.575*m*V].

8. For what voltage will the reverse current in a p-n junction Ge diode attain a value of 90 percent of its saturation value at room temperature? What will be the current if a forward voltage of the same magnitude is applied? (Ans. –60 *m*V, 9.1 I_s)

9. An ideal p-n junction germanium diode has a reverse saturation current of 30 μA at a temperature of 125°C. Find the static and the dynamic resistances of the junction at that temperature for a 0.2 volt bias in the forward direction. (Ans. 19.7Ω, 3.38 Ω)

10. The current through a p-n junction silicon diode is 55 mA at a forward bias voltage of 0.9V. Calculate r_{dc} and r_{ac} at 27^0C. (Ans. 16.36 Ω, 0.95 Ω)

11. A 12V zener diode is connected in series with a resistance of 150 Ω and load resistance of 1*k* Ω is connected across the zener diode. The minimum zener current is nearly zero and the maximum zener current should not exceed 20 *m*A. Calculate the operating rage of the input voltage. (Ans. 13.8 V to 16.8V)

Chapter 7

LASERS

7.1 OVERVIEW

LASER is the acronym for Light Amplification by Stimulated Emission of Radiation. The laser is the outgrowth of MASER which is a device that amplifies microwaves. The theory behind the maser operation was developed by the American physicist C. H. Townes and independently by the Russian physicists N. Basov and P. M. Prokhorov. The first successful maser, an ammonia maser, was built by Gordon, Zeiger, and Townes in 1954. Townes, Basov, and Prokhorov were awarded the Noble Prize in Physics in 1964 for their invention of the maser. Masers are cumbersome devices and they are now relegated to the back stage. The first successful laser was the ruby laser, a solid state laser built by the American physicist T. H. Maiman in 1960. The Helium-Neon laser, which is a gas laser, was developed by Ali Javan and his associates in 1961. Developments in semiconductor technology led to the fabrication of the solid state diode laser in 1962 by Hall and his co-workers, and at about the same time by Nathan and his associates. Solid state diode lasers are small in size, do not need any auxiliary equipment, and are nearly 60% efficient. Lasers find applications in many fields. They have brought amazing changes in many areas and caused spectacular developments in the field of communications. Lasers made optical communications possible.

7.2 INTERACTION OF RADIATION WITH MATTER-QUANTUM MECHANICAL VIEW

The understanding of the working principle of a laser (or maser) requires an appreciation of quantum processes that take place in a material when it is exposed to radiation. A material medium is composed of identical atoms of molecules each of which is characterized by a set of discrete allowed energy states. An atom can move from one energy state to another when it receives or releases an amount of energy equal to the energy difference between those two states. It is termed a quantum jump or transition. For the sake of ease in understanding, let us restrict our attention to two energy states E_1 and E_2 of an atom. E_1 is the lower energy state while E_2 is the excited state. As the constituent atoms of the medium are identical, the energy states E_1 and E_2 will be common to all atoms in the medium. Let a monochromatic radiation of frequency ν be incident on the medium. The radiation may be viewed as a stream of photons, each photon carrying an energy $h\nu$. If $h\nu = E_2 - E_1$, then the interaction of radiation with atoms leads to the following three distinct competing processes in the medium.

Process 1: Absorption

An atom residing in the lower energy state E_1 may absorb the incident photon and jump to the excited state E_2, as depicted in Figure 1(*a*). The transition is known as stimulated absorption or induced absorption or simply absorption. Corresponding to each transition made by an atom, one photon disappears from the incident beam. Schematically, it may be represented as

$$A + h\nu \rightarrow A^* \tag{1}$$

where A is an atom in the lower state and A* is an excited atom.

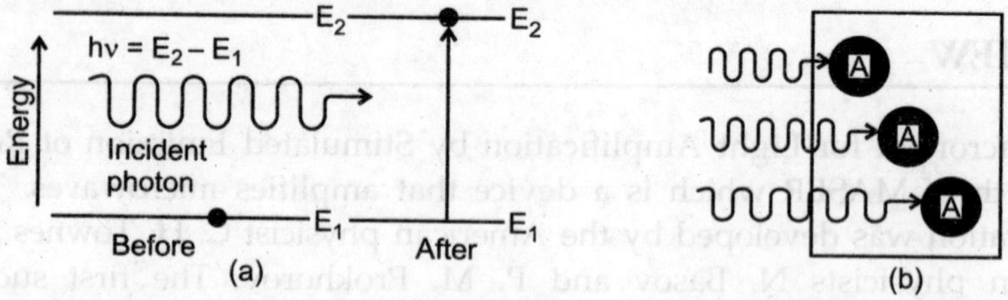

***FIGURE 1* Absorption (a) Induced absorption process. (b) Material absorbs photons.**

The number of absorption transitions occuring in the material at any instant will be proportional to the number density of atoms in the lower state E_1 and the photon density in the incident beam. The greater the number of atoms in the lower state, the more that can undergo absorption transition. Similarly, the greater the number of photons in the incident radiation, the greater the number of atoms that can be excited to a higher state. The number of atoms N_{ab} excited during the time Δt is therefore given by

$$N_{ab} = B_{12}N_1Q\Delta t \tag{2}$$

where N_1 is the number of atoms in the state E_1, Q the spectral energy density and, B_{12} is a constant of proportionality. Spectral energy density means the photon density per unit frequency range.

Normally, the number of atoms is greater in the lower energy state and the material absorbs incident energy. Therefore, the process of absorption leads to attenuation of radiation as light travels through the medium.

Process 2: Spontaneous Emission

The excited state with higher energy is inherently unstable because of a natural tendency of atoms to seek out the lowest energy configuration. Therefore, excited atoms do not stay in the excited state for a relatively longer time but tend to return to the lower state by giving up the excess energy $h\nu = E_2 - E_1$ in the form of spontaneous emission or stimulated emission.

The excited atom in the state E_2 may return to the lower state E_1 on its own out of the natural tendency to attain minimum potential energy condition. During the transition the excess energy is released as a photon of energy $h\nu = E_2 - E_1$. This type of process, in which photon emission occurs without any external impetus, is called spontaneous emission. It is depicted in Figure 2 and may be represented as

$$A^* \rightarrow A + h\nu \quad (3)$$

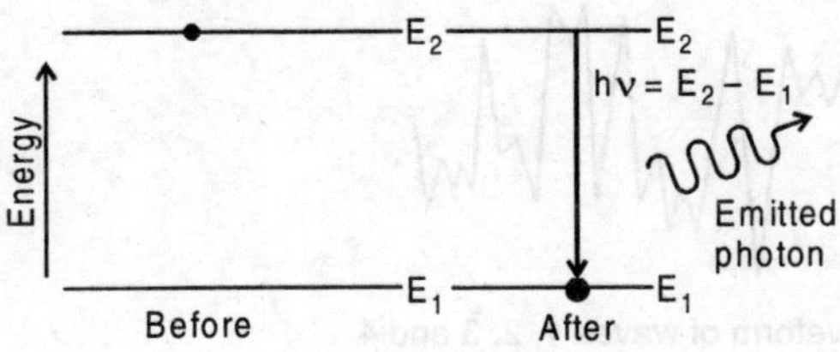

(a) Emission process

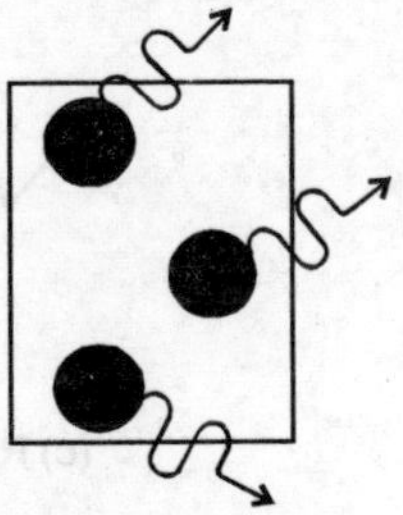

(b) Material emits photons haphazardly

FIGURE 2 **Spontaneous emission.**

A number of spontaneous transitions N_{sp} take place in the material during the time Δt is independent of the photons present in the incident radiation, and depend only on the number density of atoms N_2 lying in the excited state E_2. It is given by:

$$N_{sp} = A_{21} N_2 \Delta t \quad (4)$$

where N_2 is the number density of atoms in the state E_2 and A_{21} represents a constant of proportionality. The number of atoms making transitions per second divided by the number of atoms, is the probability of a transition occurring per second. Thus the probability for a spontaneous transition is:

$$P_{21} = \frac{1}{N_2}\left(\frac{N_{sp}}{\Delta t}\right) = A_{21}$$

The process of spontaneous emission is essentially probabilistic, that is, the atom has some probability for making the transition, and it is not amenable for control from outside. The instant of the transition, direction of emission of photon, the phase of the photon, the polarization state of the photon, are all random quantities. There will not exist any correlation among the parameters of the innumerable photons emitted spontaneously by the assembly of atoms of the medium. Therefore, the light generated by the medium will be incoherent. The light from conventional sources originates in a spontaneous emission process and is incoherent. It contains a superposition of many waves of random phases. The net intensity of such incoherent waves is proportional to the number of radiating atoms. The light is radiated in the form of short duration wave trains emitted in all directions and the intensity goes on decreasing as the wave trains travel away from the source. Each of them bears no consistent phase relationship with each other (Figure 3) nor do they share a common polarization plane. As a consequence, there is no compounding of the individual waves. The light is not monochromatic because of various line broadening processes that take place in the medium.

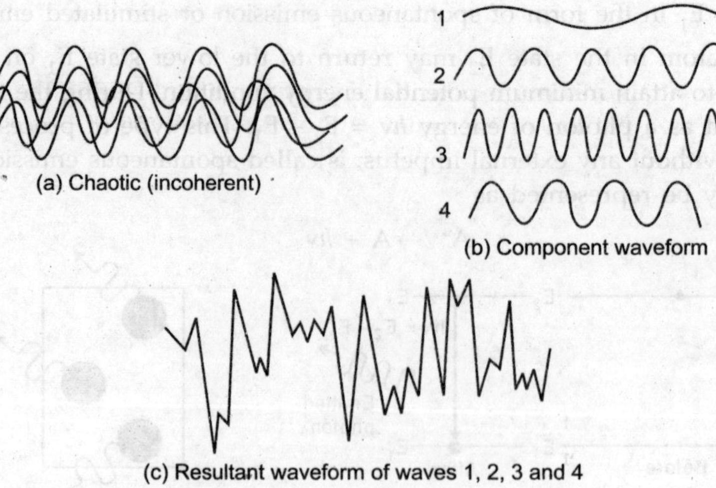

FIGURE 3 **Incoherent radiation.**

Process 3: Stimulated Emission

An atom in the excited state need not wait for a spontaneous emission to occur. There exists an alternative mechanism by which an excited atom can make a downward transition and emit light. A photon of energy $h\nu = E_2 - E_1$ can induce the excited atom to make a downward transition releasing the energy in the form of a photon. Thus, the interaction of a photon with an excited atom triggers the excited atom to drop to the lower energy state giving up a photon. The phenomenon of forced emission of photons is called induced emission or stimulated emission. The existence of this mechanism was predicted by Einstein in 1916. The process is illustrated in Figure 4 and may be represented as

$$A^* + h\nu \rightarrow A + 2h\nu \tag{5}$$

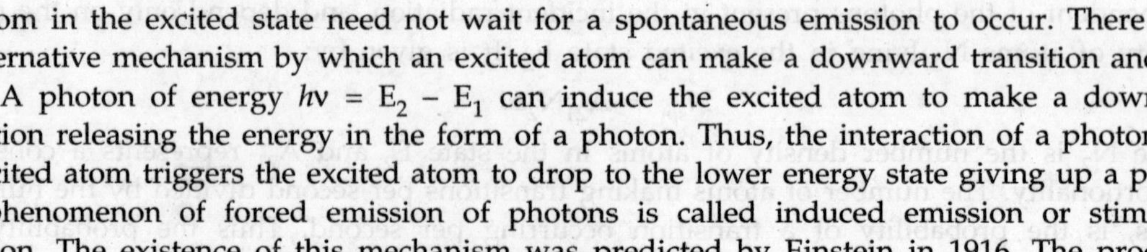

FIGURE 4 **Stimulated emission.**

The number of stimulated transitions N_a occurring in the material during the time Δt may be written as

$$N_{st} = B_{21} N_2 Q \Delta t \tag{6}$$

where B_{21} is a constant of proportionality.

The process of stimulated emission is characterized by some very interesting features:

- The emitted photon is identical to the incident photon in all respects. It has the same frequency ν as that of incident photon.

  It will be in phase with the incident photon. Both the photons travel in the same direction. They will be in the same state of polarization.
- The process is controllable from outside.
- The most important feature is that multiplication of photons takes place in the process. One photon induces an atom to emit a second photon, these two travelling along the same direction de-excite two atoms in their path, producing a total of four photons which in turn stimulate four atoms generating eight photons and so on. The number of photons builds up in an avalanche-like manner as illustrated in Figure 5. It suggests that electromagnetic waves of extremely high amplitude could be generated by the combined stimulated emissions from a large sample of atoms leading to the amplification of light. The constructive interference of many waves travelling in the same direction with a common frequency and common phase produces an intense coherent light beam, as illustrated in Figure 6. The resultant light is a single plane wave in which all of the individual waves emitted by the individual atoms have combined in phase. The net amplitude of the wave is proportional to the number of atoms that contributed to it, and the net intensity is proportional to the square of the number of atoms. Since the number of atoms in light sources is very large, coherent emission leads to an enormously high intense light rather than incoherent emission. Therefore, the process of stimulated emission is the key to the operation of a laser.

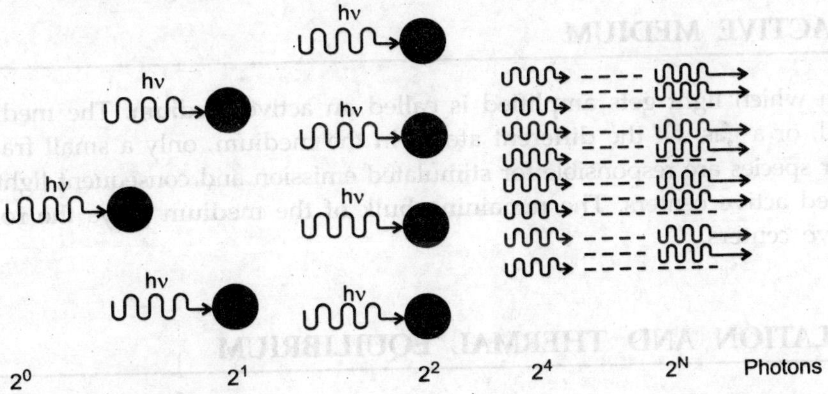

FIGURE 5 **Multiplication of stimulated photons into an avalanche.**

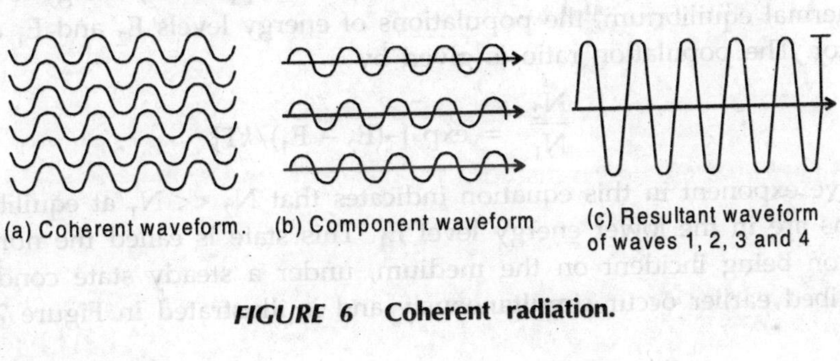

FIGURE 6 **Coherent radiation.**

7.3 THE METASTABLE STATE

An atom can be excited to a higher level by supplying energy to it. Normally, excited atoms have short lifetimes and release their energy in a matter for nanoseconds (10^{-9} s) through spontaneous emission. It means that atoms do not stay long enough at the excited state to be stimulated. As a result, even though the pumping agent continuously raises the atoms to the excited level, they undergo spontaneous transitions and rapidly return to the lower energy level. Population inversion cannot be established under such circumstances. In order to establish the condition of population inversion, the excited atoms are required to 'wait' at the upper energy level until a large number of atoms accumulate at that level. In other words, it is necessary that the excited state has a longer lifetime. A metastable state is such a state. Atoms excited to the metastable state remain excited for an appreciable time, which is of the order of 10^{-6} to 10^{-3}. This is 10^3 to 10^6 times the lifetime of the ordinary energy levels.

Therefore, the metastable state allows accumulation of a large number of excited atoms at that level. The metastable state population can exceed the population at a lower level and establish the condition of population inversion in the lasing medium. It would be impossible to create the state of population inversion without a metastable state. A metastable state can be readily obtained in a crystal system containing impure atoms. These levels lie in the forbidden band gap of the host crystal. Population inversion readily takes place, as the lifetimes of these levels are large, and there is no competition in filling these levels as they are localized levels. There could be no population inversion and hence no laser action, if metastable states do not exist.

7.4 THE ACTIVE MEDIUM

A medium in which light gets amplified is called an active medium. The medium may be a solid, a liquid, or a gas. Of the different atoms in the medium, only a small fraction of atoms of a particular species are responsible for stimulated emission and consequent light amplification. They are called active centers. The remaining bulk of the medium plays the role of host and supports active centers.

7.5 POPULATION AND THERMAL EQUILIBRIUM

The number of active atoms occupying an energy state is called the population of that state. Thus N_1 and N_2 are the populations of the lower (E_1) and upper (E_2) energy levels respectively. In a state of thermal equilibrium, the populations of energy levels E_2 and E_1 are fixed by the Boltzmann factor. The population ratio is given by

$$\frac{N_2}{N_1} = \exp\left[-(E_2 - E_1)/kT\right] \tag{7}$$

The negative exponent in this equation indicates that $N_2 << N_1$ at equilibrium. It means that more atoms are in the lower energy level E_1. This state is called the normal state. With external radiation being incident on the medium, under a steady state condition, the three processes described earlier occur simultaneously and is illustrated in Figure 7. Atoms in the lower level E_1 will occasionally absorb radiation and make a transition to upper level E_2. Atoms in the upper level will occasionally emit radiation and make a transition to the lower level. In order to maintain N_1 and N_2 constant, the number of upward transitions must be equal to the number of downward transitions. Thus

$$N_{ab} = N_{sp} + N_{st} \tag{8}$$

$$B_{12}N_1Q = A_{21}N_2 + B_{21}N_2Q \tag{9}$$

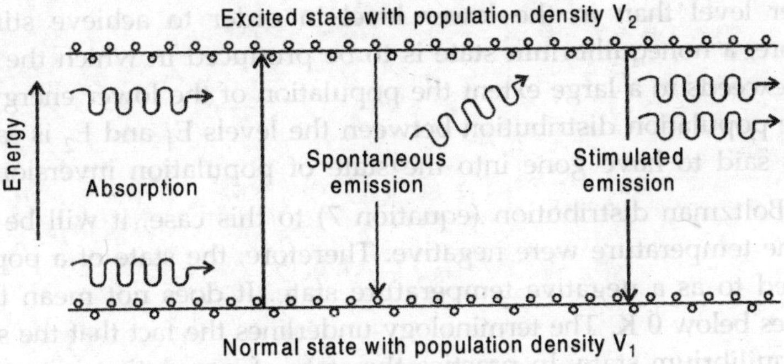

FIGURE 7 **Absorption and emission processes in steady state of the material.**

This relation was predicted by Einstein and the coefficient A_{21}, B_{12} and B_{21} are called the Einstein coefficients.

7.6 CONDITIONS FOR LIGHT AMPLIFICATION

At thermal equilibrium, the radio of the stimulated to spontaneous transitions is generally very small and the stimulated emission is negligible. The ratio is given by

$$\frac{\text{Stimulated transitions}}{\text{Spontaneous transitions}} = \frac{B_{21}N_2Q}{A_{21}N_2} = \frac{B_{21}}{A_{21}}Q \tag{10}$$

The ratio of stimulated to absorption transitions is given by

$$\frac{\text{Stimulated transitions}}{\text{Absorption transitions}} = \frac{B_{21}N_2Q}{B_{12}N_1Q} = \frac{N_2}{N_1} \tag{11}$$

$B_{21} = B_{12}$ as the probability or stimulated emission transition P_{st} must equal the probability of stimulated absorption transition P_{ab}. Hence

- Equation (10) suggests that in order to enhance the number of stimulated transitions the radiation density Q is to be made larger.
- Relation (11) indicates that stimulated emission will be larger than absorption only when $N_2 > N_1$. As long as $N_1 > N_2$, the absorption dominates stimulated emission and the medium will absorb the incident light rather than amplify it. Only when the above two conditions are fulfilled, will the medium amplify light passing through it.

7.7 POPULATION INVERSION

Laser operation requires obtaining stimulated emission almost exclusively. To achieve a high percentage of stimulated emission, an artificial situation known as *population inversion* is to be created.

It has been shown previously that in a state of thermal equilibrium there are more atoms in the lower level than in the upper level. It is seen from relation (11) that there must be more atoms in the upper level than in the lower level in order to achieve stimulated emission exclusively. Therefore, a nonequilibrium state is to be produced in which the population of the upper energy level exceeds to a large extent the population of the lower energy level. When this situation occurs, the population distribution between the levels E_1 and E_2 is said to be inverted, and the medium is said to have gone into the state of population inversion.

Extending the Boltzman distribution (equation 7) to this case, it will be seen that N_2 can exceed N_1 only if the temperature were negative. Therefore, the state of a population inversion is sometimes referred to as a negative temperature state. It does not mean that we can reach down to temperatures below 0 K. The terminology underlines the fact that the state of population inversion is a nonequilibrium state. In practice the state of population inversion is obtained at ordinary temperatures.

7.8 NEGATIVE ABSORPTION

Intensity of light decreases exponentially as it propagates through a medium. It is described by the relation

$$I = I_0 e^{-\alpha l} \tag{12}$$

where I_0 is the initial intensity, l is the length travelled by light in the medium and α is the absorption coefficient. The phenomenon of light amplification may be described by the same relation assuming α to be a negative quantity, as this process also occurs exponentially. In view of this, a medium in the state of population inversion is called a medium with a negative absorption coefficient and the light amplification is referred to as negative absorption.

7.9 PUMPING

For realising and maintaining the condition of population inversion, the atoms have to be raised continuously to an excited state. It requires energy to be supplied to the system. The process of supplying energy to the medium with a view to transfer it into the state of population inversion is known as pumping.

There are a number of techniques for pumping a collection of atoms to an inverted state. The most commonly used are optical pumping, electrical discharge, and direct conversion. In optical pumping a light source such as a flash discharge tube is used. This method is adopted in a solid state laser. In the electric discharge method, the electric field causes ionization of the medium and raises it to the excited state. This technique is used in gas lasers. In semiconductor diode lasers, a direct conversion of electrical energy into light energy takes place.

7.10 THE PRINCIPAL PUMPING SCHEMES

Atoms in general are characterized by a large number of energy levels. Among them only three or four levels will be pertinent to the pumping process. Therefore, only those levels are depicted in the pumping scheme diagrams. The transition between the two levels that generate stimulated emission is called a lasing transition. The terminal level is called the lower lasing level and the upper level as upper lasing level. The upper most level is termed as the pumping level. The upward arrow in the diagram implies the pumping transition and the downward thick arrow the lasing transition. The slant arrows represent radiative or nonradiative spontaneous transitions.

Two important pumping schemes are widely employed. They are known as three-level and four-level pumping schemes.

(a) *Three-Level Pumping Scheme:* Let us assume that an atomic system has three energy levels as shown in Figure 8. The state E_1 is the ground state and E_2 and E_3 are the excited states. In the scheme, the energy states are such that atoms are readily excited to the uppermost state E_3 (known as the pumping level) when the light of frequency ν_p (known as pump frequency) $= (E_3 - E_1)/h$ is incident on them. The pump level E_3 is not a stable state. Atoms do not stay at the E_3 level and undergo downward transitions either to E_1 or E_2 levels. The probability of spontaneous transition $E_3 \to E_1$ is comparable to that of $E_3 \to E_2$. E_2 is a metastable state, since the probability of spontaneous transition $E_2 \to E_1$ is extremely small. When the medium is exposed to radiation of frequency ν_p, a large number of atoms will be excited to the higher energy level E_3. Some of these atoms make spontaneous transitions to the lowest level E_1 through a nonradiative transition. As spontaneous transitions from E_3 to E_1 occur rarely, the atoms get trapped in E_2. The process continues because of pumping and after a short time there will be a large accumulation of atoms at E_2. When more than half of the ground state atoms accumulate at E_2 the population inversion is achieved between the states E_1 and E_2. Now a photon of energy $h\nu = (E_2 - E_1)$ can trigger stimulated emission of atoms at E_2.

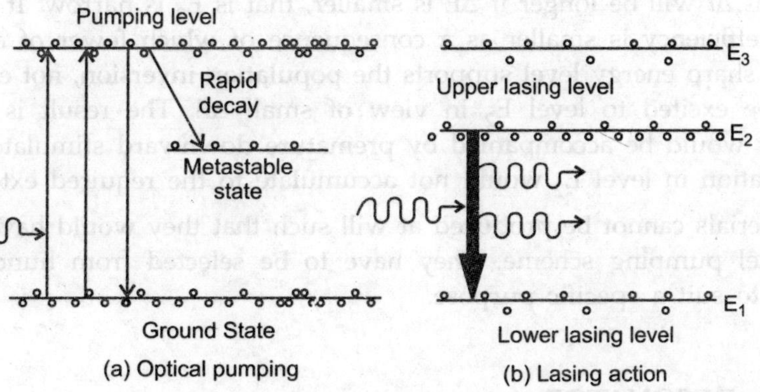

FIGURE 8 **A typical three-level pumping scheme.**

In this scheme, the terminal state of laser transition is the ground state. To achieve population inversion more than half of the ground state atoms must be pumped to the upper state. Therefore, a very high pump power is required in this type of pumping scheme.

(b) *Four-Level Pumping Scheme:* A typical four-level pumping scheme is shown in Figure 9. Pump frequency lifts the active centers from the ground level E_1 to the uppermost level E_4. From the pump level E_4 the atoms rapidly fall to the metastable state E_3. The populations at this state grow rapidly while the level E_2 is virtually empty. Therefore, population inversion is achieved between the states E_2 and E_3. A photon of energy $h\nu = E_3 - E_2$ can start a chain of stimulated emissions, bringing the atoms into the state E_2. From there the atoms undergo nonradiative transitions subsequently to the ground state E_1 and will be available once again to participate in the process.

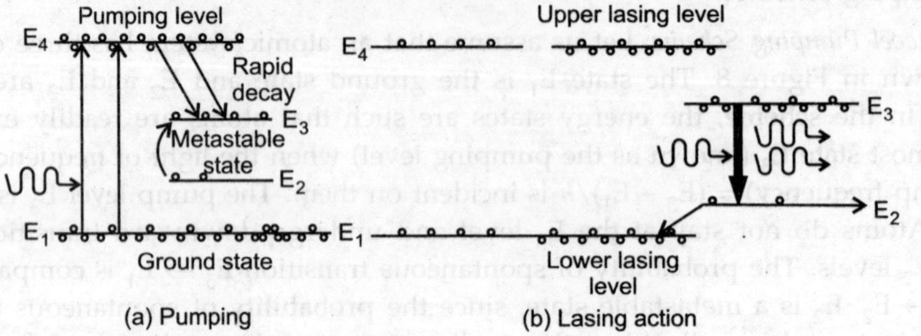

FIGURE 9 **A typical four-level pumping scheme.**

(c) *Two-Level Pumping Scheme:* The population inversion can be achieved only by the indirect routes (*a*) and (*b*) mentioned above. A two-level pumping is not suitable for obtaining populations inversion. The time span Δt, for which atoms have to stay at the upper level E_2, must be longer for achieving a population inversion condition.

According to the Heisenberg uncertainty principle $\Delta E.\Delta t \geq \hbar$.

This means Δt will be longer if ΔE is smaller, that is E_2 is narrow. If ΔE is smaller, the pumping efficiency is smaller as a consequence of which fewer of atoms are excited. Though a sharp energy level supports the population inversion, not enough population can not be excited to level E_2 in view of small ΔE. The result is that the upward transitions would be accompanied by premature downward stimulated transitions and the population in level E_2 would not accumulate to the required extent.

Laser materials cannot be produced at will such that they would have a three-level or a four-level pumping scheme. They have to be selected from hundreds of available materials to suit a specific purpose.

7.11 OPTICAL RESONATOR

Light can be amplified by an active medium and taken into the state of population inversion. Just as in the case of an electronic oscillator, the input light is generated by the medium itself. A photon spontaneously emitted by an atom acts as the input and starts stimulated emission. To sustain stimulated emission acts and to increase the light intensity in a cumulative way, a

20.61 eV above the ground state respectively. These two states are metastable states and the excited helium atoms cannot return to the ground state by spontaneously emitting photons of energy 20.61 eV and 19.81 eV. With the passage of a current through the discharge tube, more and more helium atoms accumulate in the excited states. Though a radiative transition is forbidden, the excited helium atoms can return to the ground state by transferring their energy to neon atoms through collisions. Such an energy transfer can take place when the two colliding atoms have identical energy states. The E_6 level which is at 20.66 eV, and the E_4 level which at 18.7 eV of neon atom, nearly coincide in energy with the F_3 and F_2 levels of the helium atom. Therefore, neon atoms acquire energy and go to the excited state on collision with helium atoms, which return to the ground state after transfering the energy. The kinetic energy of helium atoms provides the additional 0.05 eV for exciting the neon atom to E_6 level. This is the main pumping mechanism in He-Ne system. Neon atoms are the active centers, and the role of helium is to excite neon atoms and cause a population inversion. The probability of energy transfer from helium atoms to neon atoms is more, as there are 10 helium atoms to 1 neon atom in the mixture. For this reason, the probability of reverse transfer of energy, *i.e.*, from neon to helium atom, is extremely small. The E_6 and E_4 levels of neon atoms are also metastable states. Therefore, as the collisions go on, neon atoms accumulate in these two excited states. Successive collisions selectively increase the population of the E_6 and E_4 levels. At ordinary temperatures, the E_5 and E_3 levels of a neon atom are sparsely populated, and a state of population inversion is achieved between E_6, E_5, and E_3 levels and between E_4 and E_3 levels. Consequently, three laser transitions can occur. They are $E_6 \rightarrow E_5$, $E_6 \rightarrow E_3$, and $E_4 \rightarrow E_3$ transitions. As the terminal levels of lasing transitions are sparsely populated, the fraction of neon atoms that must be excited to the upper levels can be much less. As such the power required for pumping is low. Random photons emitted spontaneously set on stimulated emission and coherent radiation is produced.

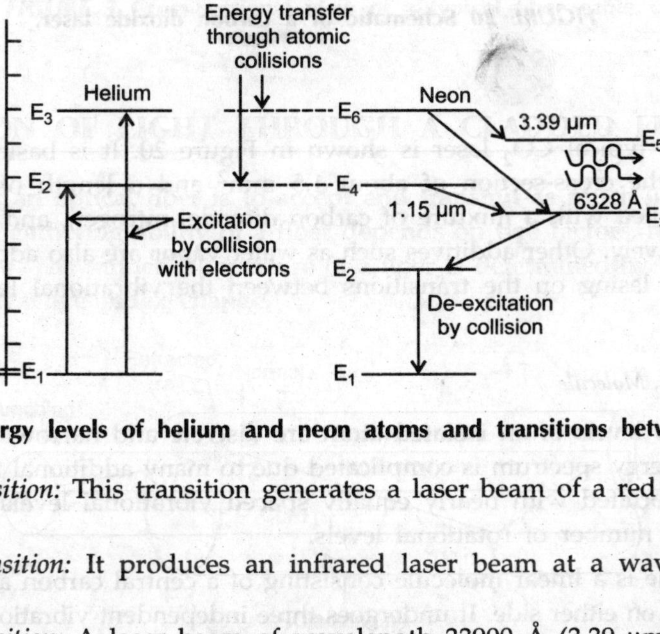

FIGURE 19 **Energy levels of helium and neon atoms and transitions between the levels.**

(*i*) *$E_6 \rightarrow E_3$ transition:* This transition generates a laser beam of a red color of wavelength 6328 Å.

(*ii*) *$E_4 \rightarrow E_3$ transition:* It produces an infrared laser beam at a wavelength of 11500 Å (1.15 µm)

(*iii*) *$E_6 \rightarrow E_5$ transition:* A laser beam of wavelength 33900 Å (3.39 µm) in the far infrared region arises due to this transition. From the terminal levels E_5 and E_3 of lasing transitions, neon atoms can make downward transitions to E_2 levels. Incoherent light is emitted due to these spontaneous transitions. As the lower lasing levels depopulate faster than the upper metastable states, it is easier to maintain the state of population inversion between the lasing levels throughout the time of laser operation.

The E_2 level is a metastable state. Therefore, neon atoms tend to accumulate at this level once again. However, when they drift toward the surrounding discharge tube wall and collide with it, they give up their energy and return to the ground state. These atoms are to be brought to the ground state quickly; otherwise the number of atoms available at the ground state will go on diminishing and the laser output decreases. To increase the probability of atomic collisions with the walls, the discharge tube is made narrow. The He-Ne laser operates in CW mode as the neon atoms are excited to upper levels continuously through collisions.

The Carbon Dioxide Laser

The carbon dioxide laser is a very useful and efficient laser. It is a four-level molecular laser and operates at 10.6 μm in far IR region.

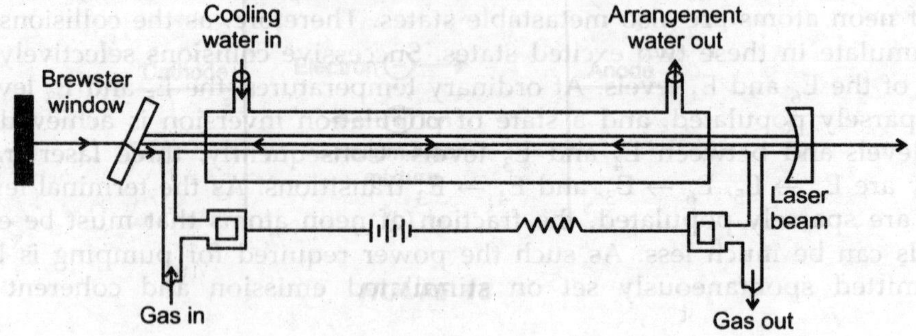

FIGURE 20 **Schematic of a carbon dioxide laser.**

Construction

The schematic of a typical CO_2 laser is shown in Figure 20. It is basically a discharge tube having a bore of the cross-section of about 1.5 mm^2 and a length of about 260 mm. The discharge tube is filled with a mixture of carbon dioxide, nitrogen, and helium gases in 1:2:3 proportions respectively. Other additives such as water vapor are also added. The active centers are CO_2 molecules lasing on the transitions between the vibrational levels of the electronic ground state.

Energy Levels of a CO_2 Molecule

The electron energy levels of an isolated atom are discrete and narrow. However, in the case of molecules the energy spectrum is complicated due to many additional features. Each electron energy level is associated with nearly equally spaced vibrational levels and each vibrational level in turn has a number of rotational levels.

A CO_2 molecule is a linear molecule consisting of a central carbon atom with two oxygen atoms attached one on either side. It undergoes three independent vibrational oscillations known as the vibrational modes. These vibrational degrees of freedom are quantized. At any one time, a CO_2 molecule can vibrate in a linear combination of three fundamental modes. The energy state of the molecule is then represented by three quantum numbers (m n q). These numbers represent the amount of energy associated with each mode. For example, the number (020) indicates that the molecule in this energy state is in the pure bending mode with two units of energy. Each vibrational state is associated with rotational states corresponding to the rotation of a CO_2 molecule about its center of mass. The separations between vibrational-rotational states are much smaller on the energy scale compared to the separations between electron energy levels. The nitrogen molecule N_2 is also characterized by similar vibrational levels. Figure 21 shows the vibrational modes and rotations of a CO_2 molecule.

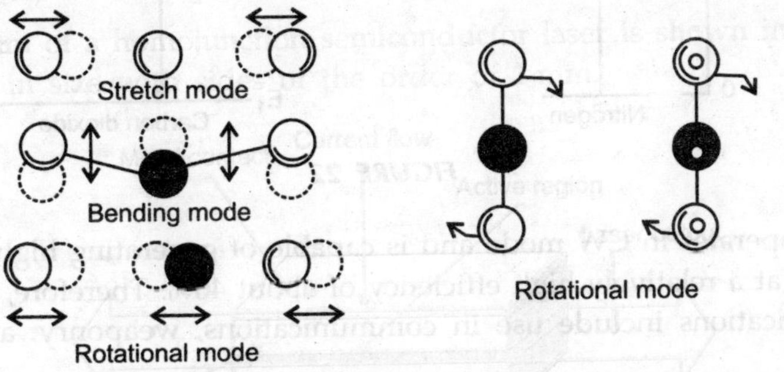

FIGURE 21 **Vibrational modes of a CO_2 molecule.**

Working

Figure 22 shows the lowest vibrational levels of the ground electron energy state of a CO_2 molecule and an N_2 molecule. The excited state of an N_2 molecule is metastable and it is identical in energy to (001) vibrational level of a CO_2 molecule, indicated as E_2 in Figure 22. As the current passes through the mixture of gases, the N_2 molecules get excited to the metastable state. The excited N_2 molecules cannot spontaneously lose their energy and consequently, the number of N_2 molecules at that level keeps on increasing. These N_2 molecules return to a ground state through inelastic collisions with the ground state CO_2 molecules. Thereby the CO_2 molecules are excited to the E_5 level. Some of the CO_2 molecules are also excited to the upper level E_5 through collisions with electrons. The excitation of CO_2 molecules through collisions with excited N_2 molecules is similar to that of helium atoms by neon atoms in a helium-neon laser. The E_5 level is the upper laser level while the (020) and (100) states marked as E_3 and E_4 levels act as the lower laser levels. As the population of CO_2 molecules build up at E_5 level, population inversion is achieved between the E_5 level and the levels at E_4 and E_3. The laser transition between the $E_5 \rightarrow E_4$ levels produces far IR radiation at the wavelength 10.6 μm (1,06,000 Å). The lasing transition between $E_5 \rightarrow E_3$ levels produces far IR radiation at 9.6 μm (96,000 Å) wavelength. E_3 and E_4 levels are also metastable states and the CO_2 molecules at these levels fall to the lower level E_2 through inelastic collisions with normal (unexcited) CO_2 molecules. This process leads to accumulations of population at the E_2 level. Also, as the gaseous mixture heats up, the E_2 level which is close to the ground state E_1, tends to be populated through thermal excitations. Thus, the de-excitation of CO_2 molecules at the lower laser level possesses a problem and inhibits the laser action. The presence of helium along with CO_2 helps to decrease the population density at the E_2 level. It de-excites CO_2 molecules through inelastic collisions and aids cooling the gaseous mixture through heat conduction.

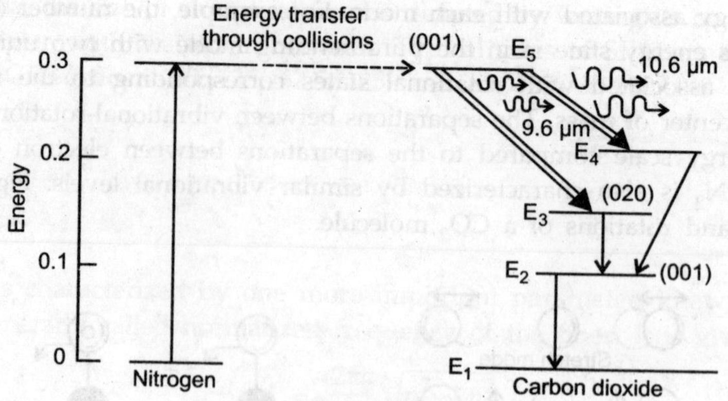

FIGURE 22

The CO_2 laser operates in CW mode and is capable of generating high powers of the order of several kilowatts at a relatively high efficiency of about 40%. Therefore, it is the most widely used laser. Its applications include use in communications, weaponry, and laser fusion.

Liquid Dye Lasers

In liquid dye lasers, an organic dye dissolved in chemicals such as ethyl alcohol, benzene, water, etc., acts as the active center. The lasers operate in the range 3000 Å to 13000 Å.

Semiconductor Diode Lasers

A solid state diode laser, also known as a semiconductor diode laser, is a specially fabricated *pn*-junction device that emits coherent light when it is forward biased. The first semiconductor lasers were made by Hall and his co-workers, and Nathan and his associates in 1962, using gallium arsenide—GaAs. The initial diode lasers are based on junctions formed on the same material and are known as homo-junction lasers. Using dissimilar semiconductors to form a *pn*-junction, heterojunction lasers are made. The present day lasers are known as double heterojunction lasers (DR lasers) in which a thin active layer of a narrow band gap semiconductor is sandwiched between a larger band gap semiconductor. The DH lasers are widely used as light sources in optical communications. We shall confine our study to a simple homojunction *pn*-junction diode laser.

Background

An intrinsic semiconductor is characterised by a valence band and a conduction band separated by a band gap E_g. In the normal state the valence band lying at a lower energy E_V is occupied by a large number of electrons and a small number of holes. The conduction band E_C located at an energy E_g above E_V is sparsely occupied by electrons. Occasionally, an electron may fall from a conduction band to a valence band and recombine with one of the holes. The energy corresponding to $(E_C - E_V) = E_g$ is released in the form of a photon $h\nu$. The electron hole recombination is the basic mechanism involved in the emission of radiation. In case of silicon and germanium the energy is released in the form of heat. Hence these materials are not suitable for fabricating lasers. In case of gallium arsenide GaAs, cadmium selenide CdSe, etc., direct band gap semiconducting materials, the energy released in the process of electron-hole recombination lies in the optical region. Therefore they are suitable materials for making lasers. GaAs is more commonly used in making semiconductor lasers.

Semiconductor Diode Laser

A semiconductor diode laser is a specially fabricated *pn*-junction device that emits coherent light when it is forward biased.

Construction

A schemetic diagram of a homojunction semiconductor laser is shown in Figure 23. The diode is extremely small in size with sides of the order of 1 mm.

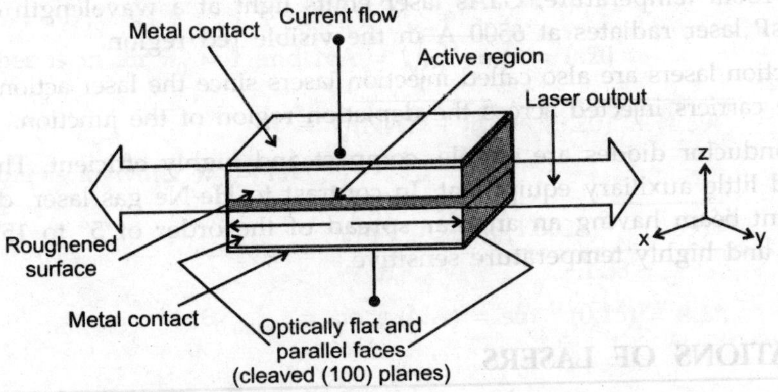

FIGURE 23 **Schematic of a semiconductor diode laser.**

The junction lies in a horizontal plane through the center. The top and bottom faces are metallized and ohmic contacts are provided to pass current through the diode. The front and rear faces are polished parallel to each other and perpendicular to the plane of the junction. The polished faces constitute the Fabry-Perot resonator. The other two opposite faces are roughened to prevent lasing action in that direction. The active region consists of a layer of about 1 μm thickness.

Working

A simple way of achieving population inversion in a semiconductor is to use it in the form of a heavily doped *pn*-junction and to forward bias the junction. The energy band diagram of such a *pn*-junction diode having highly doped *p*- and *n*-regions is shown in Figure 24. With very high doping on the *n*-side, the donor levels as well as a portion of the conduction band, are occupied by electrons and the Fermi level lies within the conduction band. Similarly, on the heavily doped *p*-side the acceptor levels are unoccupied and holes exist in the valence band and the Fermi level lies within the valence band. At thermal equilibrium, the Fermi level is uniform across the junction, as shown in Figure 24. When a forward bias is applied to the junction, the energy levels shift and the new distribution as shown in Figure 24 will be taken up. Electrons and holes are injected into the depletion region which, as a result, decreases in width. The injected electrons and holes appear in high concentrations in this transition region. At a low forward current level, the electron hole recombinations cause a spontaneous emission of photons and the junction act as an LED. The bandwidth of the emitted light will be larger. As the current is increased, the intensity of light increases linearly. When the current reaches a threshold value the carrier concentration in the depletion region will reach very high values. The region contains a large concentration of electrons within the conduction band and a large concentration of holes within the valence band, as indicated in the hatched region on Figure 24(*b*). The upper levels in the depletion region are having high population density of electrons while the lower levels in the same region are vacant. This is the state of population inversion. The narrow region where the state of population inversion is achieved is called *inversion region* or *active region*. Thus the forward bias (current) plays the role of pumping agent in a semiconductor diode laser. The photons that propagate in the junction plane induce the conduction electrons to jump into the vacant states of the valence band. The stimulated electron hole recombinations cause emission of coherent radiation of very narrow bandwidth. At room temperature, GaAs laser emits light at a wavelength of 9000 Å in IR region. A GaAsP laser radiates at 6500 Å in the visible red region.

The *pn*-junction lasers are also called injection lasers since the laser action is generated by minority charge carriers injected across the depletion region of the junction.

The semiconductor diodes are simple, compact and highly efficient. They require very little power and little auxiliary equipment. In contrast to He-Ne gas laser, diode lasers give a more divergent beam having an angular spread of the order of 5° to 15°. They are less monochromatic and highly temperature sensitive.

7.14 APPLICATIONS OF LASERS

It is not an exaggeration to say that next to computers it is the laser that is bringing revolutionary changes in our civilization. Lasers are profitably used in almost every field. They are used in fundamental research. They are applied in entertainment electronics, industrial electronics, communications, mechanical working, machining, welding, surveying, astronomy, mechanical measurements, information processing, surgery and related medical fields, and even in warfare to guide missiles to targets. The list is unending and continues to grow. Low power semiconductor lasers are used in CD players, laser printers, laser copiers, facsimile machines, etc., and office automation equipment; optical floppy discs, optical memory cards, etc., measurement instruments, and as light sources, and information transmitters in optical communications. High power gas and solid state lasers are used in machining, welding, and cutting applications. Very high power lasers are used to bring about thermonuclear reactions which would become the ultimate inexhaustible power source for human civilization. One of the most important applications of the laser is in the production of three-dimensional images of an object in a process called holography.

Lasers are profitably used in almost every field including fundamental research. The common applications of lasers are follows:

- Metallic rods can be melted and joined by means of a laser beam (laser welding).
- The laser beam is used to vaporise unwanted material during the manufacture of electronic circuits on semiconductor chips.
- CO_2 gas lasers of about 100 W output are helpful in surgery because they seal small blood vessels while cutting through tissue by vaporizing water in the path of their IR beams.

- Lasers are used to detect and destroy enemy missiles during warfare.
- Low power semiconductor lasers are used in CD (compact disc) players, laser printers, facsimile machines, etc.
- Semiconductor lasers are ideal for fiber optic transmission lines in which the electric signals that would normally be sent along copper wires are first converted into a series of pulses according to a standard code. Lasers then turn the pulses into flashes of IR light that travel along thin glass fibers and at the other end are changed back into electric signals.
- High-power lasers are used to bring about thermonuclear reactions which could become the ultimate inexhaustible power source for human civilization.
- Lasers are also being employed for separating the various isotopes of an element.
- Lasers are used in the production of three-dimensional images of an object in holography.
- Laser beams are also being used in the "inertial confinement" of plasma.
- The narrow red laser beam is used in supermarkets to read bar codes.

Holography

In conventional photography a negative is made first and using it, then a positive print is produced later. The positive print is only a two dimensional record of light intensity received from a three-dimensional object. It contains information about the square of the amplitude of the light wave that produced the image but information about the phase of the light wave is not recorded and is lost. In 1947 Dennis Gabor, the English physicist, outlined a radically new technique of photographing objects. He called this technique wavefront construction. According to this technique both the phase and intensity attributes of the wave are recorded and when viewed the photograph shows a three dimensional image of the object. This technique is named holography. In 1971 Gabor was awarded the Nobel prize in physics for this invention.

Figure 24 illustrates the principle of holography. A weak but broad beam of laser light is split into two beams, a reference beam and an object beam. The reference beam is allowed to reach the photographic plate directly while the object beam illuminates the object. Part of the

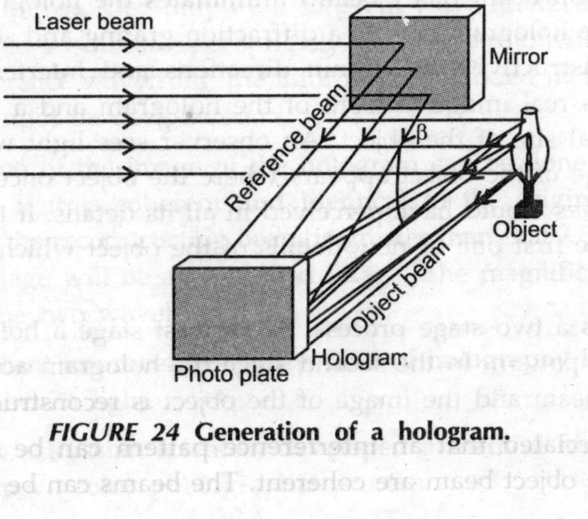

FIGURE 24 **Generation of a hologram.**

light scattered by the object travels towards the photographic plate and interfaces with the reference beam and produces an interference pattern on the photographic plate. The photographic plate carrying the interference pattern is called a hologram. "Holes" means complete in Greek and "gramma" means writing. Thus a hologram means complete recording. Like any ordinary photographic plate, a hologram is developed, fixed and stored.

A hologram does not contain a distinct image of the object. It is only a record of the interference pattern formed by the superposition of two coherent light beams. The interference pattern on a hologram consists of a complex pattern of alternate regions of dark and bright fringes. The hologram is also called a Gabor zone plate in honour of Dennis Gabor who conceived the principle of holography.

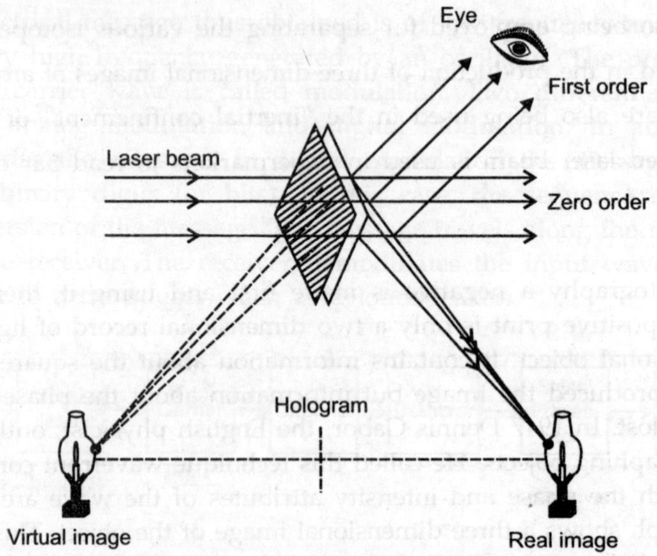

FIGURE 25 **Viewing of hologram.**

Whenever required, the object can the viewed by illuminating the hologram as shown in Figure 25. A laser beam identical to the reference beam is used for the reconstruction of the object. The reconstruction beam (laser beam) illuminates the hologram at the same angle as the reference beam. The hologram acts as a diffraction grating and secondary waves from the hologram interfere constructively in certain directions and interfere destructively in other directions. They form a real image in front of the hologram and a virtual image behind the hologram at the original site of the object. An observer sees light waves diverging from the virtual image. An image of the object appears where the object once stood and that image is identical to what our eyes would have perceived in all its details. If the observer tilts his head other objects behind the first one or new details of the object which were not noticed earlier would be observed.

Holography is thus a two stage process. In the first stage a hologram is recorded in the form of an interference pattern. In the second stage the hologram acts as a diffraction grating for the reconstruction beam and the image of the object is reconstructed from the hologram.

It should be appreciated that an interference pattern can be obtained only when the reference beam and the object beam are coherent. The beams can be coherent only when they belong to the same original wave train. As the two beams travel over a large distance it would be difficult to fulfil this condition with ordinary sources of light. An interference pattern can be obtained when the optical path difference between two beams Δ is very small compared to the length of the wave train, *i.e.*, coherence length λ_{coh}. Then, only the two parts of the wave train in the two beams will be coherent at the point of observation. For conventional monochromatic sources of light, the coherence length is found to be of the order of a few centimeters and a hologram cannot be obtained with them. Therefore, holography could not be realized in practice for a long time though the principle was formulated in 1947. The invention of lasers made holography possible. Laser beams have a coherence length of the order of a few kilometers and, therefore, the interference of split rays with a large path difference could be achieved. The first practical hologram was made using a laser beam by two American investigators, E. N. Leith and Y. Upatneiks in 1964.

Advantages and Applications of Holography

The following are the advantages and applications of Holography:

- In an ordinary photograph each region contains a separate and individual original object. Therefore, destruction of a portion of a photographic image results in irreparable loss of information corresponding to the destroyed part. On the other hand, in a hologram each part contains information about the entire object. Therefore, destruction of a part of a hologram does not cause a loss of information about the object. From even a small part of the hologram, the entire image can still be reconstructed, if only with reduced clarity and definition of the image. Therefore, a hologram is a reliable medium for data storage.
- It is not useful to record several images on a single photo film. Such a record cannot give information about any of the individual images. On the other hand, several images can be recorded on a hologram. Therefore, the information holding capacity of a hologram is extremely high. While a 6 × 9 mm photograph can hold one printed page, a hologram of the same size can store up to 300 such pages.
- On a hologram, information is recorded in the form of an interference pattern. The type of pattern obtained depends on the reference beam used to record the hologram. The information can be recorded only by a coherent wave identical to that of the wave. The reference wave can be chosen appropriately. Consequently, without the knowledge of the shape of the reference wavefront, the information encoded in the interference pattern on the hologram cannot be deciphered.
- The reconstruction of the image of the hologram can be done with a reference beam of any wavelength if it is coherent and identical to the original reference beam. If the wavelength λ of the reconstructing beam is greater than that λ_0 of the reference beam, the reconstructed image will be a magnified image. The magnification will be proportional to the ratio of the two wavelengths.
- The most significant improvement in the field of holography is that now holograms can be viewed with white light.
- Yu N. Denisyuk a Russian scientist, developed a method for making color holograms based on Lippman's color photographic process.

Acoustical holography is another area which is in a developmental stage. It is easy to produce coherent sound waves. Sound waves readily propagate in solids. Therefore, a three-dimensional acoustical hologram of an opaque object can be made. By viewing such holograms though invisible, the internal structure of an opaque object can be observed. Such techniques will be highly useful in the fields of medicine and technology.

7.15 MASER

The first successful maser was the ammonia laser built by Gordan, Zeiger, at Columbia University in 1954.

The active medium of the maser is gaseous ammonia. The ammonia molecule NH_3 has a triangular pyramidal structure Figure 26(*a*). The three hydrogen atoms lie at the corners of a triangle which forms the base of the pyramid. The nitrogen atom is located at the apex of the pyramid. The nitrogen atom flips up and down from the upper apex to the lower apex point. The two configurations have the same energy. Because of this fact the ammonia molecule has a set of energy levels. Each energy level consists of a close spaced pair of levels. The vibrational levels shown in Figure 26(*b*) corresponding to the lowest pair are used in an ammonia maser. The energy difference between the pair of levels correspond to a microwave frequency of 23 m 870 MHz. The schematic of the apparatus used is shown in Figure 27. Under normal conditions the two states are equally populated. To achieve a state of population inversion, the number of molecules in the lower state is to be reduced. By passing the beam of ammonia molecules through a strong non-homogeneous electric field, the molecules in the two states can be separated. The molecules in the upper state are subsequently led into a microwave cavity that resonates at 23,870 MHz. It corresponds to a wavelength of 1.256 cm. A feeble signal of 23,870 MHz frequency applied to the cavity causes stimulated transitions of molecules from the upper state to the lower state leading to the emission of radiation at that frequency. The output is a highly amplified form of the input signal.

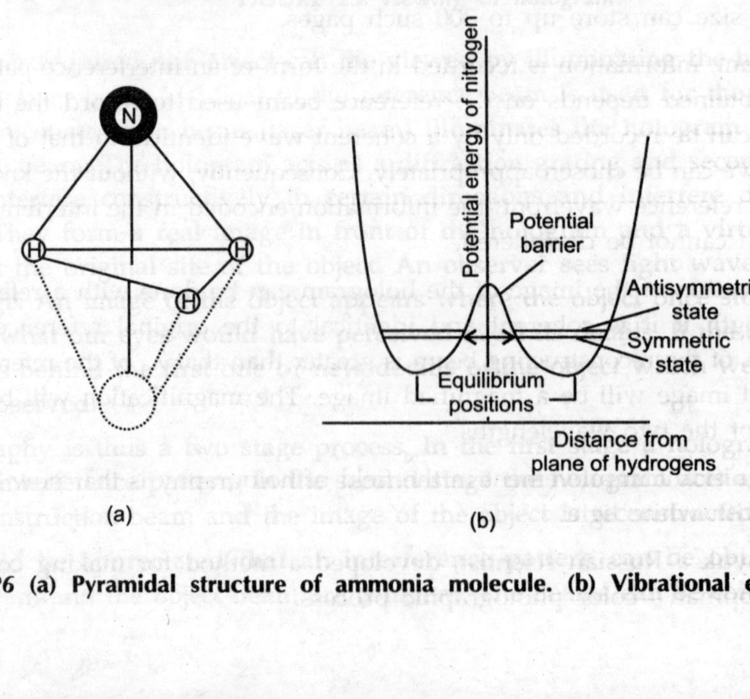

FIGURE 26 **(a) Pyramidal structure of ammonia molecule. (b) Vibrational energy levels.**

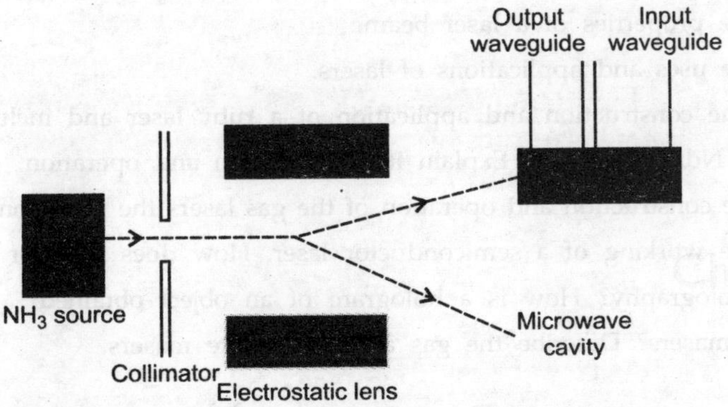

FIGURE 27 **Schematic of an ammonia maser.**

Ammonia masers have a very low power output. Their frequency of oscillation is highly stable. Hence they are useful as frequency sources of clocks. Masers are cumbersome devices and have limited applications.

EXERCISES

Short Questions for Discussion

1. What is a laser? Define spontaneous and stimulated emissions.
2. What is the basic condition in which stimulated emissions dominate?
3. Distinguish between spontaneous and stimulated emissions.
4. Define pumping and population inversion in laser action.
5. Why is population inversion often referred to as a negative temperature state?
6. What is the difference between ordinary light and a laser beam?
7. What is meant by amplification of light? How is it achieved in lasers?
8. Explain the action of an optical resonant cavity.
9. Discuss the properties of a laser beam.
10. What are the various applications of lasers?

Long Questions for Discussion

1. What is a laser? Explain the difference between ordinary light and a laser beam.
2. Differentiate between 'spontaneous emission' and 'stimulated emission' of radiation. Obtain a relation between transition probabilities of the two.
3. What is meant by active medium and active centers. Discuss the essential requirements for producing laser action.
4. Describe the working principle of an optical resonance cavity. Why is it often called a Fabry-Perot resonator?
5. Explain population inversion? How it is achieved by optical pumping?

6. What is optical pumping? Describe it in three-level and four-level lasing systems.
7. Explain the properties of a laser beam.
8. Discuss the uses and applications of lasers.
9. Describe the construction and application of a ruby laser and include necessary diagrams.
10. What is a Nd: YAG laser? Explain its construction and operation.
11. Explain the construction and operation of the gas lasers the He-Neon laser and the CO_2 laser.
12. Discuss the working of a semiconductor laser. How does it differ from other lasers?
13. What is holography? How is a hologram of an object obtained?
14. What are masers? Describe the gas and solid state masers.

Chapter 8

FIBER OPTICS

8.1 OVERVIEW

John Tyndall, a British physicist, demonstrated to the Royal Society in 1870 that light can be guided along a curved stream of water. Owing to total internal reflections light gets confined to the water stream and the stream appears luminous. A luminous water stream was the precursor of an optical fiber.

Electronic communications use radiowaves and microwaves to carry information over copper wire and coaxial cables. The information carrying capacity of these wires is highly restricted in view of their limited bandwidths and does not suffice for modern needs. The use of light waves, in place of radiowaves and microwaves, enhances tremendously the number of signals that can be transmitted simultaneously. Historically, the first technical effort to use light waves as a vehicle of communication was due to Alexander Graham Bell. He invented the photophone in 1880 which transmitted speech on a beam of light. By 1960, it had been established that light could be guided by a glass fiber but the fibers available at that time heavily attenuated light propagating through them. It was the fabrication of low-loss glass fibers by Corning Glass Works and the invention of continuous wave room temperature solid state diode lasers in 1970 that made optical communications practicable. The overwhelming advantages of optical fibers over the traditional wires and coaxial cables were soon recognized. Commercial communication systems based on optical fiber cables made their appearance by 1977. The area of photonics has been ushered in.

Apart from the use as a communication channel, optical fibers are widely employed in other areas. Sensors for detecting electrical, mechanical, and thermal energies are made using optical fibers. Inexpensive and simple display systems, copying machines, etc., utilize fiber optics. Fibroscopes are used in a variety of forms in medical diagnostics.

8.2 OPTICAL FIBERS

Optical fibers are glass or plastic conduits as thin as a human hair, designed to guide light waves along their length. An optical fiber works on the principle of total internal reflection. When light enters one end of the fiber, it undergoes successive total internal reflections from side walls and travels down the length of the fiber along a zigzag path as shown in Figure 1(*a*). A small fraction of light may escape through side walls but a major fraction emerges out from the other end of the fiber, as illustrated in Figure 1(*b*).

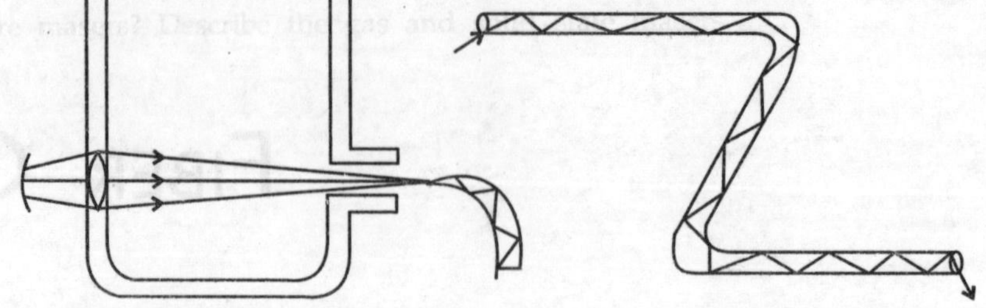

FIGURE 1 **(a) Illustration of a water stream conducting light along its path. (b) Illustration of a transparent fiber guiding light along its length.**

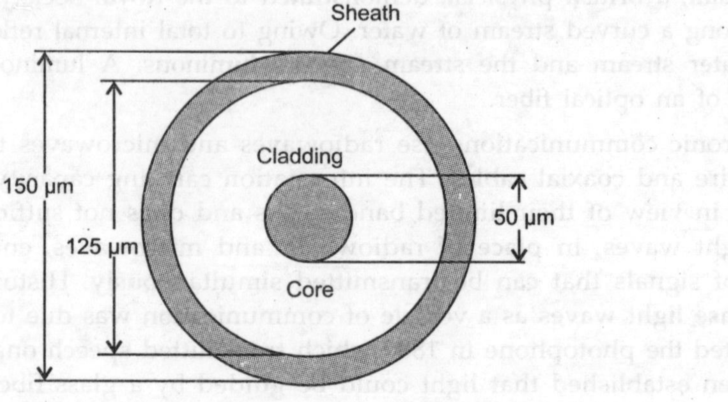

FIGURE 2 **Cross-sectional view of an optical fiber.**

A practical optical fiber has in general three coaxial regions (Figure 2). The inner most region is the light guiding region known as the core. It is surrounded by a coaxial middle region known as the cladding. The outermost region is called a sheath. The refractive index of cladding is always lower than that of the core. The purpose of the cladding is to make the light be confined to the core. Light launched into the core and striking the core-to-cladding interface at greater than critical angle will be reflected back into the core. Since the angles of incidence and reflection are equal, the light will continue to rebound and propagate through the fiber. The sheath protects the cladding and the core from abrasions, contamination and the harmful influence of moisture. In addition, it increases the mechanical strength of the fiber.

Optical fibers are constructed, either as a single fiber or a flexible bundle or cable. A fiber bundle is a number of fibers in a single jacket. Each fiber carries light independently. The cross-sectional view of a typical telecommunication cable is shown in Figure 3. It contains six fibers and has an insulated steel cable at the center for providing tensile strength. Each optical fiber consists of a core surrounded by a cladding which in turn is coated with an insulating jacket. The fibers are thus individually buffered and strengthened. Six insulated copper wires are distributed in the space between the fibers. They are used for electrical transmission, if required. The fibers are wrapped with mylar tape to bind the assembly. The assembly is then fitted in a corrugated aluminum sheath which acts as a shield. A polyethylene jacket is applied over the top.

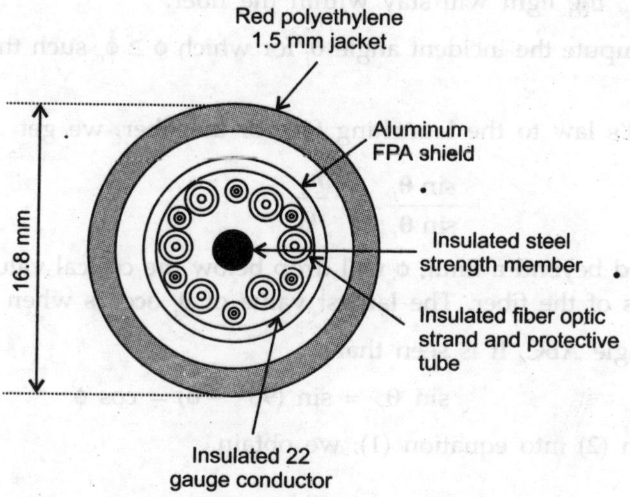

FIGURE 3 **Cross-sectional view of a typical fiber cable.**

8.3 PROPAGATION OF LIGHT THROUGH A CLADDED FIBER

The main function of an optical fiber is to accept and transmit as much light from the source as possible. The light gathering ability of a fiber depends on two factors, namely core size and the numerical aperture. The numerical aperture of a fiber is determined by the acceptance angle and the fractional refractive index change.

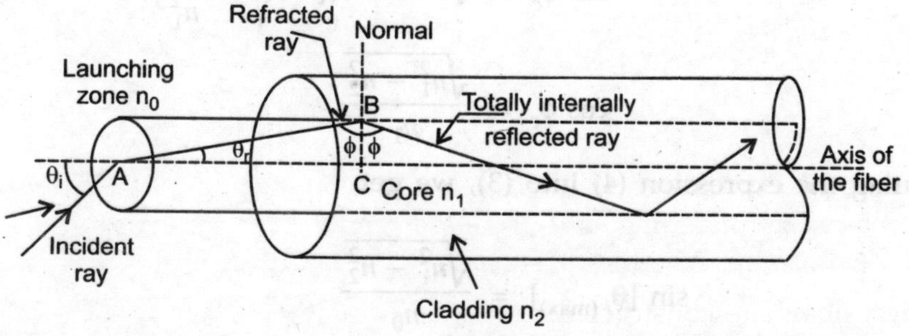

FIGURE 4 **Illustration of the path of a light ray incident on the end of an optical fiber at an angle θ_i to the fiber axis.**

Acceptance Angle and Acceptance Cone

Let us consider an optical fiber into which light is launched. The end at which light enters the fiber is called the launching end. Figure 4 depicts the conditions at the launching end. Let the refractive index of the core be n_1 and the refractive index of the cladding be n_2. n_2 is less than n_1. Let n_0 be the refractive index of the medium from which light is launched into the fiber. Let a light ray enter the fiber at an angle θ_i to the axis of the fiber. The ray refracts at an angle θ_r and strikes total internal reflection at the angle ϕ. If ϕ is greater than the critical angle ϕ_c, the ray undergoes total internal reflection at the interface, since $n_1 > n_2$. As long as the angle ϕ is greater than ϕ_c, the light will stay within the fiber.

Let us now compute the incident angle θ_i for which $\phi \geq \phi_c$ such that light rebounds within the fiber.

Applying Snell's law to the launching face of the fiber, we get

$$\frac{\sin \theta_i}{\sin \theta_r} = \frac{n_1}{n_0} \tag{1}$$

If θ_i is increased beyond a limit, ϕ will drop below the critical value ϕ_c and the ray escapes from the side walls of the fiber. The largest value of θ_i occurs when $\phi = \phi_c$.

From the triangle ABC, it is seen that

$$\sin \theta_r = \sin (90° - \phi) = \cos \phi \tag{2}$$

Using equation (2) into equation (1), we obtain

$$\sin \theta_i = \frac{n_1}{n_0} \cos \phi$$

When $\phi = \phi_c$ (Critical angle) Then $\theta_i = \theta_{i\,(\max)}$

$$\sin [\theta_{i\,(\max)}] = \frac{n_1}{n_0} \cos \phi_c \tag{3}$$

But $\sin \phi_c = \dfrac{n_2}{n_1}$

$$\cos \phi_c = \sqrt{1 - \sin^2 \phi_c} = \sqrt{1 - \frac{n_2^2}{n_1^2}}$$

$$\therefore \quad \cos \phi_c = \frac{\sqrt{n_1^2 - n_2^2}}{n_1} \tag{4}$$

Substituting the expression (4) into (3), we get

$$\sin [\theta_{i\,(\max)}] = \frac{\sqrt{n_1^2 - n_2^2}}{n_0} \tag{5}$$

Quite often the incident ray is launched from an air medium, for which $n_0 = 1$.

Designating $\theta_{i\,(max)} = \theta_0$, equation (5) may be simplified to

$$\sin\theta_0 = \sqrt{n_1^2 - n_2^2} \tag{6}$$

$$\therefore \quad \theta_0 = \sin^{-1}\left[\sqrt{n_1^2 - n_2^2}\right] \tag{7}$$

The angle θ_0 is called the acceptance angle of the fiber. The acceptance angle may be defined as the maximum angle that a light ray can have relative to the axis of the fiber and propagate down the fiber. The light rays contained within the cone having a full angle $2\theta_0$ are accepted and transmitted along the fiber. Therefore, the cone is called the acceptance cone (Figure 5). Light incident at an angle beyond θ_0 refracts through the cladding and the corresponding optical energy is lost. It is obvious that the larger the diameter of the core, the larger the acceptance angle.

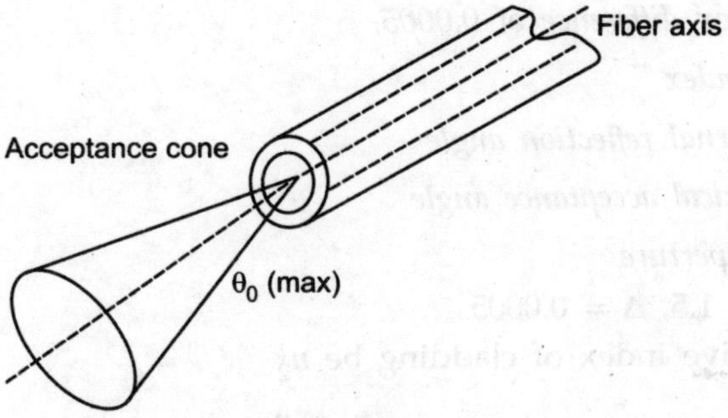

FIGURE 5 **An optical fiber accepts only those rays which are incident within a cone having a semiangle θ_0.**

Fractional Refractive Index Change

The fractional difference Δ between the refractive indices of the core and the cladding is known as the fractional refractive index change. It is expressed as

$$\Delta = \frac{n_1 - n_2}{n_1} \tag{8}$$

This parameter is always positive because n_1 must be larger than n_2 for the total internal reflection condition. In order to guide light rays effectively through a fiber, $\Delta \ll 1$. Typically, Δ is of the order of 0.01.

Numerical Aperture

The Numerical Aperture (NA) is defined as the sine of the acceptance angle. Thus,

$$NA = \sin\theta_0$$

$$\therefore \quad NA = \sqrt{n_1^2 - n_2^2} \tag{9}$$

$$n_1^2 - n_2^2 = (n_1 + n_2)(n_1 - n_2)$$

$$= \left(\frac{n_1 + n_2}{2}\right)\left(\frac{n_1 - n_2}{n_1}\right) 2n_1$$

Approximating $\frac{n_1 + n_2}{2} \approx n_1$, we can express the above relation as

$$(n_1^2 - n_2^2) = 2n_1^2 \Delta$$

$$\therefore \quad \text{NA} = n_1 \sqrt{2\Delta} \quad (10$$

The numerical aperture determines the light gathering ability of the fiber. It is a measur of the amount of light that can be accepted by a fiber. It is seen from equation (9) that NA i dependent only on the refractive indices of the core and cladding materials. Its value range from 0.13 to 0.50. A large NA implies that a fiber will accept a large amount of light from th source.

Example 1. *A glass clad fiber is made with core glass of refractive index 1.5 and the cladding is dope to give a fractional index difference of 0.0005.*

Find (a) the cladding index

(b) the critical internal reflection angle

(c) the external critical acceptance angle

(d) the numerical aperture

Solution. Given: $n_1 = 1.5$, $\Delta = 0.0005$

(*a*) Let the refractive index of cladding be n_2.

$$\Delta = \frac{n_1 - n_2}{n_1}$$

$$\therefore \quad 0.0005 = \frac{1.5 - n_2}{1.5}$$

$$\therefore \quad n_2 = 1.5 - 1.5 \times 0.0005 = 1.49925$$

(*b*) Let the critical internal reflection angle be ϕ_c.

$$\sin \phi_c = \frac{n_2}{n_1}$$

$$\therefore \quad \phi_c = \sin^{-1}\left[\frac{n_2}{n_1}\right]$$

$$= \sin^{-1}\left[\frac{1.49925}{1.5}\right]$$

$$= \sin^{-1}(0.9995) = 88.2°$$

(*c*) Let the external critical acceptance angle be θ_0.

$$\sin \theta_0 = \frac{\sqrt{n_1^2 - n_2^2}}{n_0}; \quad \text{where } n_0 = 1$$

$$\therefore \quad \theta_0 = \sin^{-1}\left[\sqrt{n_1^2 - n_2^2}\right]$$

$$= \sin^{-1}(1.5^2 - 1.49925^2)^{1/2}$$

$$= \sin^{-1} (2.25 - 2.24775)^{1/2}$$
$$= \sin^{-1} (0.0474)$$
$$= 2.72°$$

(*d*) Numerical Aperture $\quad \text{NA} = n_1\sqrt{2\Delta}$

$\therefore \quad \text{NA} = 1.5 \sqrt{2 \times 0.0005} = 1.5\ (0.03162) = \mathbf{0.0474.}$

xample 2. *An optical fiber has a NA of 0.20 and a cladding refractive index of 1.59. Determine the ceptance angle for the fiber in water which has a refractive index of 1.33.*

olution. $\text{NA} = \dfrac{\sqrt{n_1^2 - n_2^2}}{n_0}$

When the fiber is in air $n_0 = 1$ and $\text{NA} = \sqrt{n_1^2 - n_2^2} = 0.20$

$$\therefore \quad n_1 = \sqrt{(\text{NA})^2 + n_2^2} = \sqrt{(0.20)^2 + (1.59)^2} = 1.6025$$

When the fiber is in water $n_0 = 1.33$

$$\therefore \quad \text{NA} = \frac{\sqrt{n_1^2 - n_2^2}}{n_0} = \frac{\sqrt{(1.6025)^2 - (1.59)^2}}{1.33} = 0.15$$

$$\therefore \quad \theta_{0\,(max)} = \sin^{-1} (\text{NA}) = \sin^{-1} (0.15) = \mathbf{8.6°.}$$

4 MODES OF PROPAGATION

ght propagates as an electromagnetic wave through an optical fiber. It is true that all waves, ving ray directions above the critical angle, will be trapped within the fiber due to total ternal reflection. But it is not true that all such waves propagate along the fiber. In reality, ly certain ray directions are allowed to propagate. The allowed directions correspond to the odes of the fiber. In simple terms, modes can be visualized as the possible number of paths of ght in an optical fiber. The paths are all zigzag paths excepting the axial direction. Accordingly, ght rays travelling through a fiber are classified as axial rays and zigzag rays. As a ray gets peatedly reflected at the walls of the fiber, phase shift occurs. Consequently, the waves travelling ong certain zigzag paths will be in phase and intensified while the waves coursing along ertain other paths will be out of phase and diminish due to destructive interference. The light y paths along which the waves are in phase inside the fiber are known as modes. The number f modes that a fiber will support depends on the ratio d/λ, where d is the diameter of the core nd λ is the wavelength of the wave being transmitted (Figure 6).

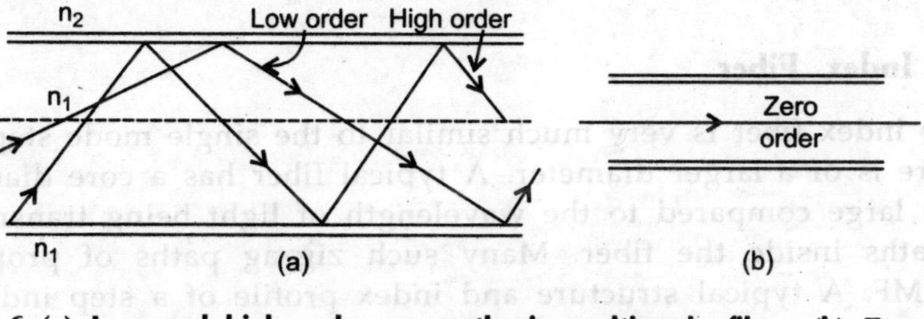

FIGURE 6 **(a) Low and high order ray paths in multimode fiber. (b) Zero order ray path (axial ray) in a single mode fiber.**

Modes are designated by an order number *m*. In a fiber of fixed thickness, the higher order modes propagate at smaller angles than the lower order modes. The zero order ray travels along the axis and is known as an axial ray.

8.5 TYPES OF OPTICAL FIBERS

Optical fibers are in general of two types: namely the single mode fiber (SMF) and the multimode fiber (MMF). A single mode fiber has a smaller core diameter and can support only one mode of propagation. On the other hand, a multimode fiber has a larger core diameter and supports a number of modes. Multimode fibers (MMF) are further distinguished on the basis of index profile. An index profile is a plot of refractive index drawn on a horizontal axis versus the distance from the core axis drawn on the vertical axis. The index profile of a MMF can be either a step index (SI) type or a graded index (GRIN) type. The index profile of a SMF is usually a step index (SI) type.

Single Mode Step Index Fiber

A single mode step index fiber has a very fine thin core of a uniform refractive index of a higher value which is surrounded by a cladding of a lower refractive index. The refractive index changes abruptly at the core-cladding boundary, as illustrated in Figure 7, because of which it is known as a step index fiber.

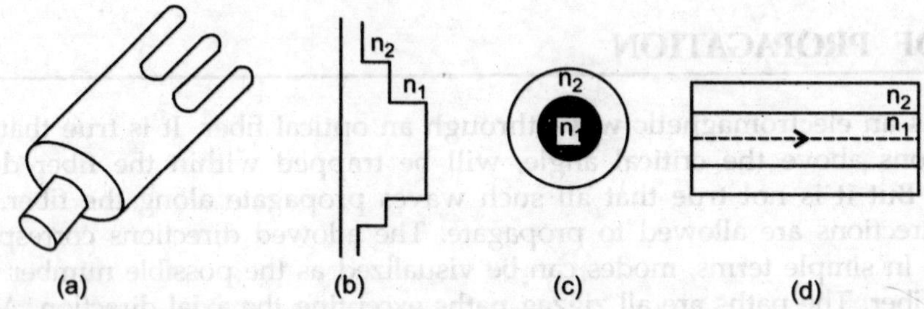

FIGURE 7 **Single mode step index fiber (a) Cable (b) Refractive index profile (c) Cross-section (d) Side view showing mode of propagation**

The fiber is surrounded by an opaque protective sheath. A typical SMF has a core diameter of 4 μm, which is of the order of a few wavelengths of light. Light travels in the SMF along a single path, that is along the axis. Obviously, it is the zero order mode that is supported by a SMF. A SMF is characterized by a very small value of Δ. It is of the order of 0.002.

Multimode Step Index Fiber

A multimode step index fiber is very much similar to the single mode step index fiber except that its core is of a larger diameter. A typical fiber has a core diameter of 100 μm which is very large compared to the wavelength of light being transmitted. Light follows zigzag paths inside the fiber. Many such zigzag paths of propagation are permitted in a MMF. A typical structure and index profile of a step index MMF are shown in Figure 8. The NA of a MMF is larger as the core diameter of the fiber is larger. It is of the order of 0.3.

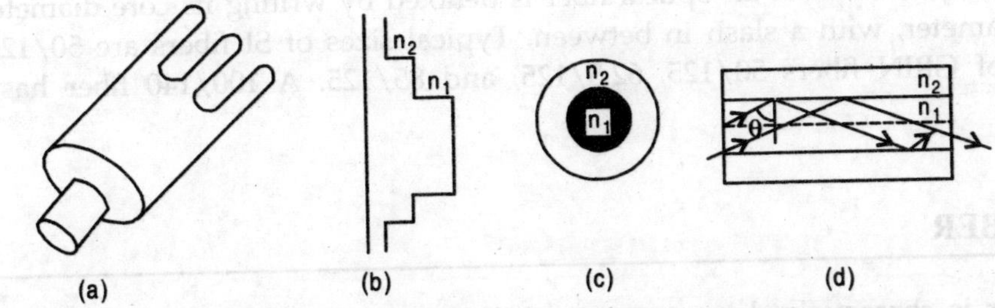

FIGURE 8 **Multimode step index fiber. (a) Cable (b) Refractive index profile (c) Cross-section (d) Side view showing modes of propagation.**

Graded Index (GRIN) Fiber

A graded index fiber is a multimode fiber with a core consisting of concentric layers of different refractive indices. Therefore, the refractive index of the core varies with the distance from the fiber axis. It has a high value at the center and falls off with increasing radial distance from the axis. A typical structure and its index profile are shown in Figure 9. Such a profile causes a periodic focussing of the light propagating through the fiber.

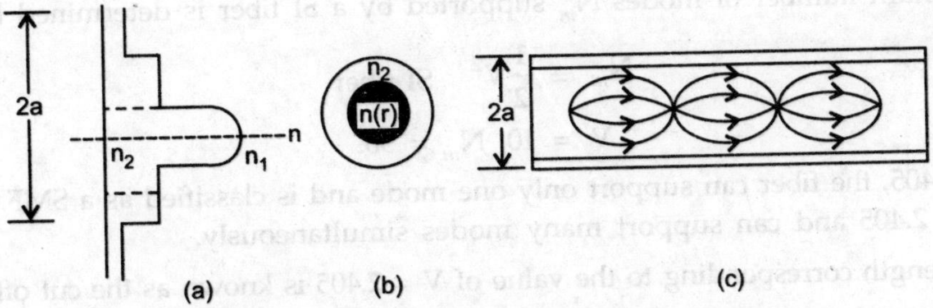

FIGURE 9 **Multimode GRIN fiber. (a) Refractive index profile (b) Cross-section (c) Side view showing typical ray paths along a GRIN fiber.**

In the case of GRIN fibers, the acceptance angle and numerical aperture decrease with the radial distance from the axis (Figure 10). For fibers of parabolic index profile, the numerical aperture is given by

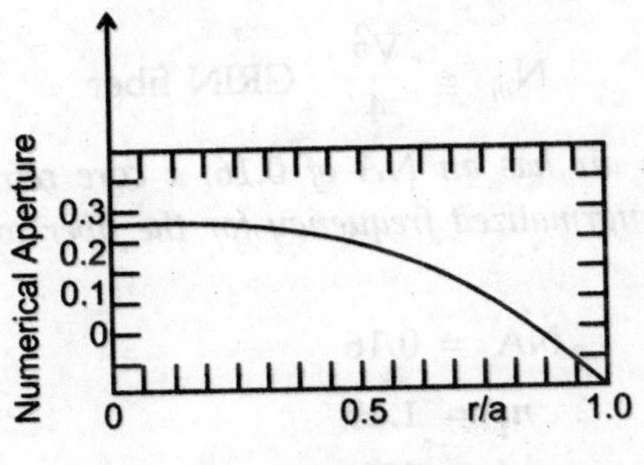

FIGURE 10 **Illustration of the variation of a numerical aperture of a parabolic index fiber.**

$$\text{NA} = n_1 (2\Delta)^{1/2} \sqrt{1 - (r/a)^2} \tag{11}$$

Conventionally, the size of an optical fiber is denoted by writing its core diameter and then its cladding diameter, with a slash in between. Typical sizes of SI fibers are 50/125, 100/140, 200/230, and of GRIN fibers 50/125, 62.5/125, and 85/125. A 100/140 fiber has a 100 μm cladding.

8.6 V-NUMBER

An optical fiber is characterized by one more important parameter, known as the V-number, which is more generally called normalized frequency of the fiber. It is given by the relation

$$V = \frac{2\pi a}{\lambda}\sqrt{n_1^2 - n_2^2} \tag{12}$$

where a is the radius of the core and λ is the free space wavelength. Using equations (9) and (10), we can express V as

$$V = \frac{2\pi a}{\lambda}(\text{NA}) \tag{13}$$

and

$$V = \frac{2\pi a}{\lambda} n_1\sqrt{2\Delta} \tag{14}$$

The maximum number of modes N_m supported by a SI fiber is determined by

$$N_m \cong \frac{1}{2}V^2 \quad \text{SI fiber} \tag{15}$$

Thus for $V = 10,\ N_m = 50.$

For $V < 2.405$, the fiber can support only one mode and is classified as a SMF. MMFs have values of $V > 2.405$ and can support many modes simultaneously.

The wavelength corresponding to the value of $V = 2.405$ is known as the cut off wavelength λ_C of the fiber.

$$\lambda_C = \frac{\lambda V}{2.405} \tag{16}$$

It may be deduced from equation (14) that single mode properties can be realized by decreasing the core diameter.

In case of GRIN fibers, for large values of V,

$$N_m \cong \frac{V^2}{4} \quad \text{GRIN fiber} \tag{17}$$

Example 3. *A step index fiber in air has an NA of 0.16, a core refractive index of 1.45, and a core diameter of 60 cm. Determine the normalized frequency for the fiber when light at a wavelength of 0.9 μm is trasmitted.*

Solution. Given

$$\text{NA} = 0.16$$
$$n_1 = 1.45$$
$$d = 60 \text{ cm} = 0.6 \text{ m}$$
$$\lambda_0 = 0.9\ \mu\text{m} = 9 \times 10^{-7} \text{ m}.$$

The normalized frequency is known as V-number.

$$V = \frac{\pi d}{\lambda_0}\sqrt{n_1^2 - n_2^2} = \frac{3.143 \times 0.060 \text{ m}}{9 \times 10^{-7} \text{ m}} \times 0.16$$

$$V = 335103.22 = \mathbf{3.35 \times 10^5}$$

xample 4. *A step index fiber is made with a core of index 1.52, a diameter of 29 μm, and a fractional ifference index of 0.0007. It is operated at a wavelength of 1.3 μm.*

ind: (a) the fiber V-number and (b) the number of modes the fiber will support.

olution. Given: Core refractive index $n_1 = 1.52$

$$d = 29 \text{ μm} = 29 \times 10^{-6} \text{ m}$$

$$\Delta = 0.0007$$

$$\lambda_0 = 1.3 \text{ μm} = 1.3 \times 10^{-6} \text{ m}$$

$$\Delta = \frac{n_1 - n_2}{n_1}$$

$$0.0007 = \frac{1.52 - n_2}{1.52}$$

$$n_2 = 1.52 - 1.52 \times 0.0007 = 1.5189$$

(*a*)

$$V = \frac{\pi d}{\lambda_0}\sqrt{n_1^2 - n_2^2}$$

$$V = \frac{3.143 \times 29 \times 10^{-6}}{1.3 \times 10^{-6}} [1.52^2 - 1.5189^2]^{1/2}$$

$$= 4.049$$

(*b*) Number of modes, $N = \frac{1}{2} V^2 = \frac{1}{2} (4.049)^2 = 8.19 = \mathbf{8 \text{ modes.}}$

xample 5. *Find the core radius necessary for a single mode operation at 850 nm in step index fiber vith $n_1 = 1.480$ and $n_2 = 1.47$. What is the numerical aperture and maximum acceptance angle of this iber?*

olution. (*i*) For single mode propagation, V-number is 2.405.

$$V = \frac{\pi d}{\lambda_0}\sqrt{n_1^2 - n_2^2}$$

$$\therefore \quad 2.405 = \frac{\pi d}{450 \times 10^{-9} \text{ m}} \times 0.1717$$

$$d = \frac{2.405 \times 850 \times 10^{-9}}{3.143 \times 0.1717} \text{ m} = 3.79 \text{ μm}$$

$$\therefore \quad r = 1.89 \text{ μm}$$

(*ii*)

$$NA = \sqrt{n_1^2 - n_2^2} = \sqrt{1.48^2 - 1.47^2} = 0.1717$$

(*iii*)

$$\sin \theta_0 = NA$$

$$\theta_0 = \sin^{-1} (NA) = \sin^{-1} [n_1^2 - n_2^2]^{1/2}$$

$$= \sin^{-1} (0.1717) = \mathbf{9°53'12''}.$$

8.7 APPLICATIONS OF OPTICAL FIBERS

Optical fibers were developed initially for use in the communications industry. Subsequently, a large number of applications have been found in other areas where fibers can out-perform over the existing devices. Fibers are extensively used in medicine, modern warfare, in communications of different types, in sensor applications, entertainment electronics, etc. A few representative applications are briefly discussed here to indicate the scope of fiber applications.

Medical Applications

One of the important applications of optical fibers is in the field of medicine. A bundle of fibers (MMF) is used to illuminate the areas in the human body which are otherwise inaccessible. A second bundle is used to collect the reflected light. An incandescent bulb can be used as the source of light in this application. Such fiberoscopes are employed widely in endoscopic applications.

In ophthalmology, a laser beam guided by the fibers is used to re-attach detached retinas and to correct defects in vision. In cardiology, laser angioplasty is expected to replace balloon angioplasty and bypass surgery. A special catheter is developed for the removal of blockages in the veins. The catheter consists of three channels—for visualization, for delivering laser power, and an open tube to flush or suck the debris.

Military Applications

An aircraft, a ship, or a tank needs tons of copper wire for the wiring of the communication equipment, control mechanisms, instrument panel illumination, etc. Use of fibers in place of copper wires reduces weight and also maintains communications silence.

Fiber guided missiles have been pressed into service during recent wars. Sensors are mounted on the missile which transmit video information through the fiber to a ground control van, and can receive commands from the van repeatedly. The control van continuously monitors the course of the missile, and if necessary, corrects it to ensure that the missile precisely hits the target.

Entertainment Applications

A coherent optical fiber bundle is used to enlarge the image displayed on a TV screen. Conventional optical projection system, are bulky and expensive.

Optical Fiber Sensors

In this particular application, the inherent physical property of the fiber material is utilized. The variation in the refractive index of the fiber under the influence of external forces leads to the possibility of an optical fiber being used as a transducer.

If the fiber is subjected to heating, the temperature causes a change in the refractive index of the fiber. As temperature increases, the difference between the refractive indices of core and cladding reduces, leading to the leakage of light into the cladding. A simple thermometer can be built by using a LED as a light source, a coil of fiber as a heat sensing element and a photodetector to measure the intensity of light. Temperatures in the range from 80° to 700° are measured using such a thermometer.

A smoke detector and pollution detector can be built using fibers. A beam of light radiating from one end of a fiber can be collected by another fiber. If foreign particles are present, they scatter light and the variation in intensity of the collected light reveals their presence.

A loop of fiber can be used to determine the level of liquid in a container. A part of the cladding is scraped and the loop is suspended above the liquid level. Light is directed to pass through the fiber and its intensity is measured at the output. A bare core loses more light when it is immersed in liquid then when it is exposed to air. A sudden change of light intensity indicates the liquid level. A LED source, a MMF, and photodetector are used in building such a liquid level sensor. These sensors are used to monitor the filling of petroleum tanks.

Double refraction is induced in the optical fiber when external forces are applied. This property of induced birefringence is exploited in fabrication of various sensors. A pressure sensor using stress induced birefringence, a voltage sensor using Pockel's effect, a current sensor based on Faraday rotation, and magnetic field sensors based on magnetostriction are built.

Communications Applications

Optical fiber communications systems can be broadly classified into two groups, (*i*) local and intermediate range systems where the distances involved are small and (*ii*) long-haul systems where cables span large distances.

Smaller systems cover communication networks in a small community, an industry, or a bank etc. In many organizations, a LAN distributes information to several stations within the organization. A number of computer terminals are interconnected over a common channel to keep track of the flow of the data and to process the data. An optical fiber data bus offers a great reduction in cost and enormously increases the information handling capacity compared to a parallel multiwire data bus.

The wired-city concept was tried in Japan using an interactive optical fiber system. The Higashi-Ikoma Optical Visual Information System (Hi-OVIS) is an example in which a two-way interactive cable TV service was provided for the suburban city of Osaka. The system consists of computer-controlled video services to 158 home subscribers and 8 local studio terminals. Subscribers may select any TV station and key-in any desired program. A two-way video phone provides shopping opportunities. Plastic glass fibers with NA = 0.25 and LEDs operating at 0.82 μm are used in this system.

Telephone cables connecting various countries come under the category of the long-haul system. A 6600 km long submarine transatlantic telephone cable was commissioned under international cooperation in 1980. Many such projects have been completed since then.

Optical Waves in Communications

The most important application of optical fibers is in the field of communications as an information channel or transmission medium. Traditionally, telecommunications are carried out between distance transmitters and receivers with the aid of radio frequency and microwaves. Wires or coaxial cables are used to transmit information signals. In a wireless system, antennas are used for transmitting and receiving the signal.

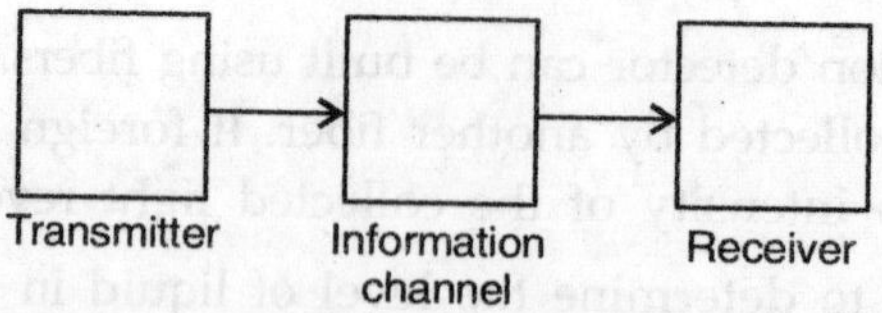

FIGURE 11 **A basic communications system.**

A basic communications system consists of a transmitter, a receiver and an information channel (Figure 11). Normally, the information to be communicated is a non-electrical message which is to be converted first into an electrical form. A transducer performs this task. For example, a microphone converts sound waves into currents and a video camera converts images into currents. The electrical message thus obtained is of low frequency and is impressed onto a carrier wave of very high frequency generated by an oscillator. The process of imposing a message signal on a carrier wave is called modulation. Two different techniques exist for modulation, namely analog modulation and digital modulation. In analog modulation a continuous wave carries the message. In digital modulation the message is transmitted in discrete form using binary digits (or bits). In this case, the sequence of on or off pulses constitutes a coded version of the message. The message travels along the transmission channel and is received at the receiver. The receiver demodulates the input wave, separates out the message, and feeds it to a transducer, such as a loudspeaker.

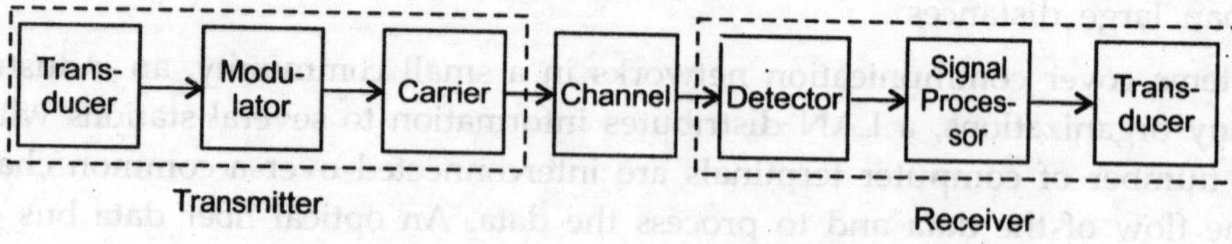

FIGURE 12 **A generalized communication system.**

The number of messages that can be simultaneously transmitted on an information channel is determined by (*i*) the bandwidth requirement of the message and (*ii*) the bandwidth of the carrier. For example, a bandwidth of 4 kHz is required for voice transmission, 10 kHz for music transmission, and 6 MHz for television transmission. When the signals are transmitted in analog form, a channel having a width equal to double that of the above has to be provided. Normal TV communication has a bandwidth of 250 MHz. Therefore 250 MHz ÷ 2[6 MHz] = 20 TV programs which can be simultaneously transmitted. When optical waves are used, the bandwidth comes to 7.5×10^{14} Hz $- 4 \times 10^{14}$ Hz $= 350 \times 10^{6}$ MHz, and the optical band can accommodate about 30×10^{6} TV programs at a time. Similarly, the band can carry 4×10^{10} telephone calls simultaneously. Thus, the use of light waves expands tremendously our communications capabilities.

When the signals are transmitted in digital form the data rate is to be considered. The data rate is the number of bits per second (bps) transmitted. An analog signal such as a voice can be faithfully transmitted if it is sampled at a rate of twice the highest frequency contained in that signal. Further, a coding scheme is employed which, uses 8 bits. Therefore, a single voice message is constituted by 64 kbps. The data rate necessary for digital transmission of a television broadcast comes to 81.24 Mbps, using a video bandwidth of 4.5 MHz encoded and using 9 bits per sample, and the audio of 15 kHz encoded using 8 bits per sample.

In view of the increasing demand on simultaneous transmission of more messages, optical communications are preferred. Nationally and internationally, optical fiber cables of thousands of kilometers are installed for transmission of voice signals.

8.8 ADVANTAGES OF OPTICAL FIBERS

Optical fibers have many advantageous features which are not found in conducting wires.

(*i*) *Optical fibers are cheaper:* Optical fibers are made from silica (SiO_2), one of the most abundant materials on the earth.

(*ii*) *Optical fibers are small in size, light in weight, flexible, and mechanically strong:* The cross-section of an optical fiber is about a few hundred micrometers (μm) whereas wires are bigger in size and bulkier in weight. Typically, an RG-19/U coaxial cable weighs about 1100 kg/km whereas a PCS fiber cable weighs 6 kg/km. Therefore, fiber cables are easier to transport and install than metal cables. Fibers are quite flexible and strong.

(*iii*) *Optical fibers are not hazardous:* A wire communication link could, accidentally, short circuit high voltage lines and the sparking occurring thereby could ignite combustible gases in the area leading to great damage. Such accidents cannot occur with fiber links because of their insulating nature.

(*iv*) *Optical fibers are immune to EMI and RFI:* In optical fibers, information is carried by photons. Photons are electrically neutral and cannot be disturbed by high-voltage fields, lightning, etc., external influences so common to wire and wireless transmission. Fibers are therefore immune to externally caused background noise generated through electromagnetic interference (EMI) and radio frequency interference (RFI).

(*v*) *Optical fibers reduce cross-talk possibility:* The light waves propagating along the fibers are completely trapped within the fiber and cannot leak out. Similarly, light cannot couple into the fiber from its sides. Because of this feature, cross-talk susceptibility is greatly reduced.

(*vi*) *Optical fibers have a wider bandwidth:* While a telephone cable composed of 900 pairs of wire can handle 10,000 calls, 1 mm fiber cable can transmit 50,000 calls. Thus, fibers have the ability to carry large amounts of information.

(*vii*) *Optical fibers have low loss per unit length:* The transmission loss per unit length of an optical fiber is about 4 dB/km. Therefore, longer cable runs between repeaters are feasible. The spacing of repeaters are about 2 km for copper cables whereas it can range from 30 km to 100 km in the case of optical fibers.

A length of optical fiber serves as a transmission line that guides the message-carrying light wave to the receiver. The transmission of the light wave along the fiber suffers transmission dispersion.

8.9 ATTENUATION

An optical signal propagating through a fiber will get progressively attenuated. The signal attenuation is defined as the ratio of the optical output power from a fiber of the length L to the input optical power. It is expressed in decibel per kilometer dB/km.

$$\alpha = \frac{10}{L} \log \frac{P_i}{P_0} \tag{18}$$

where P_i is the power of the optical signal launched at one end of the fiber and P_0 is the power of the optical signal emerging from the other end of the fiber. In case of an ideal fiber $P_0 = P_i$ the attenuation would be 0 dB/km. A typical attenuation of 3 dB/km is usual in practice. The attenuation is wavelength dependent, and therefore the wavelength should also be specified.

Example 6. *A certain optical fiber has an attenuation of 3.5 dB/km at 850 mm. If 0.5 mW of optical power is initially launched into the fiber, what is the power level in μW after 4 km?*

Solution. Given: Attenuation $\alpha = 3.5$ dB/km

Initial power level, $P_i = 0.5$ mW

Length of the cable, $L = 4$ km.

$$\alpha = \frac{10}{L} \log \frac{P_i}{P_0} = 3.5 \text{ dB/km.}$$

$$\therefore \quad 10 \log \frac{0.5 \text{ mW}}{P_0} = (3.5 \text{ dB/km})\,(4 \text{ km}) = 14 \text{ dB.}$$

$$\therefore \quad \log \frac{0.5}{P_0} = 1.4$$

$$\therefore \quad P_0 = \left(\frac{0.5 \text{ mW}}{25.11}\right) = \mathbf{19.9\ \mu W.}$$

Example 7. *An optical signal has lost 85% its power after traversing 500 m of fiber. What is the loss in dB/km of this fiber?*

Solution. Let the initial power be

$$P_i = 1 \text{ mW}$$

$$P_0 = 85\% \text{ of } P_i = 0.85 \text{ mW}$$

$$L = 500 \text{ m} = 0.5 \text{ km}$$

$$\alpha = \frac{10}{L} \log \frac{P_i}{P_0} = \frac{10}{0.5} \log \frac{1}{0.85} = \mathbf{1.41\ dB/km.}$$

8.10 FIBER LOSSES

The losses occurring in optical fiber (glass fibers) may be attributed to three mechanisms: absorption, Rayleigh scattering, and geometric effects.

Absorption

Even a highly pure glass absorbs light in specific wavelength regions. Strong electronic absorption occurs at UV wavelengths, while vibrational absorption occurs at IR wavelengths 7 to 12 μm. These absorption losses are the inherent property of the glass itself and is called intrinsic absorption. However, intrinsic losses are insignificant where fiber systems operate at present.

Impurities are a major source of losses in fibers. Hydroxyl radical ions (OH), and transition metals such as copper, nickel, chromium, vanadium, and manganese have electronic absorption in and near the visible part of the spectrum. Their presence causes heavy losses. Metal ions must be kept to less than a few parts per billion and OH impurity to less than a few parts per million. The absorption of light either through intrinsic or impurity processes constitutes a transmission loss because that much energy is subtracted from the light propagating through the fiber. The absorption losses are found to be a minimum at around 1.3 μm.

Rayleigh Scattering

Glass is a disordered structure having local microscopic variations in density which in turn cause local variations in refractive index. Light propagating through such a structure suffers scattering losses. It is known as the Rayleigh scattering loss. Because Rayleigh scattering is proportional to λ^{-4}, it becomes important at lower wavelengths. Thus, Rayleigh scattering sets a lower limit, on the wavelengths that can be transmitted by a glass fiber at 0.8 μm below which the scattering loss is very high.

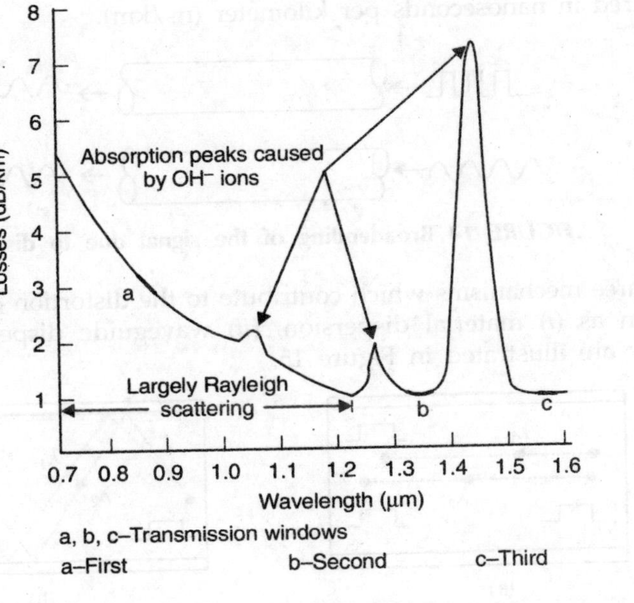

FIGURE 13 **A typical plot of fiber attenuation versus wavelength for a silica based optical fiber.**

Geometric Effects

These are the fiber losses introduced during manufacturing processes. Irregularities in fiber dimensions may arise in the fiber drawing process, in the coating and cabling process, or in the installation process. An attenuation curve combing the losses due to these effects is illustrated in the figure.

8.11 OPTICAL WINDOWS

It may be observed from the attenuation curve that it has a minimum at around a particular band of optical wavelengths. The band of wavelengths at which the attenuation is minimum is called an optical window or transmission window. It means that optical signals having wavelengths in this region travel through the fiber with minimum attenuation. Such windows are selected for information transmission.

Two windows are used in fiber optic communication:

- The first window lies between 800 nm to 900 nm (0.8 μm to 0.9 μm).
- The second window lies between 1300 nm to 1600 nm (1.3 μm to 1.6 μm).

8.12 DISPERSION

A light pulse launched into a fiber decreases in amplitude as it travels along the fiber, due to losses in the fiber. It also spreads during its travel. The pulse received at the output is wider than the input pulse, as shown in Figure 14. It means that the pulse becomes distorted as it propagates through the fiber. Such a distortion arises due to dispersion effects. Dispersion is typically measured in nanoseconds per kilometer (ns/km).

FIGURE 14 **Broadending of the signal due to dispersion.**

There are three mechanisms which contribute to the distortion of the light pulse in a fiber. They are known as (*i*) material dispersion, (*ii*) waveguide dispersion, and (*iii*) intermodal dispersion. They are illustrated in Figure 15.

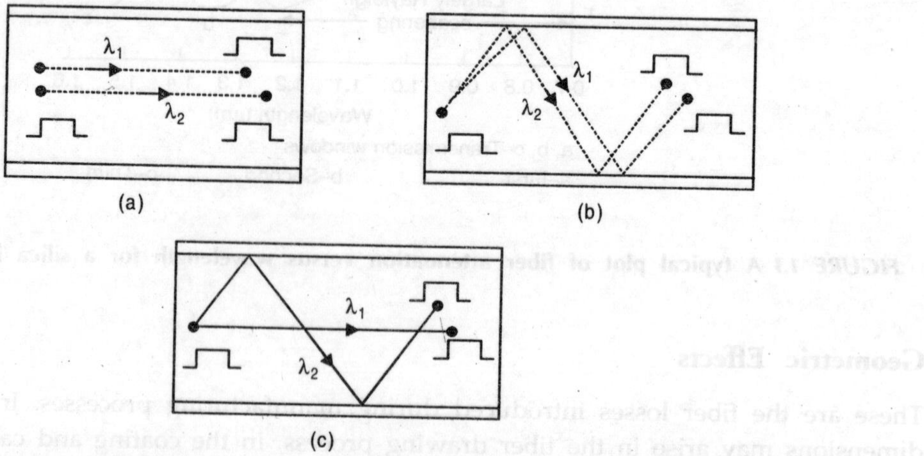

FIGURE 15 **An illustration of signal spreading due to various dispersion processes. (a) Material dispersion pulses at different wavelengths travel with different velocities. (b) Waveguide dispersion-pulses at different wavelengths, though propagating in the same mode, travel at slightly different angles. (c) Modal dispersion of a pulse at a single wavelength divides itself into modes which travel at different axis velocities.**

Material Dispersion

Light waves of different wavelengths travel at different speeds in a medium. The short wavelength waves travel slower than long wavelength waves. Consequently, narrow pulses of light tend to broaden as they travel down the optical fiber. This is known as material dispersion.

Obviously, the spectral width of the source determines the extent of the material dispersion. The material dispersion is given by the equation

$$D_m = \frac{\lambda(\Delta\lambda)}{c} L \frac{d^2 n}{d\lambda^2} \tag{19}$$

where λ is the peak wavelength, $\Delta\lambda$ is the spectral width, L is the length of the core, and n is the refractive index of the core.

To cite an example, an LED operating at 820 nm and having a spectral width of 38 nm results in a dispersion of about 3 ns/km in a certain fiber. In the same fiber, dispersion can be reduced to 0.3 ns/km using a laser diode operating at 1140 nm and having a 3 nm spectral width. Thus, material dispersion can be reduced using a more monochromatic source.

Waveguide Dispersion

Waveguide dispersion arises from the guiding properties of the fiber. The effective refractive index for any mode varies with the wavelength, which causes pulse spreading just like the variation in a refractive index does. This is known as waveguide dispersion. The amount of waveguide dispersion is governed by an equation similar to equation (19), with the material refractive index being replaced by the effective refractive index.

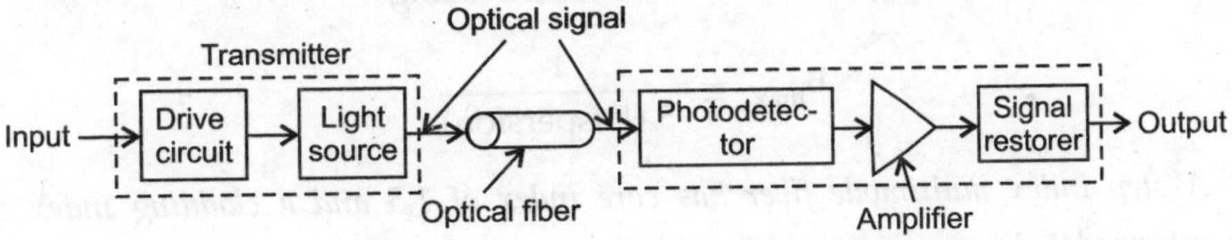

FIGURE 16 **Illustration of a typical fiber optic communication link.**

Intermodal Dispersion

A ray of light launched into a fiber follows different zigzag paths. When numerous modes are propagating in a fiber, they travel with different net velocities with respect to the fiber axis. Parts of the wave arrive at the output before other parts leading to a spread of the input pulse. This is known as intermodal dispersion. It does not depend on the spectral width of the source. A light pulse from an ideal monochromatic source ($\Delta\lambda$ = 0) would still exhibit spreading.

In a MMF, all three pulse spreading mechanisms exist simultaneously whereas in a SMF, only material and waveguide dispersions exist. Low NA fibers exhibit a smaller dispersion. In fibers with a large NA, there exist more modes leading to larger dispersion. Dispersion may be restricted by a careful selection of low NA fiber and a narrow spectral width source. A solution to the dispersion problem is to use GRIN fiber which produce less distortion than SI fibers. Typically, they have pulse spreads of only about a few ns/km which is much smaller than the pulse spreads in SI fibers. However, GRIN fibers are more expensive to manufacture. In a short length of fiber, dispersion is proportional to length but may not be that much of a problem. For longer lengths, it becomes proportional to the square root of the fiber length and is a limiting factor to the rate of data transmission.

Dispersion limits the bandwidth of a fiber. The information flow, *i.e.*, pulse rates, must be slow enough that dispersion does not cause adjacent pulses to overlap. If the distortion due to dispersion is large, the pulses may overlap and appear at the output as a single pulse with a number of humps. The detector cannot distinguish between the individual pulses (Figure 16).

8.13 BANDWIDTH DISTANCE PRODUCT

The bandwidth of optical fiber is obtained from a knowledge of dispersion, with the aid of the following relation:

$$\text{Bandwidth (MHz. km)} = \frac{310}{\text{Dispersion (ns/km)}} \tag{20}$$

The bandwidth is a very important characteristic of fiber cable. Obviously, low NA cables have the widest bandwidth. The bandwidth-distance product specifies the information capacity of an optical fiber. The attenuation per kilometer and the bandwidth-kilometer products are the important characteristics of fiber cables. They are determined by taking values of measurements on a cable length of 1 km. For longer cables, these results are converted to equivalent values from figures for 1 km length.

The maximum bit rate B_{max} allowed is estimated using

$$B_{max} \cong \frac{1}{5\,(\text{Dispersion})} \tag{21}$$

Example 8. *A step index multimode fiber has core index of 1.5 and a cladding index of 1.498. Find: (i) the intermodal dispersion factor for the fiber*

(ii) the total dispersion in a 18 km length

(iii) the maximum bit rate allowed assuming dispersion limiting

Solution. Given: $n_1 = 1.5$, $n_2 = 1.498$, $l = 18$ km.

(*i*)
$$\Delta = \frac{n_1 - n_2}{n_1} = \frac{1.5 - 1.498}{1.5} = 0.00133$$

For a 1 km length fiber,

$$\Delta = \frac{n_1 l}{c}\left[\frac{D}{1-D}\right] = \frac{1.5 \times 1000 \text{ m}}{3 \times 10^8 \text{ m/s}} \cdot \frac{0.00133}{1 - 0.00133}$$
$$= 0.6659 \times 10^{-8} \text{ s} = 6.66 \text{ ns/km}$$

(*ii*) Total dispersion in 18 km length

$$\Delta_t \text{ (Total)} = (\Delta_t \text{ per km})\ (1 \text{ km})$$
$$= (6.66 \text{ ns/km})\ (18 \text{ km}) = 119.9 \text{ ns}$$

(*iii*) Maximum bit rate $B_{max} = \frac{1}{5\Delta_t \text{ (Total)}} = \frac{1}{5(119.9 \text{ ns})}$

$= 0.00166 \times 10^9$ bits/sec = **1.66 M bits/s.**

Example 9. *A 15 km optical fiber link uses fiber with a loss of 1.5 dB/km. The fiber is jointed every kilometer with connectors which give an attenuation of 0.8 dB each. Determine the minimum mean optical power which must be launched into the fiber in order to maintain a mean optical power level of 0.3 mW at the detector.*

Solution. (*i*) Connector loss = 0.8 dB/km

∴ For 15 km length the loss = 0.8 dB/km (15 km) = 12 dB

(*ii*) Fiber loss = 1.5 dB/km

∴ For a length of 15 km, the loss = (1.5 dB/km) (15 km) = 22.5 dB

Total loss = 12 dB + 22.5 dB = 34.5 dB

Optical power level at the detector $P_2 = 0.3\ \mu W$

Optical power to be launched into the fiber P_1.

$$x\text{dB} = 10 \log \frac{P_1}{P_2}$$

$$\therefore \quad 34.5 = 10 \log \frac{P_1}{0.3\ \mu W}; \quad \log \frac{P_1}{0.3\ \mu W} = 3.45$$

$$\therefore \quad \frac{P_1}{0.3\ \mu W} = 2818.3; \quad P_1 = 8.455 \times 10^{-4} \text{ W} = \mathbf{0.85\ mW.}$$

8.14 FIBER OPTIC COMMUNICATIONS

The fiber optic communication system is very similar to that of a traditional communication system. A transducer converts a nonelectrical message into an electrical signal. In case of fiber communications, light waves are carrier waves. They are generated by a light emitting diode (LED) or a laser diode. The intensity of light radiated by these sources is proportional to the current passing through them. When the message in electrical form is fed to the optic source, the light output from it follows the variations of the message. Digital modulation is more frequently used in optical communications, as it allows greater transmission distance than analog modulation with the same power. The intensity modulated carrier light wave is then fed to the transmission channel, the fiber link. In order to have efficient transmission through the optical fiber, the frequency of operation of LEDs and lasers are chosen to lie in the optical window region of the fiber.

The main difficulty arises in coupling light into the fibers which have diameters in the order of a few μm. The optical signal travelling through the fiber will become progressively attenuated and distorted because of various losses and dispersion occurring in the fiber. Therefore, repeaters are used in the transmission line to amplify and reshape the signal. At the receiver end, an output coupler directs the light emerging from the fiber channel onto a photodetector.

The photodetector converts the light wave into an electric current. The message is contained in the detector output current. In the next step, the detector output current is filtered to extract the message and is amplified if needed. It is then fed to a suitable transducer to convert into an audio or video form.

EXERCISES

Short Questions for Discussion

1. What are optical fibers? Explain how glass fiber guides light from one end to the other.
2. Describe the structure of typical optical fibers used in practice.
3. Define the acceptance angle and the acceptance cone in fiber optics.
4. Define the relative refractive index change of an optical fiber. Show how is it related to numerical aperture.
5. Explain the importance of total internal reflection in optical fibers.
6. What are modes in optical communication? Explain single mode and multimode fibers.
7. What is the numerical aperture (NA) of an optical fiber? What does it signify?
8. Explain how energy is lost at a sharp bend in a fiber.
9. What is pulse dispersion in optical fiber?
10. Write various applications of optical fibers.

Long Questions for Discussion

1. Explain how light propagates through an optical fiber from one end to the other.
2. What is meant by the acceptance angle for an optical fiber? Show how it is related to the numerical aperture.
3. Derive the expression for the numerical aperture of the step index fiber. Show that it does not depend on the physical dimensions of the fiber.
4. Define the relative refractive index difference of an optical fiber. Show how it is related to the numerical aperture.
5. Explain how optical fibers are classified. Discuss their characteristic features.
6. Explain the difference between a step index fiber and graded index fiber.
7. Explain why and how a light signal gets attenuated in an optical fiber.
8. What is meant by modes? Compare a single mode and multimode fiber.
9. What is dispersion in optical fibers? Explain the different mechanisms that contribute to dispersion.
10. What are the advantages of an optical fiber communication system over the conventional ones?
11. Draw a block diagram of an optical fiber communications system and explain the function of each block.

INDEX

F

H

I

L

M

N

O

P

Q

S

T

V

W

Y